# GENUS MEDICAGO

# GENUS MEDICAGO (LEGUMINOSAE)

## A TAXOGENETIC STUDY

KARLIS ADOLFS LESINS & IRMA LESINS

Dr. W. Junk bv Publishers The Hague-Boston-London 1979

The distribution of this book is handled by the following team of publishers:

*for the United States and Canada*

Kluwer Boston, Inc.
160 Old Derby Street
Hingham, MA 02043
USA

*for all other countries*

Kluwer Academic Publishers Group
Distribution Center
P.O. Box 322
3300 AH Dordrecht
The Netherlands

Library of Congress Cataloging in Publication Data     C̄ĪP̄

Lesins, Karlis Adolfs.
   Genus Medicago: (Leguminosae): a taxogenetic study.

   Bibliography: p.
   Includes index.
   1. Medicago. 2. Botany-Classification. 3. Plant genetics. I. Lesins, Irma,
joint author. II. Title.
QK495.L52L42  583'.322  79-18960
ISBN-13:978-94-009-9636-6    e-ISBN-13:978-94-009-9634-2
DOI: 10.1007/978-94-009-9634-2

Cover design Max Velthuijs

# CONTENTS

\* In General Part italics refer to figures, or pages facing color plates

# INTRODUCTION

In introducing ourselves it should be told that in our native Latvian language our name is written Lesiņš. In most English publications, as in this work, the writing has been simplified to Lesins, and often only the first initial has been used.

Our interest in *Medicago* was first aroused during 1936-38, while employed as teachers in the agricultural and home economics school at Bebrene, Upper Zemgale, Latvia. Some plants of alfalfa (*M. varia, M. media*), locally called 'lucerna', were found growing wild along roadsides in that area, though no alfalfa fields had been seen in the vicinity within the memory of local farmers. Some roadside plants were dug out and transplanted to the garden, but their seedset was poor. During the next few years we paid only slight attention to alfalfa, the reason being that Latvia is a country with Atlantic climatic features (annual precipitation 600-700 mm; mild winters for its 56°-58° N. Lat., with January isotherms between −3° and −7° C; moderately warm summers, with July isotherms between 16° and 18° C), which together with its soils, mostly of acidic, podzolic type, is not well-suited for alfalfa production. It was not until 1945 in Sweden that work on alfalfa came to the foreground, when the senior author was assigned investigations on alfalfa seed setting by Dr. Erik Åkerberg, then director of the Swedish Seed Association branch station at Ultuna. The junior author, Irma Lesins, soon joined the project. Conditions at Ultuna were more suitable for alfalfa growing, as the soils were neutral. Moreover, alfalfa seed setting investigations were and still are a tradition at that institution as a continuation of studies by its former director, R. Torssell (1936, 1943, 1948). Our results confirmed the earlier findings that good seedset in alfalfa required cross-pollination, that honey bees were not effective cross-pollinators, and that there were too few wild bees to provide cross-pollination necessary for a good seedset (Lesins, 1950). An intensive search for means to improve the seed setting situation followed. In one direction we looked for related taxa that were self-fertile and could be hybridized with alfalfa, thus transferring self-fertility to the cultivated crop. In another direction, noting that self-opening of florets (self-tripping) in alfalfa under certain conditions was quite common, we searched for plants that, after self-tripping and, consequently, after self-pollination and self-fertilization, would not suffer drastically from inbreeding depression.

Following the first approach, i.e., hybridization, a prerequisite for further work was the acquisition of as many different *Medicago* species as possible, and also of some *Trigonella* and *Melilotus*. In addition, techniques for determining chromosome number, duplicating chromosome sets, as well as embryo culture, grafting and hybridization had to be adjusted for our material or developed anew. In the other direction, the search for inbreeding-tolerant plant material was encouraged following the observation that progenies from certain plants suffered much less from inbreeding depression than others. Studies on inheritance of some plant characters were also started at Ultuna.

Since 1951, at the University of Alberta (Edmonton, Canada), the abovementioned studies have been continued and extended. The aim was set to acquire all known *Medicago* species and as many varieties as possible. Unfortunately, from botanical institutions and gardens only the most common species were available and even these under confusingly different names. It soon became obvious that without field expeditions we would never come close to achieving a complete collection of all *Medicago* species.

After the first expedition to Italy in 1959, the idea of systematically investigating and compiling the results in a *Medicago* monograph was first conceived. However, a number of countries or districts harboring certain rare endemic species were out of our reach. We are greatly indebted to those botanists (only a few of whom are mentioned in the acknowledgements) who have supplied us with live seeds from those areas. With the accumulation of live *Medicago* material, and an aim set for its systematization, all aspects concerning different *Medicago* taxa became of interest. Gradually much of the time and effort invested in improvement of cultivated alfalfa shifted to systematization of accumulated *Medicago* material, though interest in alfalfa improvement was never left out of sight. As time went on some specialization in duties between the two authors was worked out. Morphological notes with the thousands of necessary measurements, together with some of the chromatography and cytology, fell to the junior author, who had completed about nine-tenths of the morphological descriptions when she died in 1966. The senior author's responsibility was hybridization and genetic studies together with the alfalfa breeding work. The results, of which this book represents the final outcome, have appeared in part in a number of co-authored papers. Throughout this book the term 'we' is used to express the fact that ideas and items were discussed by both authors from the early stages. The final formulations are those of the senior author.

# ACKNOWLEDGEMENTS

The authors are indebted to both the Swedish Seed Association and the University of Alberta for facilities provided during the different phases of this work from 1945 to the present (1979). Grants made either directly to the authors or through directors of the institutions concerned have been received from the Wenner-Grens, Helge Axelsson-Johnsson, and the Harvie foundations, the National Research Council of Canada, the Alberta Wheat Pool, the Alfalfa Processors Co-op Association (Alberta), the Canada Department of Agriculture, the Alberta Agricultural Research Trust, and the Scientific Division of NATO. This latter support was the most important one since it allowed us to make the expeditions to Mediterranean countries in 1962 and 1963, during which most of our *Medicago* material was collected. These collections, in fact, were the foundation of our present *Medicago* seed collection, which now comprises about 3000 accessions.

Individual botanists who have supplied us with live *Medicago* seeds are many. We have first to thank Dr. I. T. Vassilczenko who sent us some seeds as early as 1948. Thanks are also due to respected botanists who have taken pods from their herbaria and sent them to us; thus, we have received seeds from Drs. K. H. Rechinger, P. Aellen, I. J. Latschaschvili, B. Stefanoff, I. V. Kriukova and S. Kožuharov. In response to our request for seeds of some rare species, special field trips had been undertaken by workers of the Nikita Botanical Gardens and of the Tadzik Botanical Institute. We appreciate their help very much. Technical assistance in the field, greenhouses, and laboratory work, and in the photographic and office work has been received from a great number of assistants. Their input in this work is considerable and we wish to express our sincere thanks to them.

The editing of the manuscript was done by Professor Dr. C. C. Heyn at the University of Jerusalem. The senior author highly appreciates her thorough editing, and her generous professional help in improving the book's value. For editorial suggestions and valuable discussions concerning general matters, the authors are indebted to Dr. G. W. R. Walker at the University of Alberta. For what is said and illustrated in this book the senior author takes full responsibility.

# GENERAL PART

# PROCEDURAL

Undoubtedly, the authors' background and interests have influenced their approach to the study of relationships within the genus *Medicago*, which differs considerably from that adopted in the orthodox plant taxonomy. We consider our work as an attempt at taxogenetic systematization of the genus *Medicago*, rather than as a strict taxonomic treatment which, as regulated by the International Code of Botanical Nomenclature and applied by herbarium taxonomists, is a rather formal discipline, to say the least.

Our use of herbaria consisted almost exclusively of noting where a species or a variety of interest to us had been collected, keeping in mind the possibility that we might include the particular place in our itinerary. Altogether, during the eight collection trips undertaken to the Mediterranean countries and islands, we tried to see all herbaria on our way and gather as much new information as possible on additional variants and their whereabouts. Looking back on the difficulties we had in securing the live material, and looking ahead to what may be the needs for plant taxonomists, scientists, and users in general, it is the authors' view that plant taxonomy, as based on herbaria, should be reorganized. It is, of course, very satisfying to look at an herbarium specimen which a taxon's author has considered representative of his taxon. It should be noted, however, that the early authors have not attached that singular importance to their herbarium specimens and have placed somewhat different plants on the same sheet, therewith indicating that there are variations in nature covered by the same taxonomic name. This is at variance with the current typification cult.

It is rather frustrating to discover that a particular taxon named and herbarized may not exist any more in nature. Increased areas under agricultural crops, urbanization, and overgrazing of non-tillable land have overrun and destroyed the natural habitats from whence many taxa have been collected and described. Hills around the city of Brno, for example, on which a number of *M. falcata* variations grew and have been described, are now part of a residential area, hospital and observatory grounds. The environs of the city of Montpellier, where a number of variations of annual and some perennial species were growing, have long ago been 'developed'. On Mt. Pashtrik, Albania, from where *M. pseudorupestris* was collected and described, such plants have not been seen any more (Dr. F. Markgraf, Zurich, in a letter). The more deplorable, as this taxon, due to its thick hair cover all over, as described and seen on herbarium specimens, would now be of great interest as a possible source of resistance to some insect pests. Similarly, any other extinct taxon is an irreplaceable loss to science and possibly to the economy. For that reason a reorganization of taxonomic procedures would in-

3

volve, along with the herbarium type specimen, the deposition of an amount of viable seed in seed banks. With the presently available seed preservation techniques (Roberts, 1975), seeds of the vast majority of species may be preserved alive for centuries. Among other uses, live seeds would permit taxonomists to clarify the status of a great many hybrids described as separate species. Further, in regard to present taxonomy based on herbaria, its most serious weakness is the preclusion of testing materials as to their relationships based on hybridization. Without the possibility of hybridization tests we would never have been able to get an insight into the closeness between some species (e.g., *M. suffruticosa* — *M. hybrida*), and the rather distant relationship between some others (e.g., *M. falcata* — *M. platycarpa*). Further basic species-specific characteristics are chromosome number and life span, neither of which can be tested on herbarium specimens. From our experience we, in fact, can mention a number of other shortcomings of herbarium specimens: Flowers, especially those containing anthocyanins, become faded; seeds lose their specific color, turning brown, and leaves lose their distinctive lighter and deeper shades of green (e.g., *M. disciformis* vs. *M. tenoreana*). Early stages of seedling development (e.g., first trifoliate leaflets) which may be of use in distinguishing taxa, are usually not found on herbarium sheets; also, other plant parts are often poorly represented or lost. Pollen often becomes distorted in shape due to high moisture conditions during preparation. The more common physiological characteristics, which can be determined only from living plants, include need for winterrest (e.g., in *M. carstiensis*), the time taken to reach reproductive stages (e.g., forms in *M. minima*), the day-length requirements, the length of the afterripening period for seeds (most annual vs. perennial species), and the vulnerability to excessive soil moisture (e.g., *M. doliata*). Also, such important characteristics as susceptibility to diseases and insect pests can rarely be determined from herbarium specimens since collectors usually have put the least damaged plants on sheets. Many of these physiological characteristics are more taxon-specific than morphological differences. Last but not least, to study type specimens which are widely scattered among many institutions, and often not available for loan, necessitates expensive investment of time and travel. Moreover, sometimes, for one reason or other, the herbaria with type specimens are not accessible.

In *Medicago* the situation is such that most of the original type specimens are not in existence. Heyn (1963), in describing 50 annual taxa, could indicate only 16 which had as holotypes herbarium specimens. As substitutes, i.e., lectotypes, had to be chosen illustrations, or descriptions, or one out of several herbarium specimens. In some cases, selection of a substitute was put off because of the possibility that the original specimen may exist deposited in some unknown herbarium. Thus, *Medicago* typification is on a rather artificial basis, and in most cases not strictly tied with the original, name-providing plants. For that reason we felt that to avoid ambiguity we had to provide descriptions of the discussed *Medicago* material with as clear illustrations as we could possibly prepare.

4

Having live material, for illustrations we used photographs exclusively rather than drawings. True, with drawings certain features may be emphasized. However, certain other features not considered of value at the moment may be omitted, though later these may turn out to be of importance. For photographing we used fresh plants, usually branches. These were mounted with strips of transparent tape on a sheet of clear glass, then put on top of a frosted glass evenly illuminated from below. The specimen was weighted down with a plate of clear glass, providing additional flattening and avoiding glaring reflection from mounting strips. Such specimens remained fresh for several hours allowing rephotographing if necessary. For high quality negatives we used $4'' \times 5''$ sheet films, although narrow roll films now available on the market may be equally suitable. We have learned recently that photography using soft X-rays can give good pictures (Chmelar, 1975).

In our illustrations, branches and other vegetative parts are shown so that as many details as possible can be seen on the pagespace available. Magnifications of up to 3 times are not indicated except where vegetative parts of different magnification are shown in the same figure.

Some of the pod photographs were published in 'Alfalfa Science and Technology, ASA monograph 15, pp. 53-86, 1972'; permission for reproduction by the American Society of Agronomy is herewith acknowledged.

The methods used in studying pollen grains have been described in our previous publication (Lesins & Lesins, 1963b). The technique of autoclaving pollen (at 15 lbs. pressure for 12 minutes) in the mixture of Cummings et al. (1936), stain and glycerin (1:1), disclosed very clearly the number and locations of germinal apertures (colpi). Observations on the shape of expanded pollen grains immersed in Cummings (l.c.) stain were useful, since differences not seen in dry pollen could be detected.

The poor staining capacity of *Medicago* chromosomes required a great deal of experimentation in order to achieve good preparations. The schedule used for root tips at present is as follows:
1) From growing plants take young root tips approx. 1 cm long, put them into fixative consisting of 3 parts alcohol (ethanol or methanol) and 1 part glacial acetic acid.
2) Fix under vacuum until no air bubbles appear and root tips settle down. This would not take longer than 15 minutes.
3) Transfer root tips to a mordant consisting of 2 parts 95% ethanol and 1 part iron alum (preparation: dissolve 15 g of violet crystals of $FeNH_4(SO_4)_2 \cdot 12 H_2O$ in 500 ml water, add 0.6 ml conc. $H_2SO_4$, followed by 5 ml glacial acetic acid). Mordant root tips for 1 hour at 60-64° C.
4) Cut off the yellow-colored tips, clean them by rolling on a mordant-wetted cloth or chamois, transfer to aceto-carmine (preparation: 2 g of carmine in 100 ml of 45% aqueous acetic acid, boiled gently for several minutes with reflux, cooled and filtered). Wash once or twice with aceto-carmine, using a finely drawn glass pipette (orifice less than the diameter of root tips).

5

5) Stain in aceto-carmine for 1 hour at 60-64° C.

6) Suck off aceto-carmine with the pipette. Wash 2-3 times with 45% aqueous acetic acid.

7) Transfer to a drop of 45% acetic acid on a microscope slide. Under dissecting binoculars remove all debris such as lint or other impurities, using a thin insect pin with handle rather than a preparation needle. Macerate in bulk or, preferably, arrange single rootlets in two rows evenly spaced between rows and rootlets.

8) Put on coverslip (thin, No. 1). Press slightly with filter paper to remove excess fluid; some of it should remain as lubricant. Tap slightly with the end of a pencil or pinhandle on each root tip to spread it in an even patch. Replace the filter paper and press hard to remove as much fluid as possible; then tap hard with the eraser end of a pencil on each root tip to spread the cells in a single layer. An even color of the patch indicates that the preparation has been successful.

9) Surround the edges of the coverglass with Cellodal I (Bayer) diluted with 2-3 parts of 45% acetic acid plus 1 part of aceto-carmine staining solution. After a few hours the preparation may be examined under the microscope. [Note: Cellodal is no longer produced by Bayer, but some similar product under a different name is probably available. Enclosure of preparations in Cellodal made them semi-permanent.]

Two of the above points may be stressed. First, the rapid fixation under vacuum, and the brief mordanting and staining at the indicated temperature seem to preserve the softness of the root tips, which is very important in spreading the cells in a thin layer. Second, in the preparation of the staining solution, an excess of carmine (2 g) seems to result in the extraction of certain compounds (carmine is not a uniform chemical substance) which stain chromosomes more intensely. It may be added that washing out the fixing solution with mordant and washing in aceto-carmine before staining helps to avoid undesirable precipitates.

If contraction of chromosomes is desired, the root tips are put before fixation into a saturated aqueous solution of paradichlorobenzene for about 3 hours in the refrigerator (temp. 5-13° C). On some occasions it was felt that, by keeping in cold paradichlorobenzene for only $1\frac{1}{2}$ hours, the chromosome movements to the poles were arrested without causing noticeable chromosome contraction.

Techniques used by us in hybridization have been described previously (Lesins, 1955; Lesins & Erac, 1968a). Currently, emasculation of florets in the bud stage is done using very fine tweezers with slightly crossed tips. The front of the standard petal toward the base of the calyx teeth is grasped between the tips of the tweezers, which in closing cut through the petal like scissors. By moving the tweezer tips up, the front upper part of the standard is separated and removed.

In the chromatography of flavonoid flower pigments the use of circular filter papers gave very showy results, especially in preliminary investigations of anthocyanins (Lesins & Lesins, 1958). Techniques for the analysis of carotenoids were later worked out in cooperation with a chemist, T. Ignasiak, and have been published (T. Ignasiak & K. Lesins, 1972, 1973).

Regarding the use of numerical taxonomy in our *Medicago* work, we had some reservations. In the first place, in assessing relationship between taxa, we tried to obtain information on such characteristics as crossability, mode of inheritance of characters, chromosome complements and meiotic processes. These features can not be readily incorporated into systems of other operational taxonomic units consisting mainly of morphological attributes. Thus, the requirement of different ploidy levels for successful hybridization between *M. papillosa* and *M. sativa* is a rare character not only in *Medicago* but in higher plants in general, hence requiring very exceptional weight. Secondly, numerical taxonomy is rather time consuming. Many characters (suggested number of at least 40-50) have to be taken, their actual attributes found, evaluated often by measurements, and coded. Our *M. constricta* and *M. murex* accessions were worked out and analyzed by J. L. Fyfe (see under *M. constricta*). True, some neatly delimited clusters of *M. constricta* did appear, representing ecologically close material. On the other hand, boundaries between *M. constricta* and *M. murex*, and also between chromosome groups within *M. murex*, could be determined much easier and faster by cytological and hybridization methods.

Plant materials worked on were those accumulated in our *Medicago* seed collection. Individual accessions mentioned in the text are denoted UAG (University of Alberta Department of Genetics) with their current accession number, often omitting UAG. Herbarium specimens from such individually noted accessions and types have been prepared and are deposited as vouchers in the Department's collection. Plants for study were grown in the growth chambers at a 14-18 hr light period under illumination of approximately 400 foot-candles provided by incandescent and 2000 f.c. by fluorescent lamps, as measured at the level of plant tops. During the summer the plants were usually kept outdoors, and perennials often were transplanted in the field and left over winter. Details on outdoor climatic conditions are given in the chapter on agricultural value.

Some terms used may need mentioning: 'strain', 'type', 'variation', 'race' and 'form' are used somewhat loosely to denote deviation from the main entity. Taxa are used for plant groups that have, or in the authors' opinion deserve, a formal rank. Latin abbreviations: e.g., i.e., l.c., and others are used, especially in parenthetical text, interchangeably with their English counterparts.

Some statements made in the General Part are repeated in the Specific Part. This is because matters discussed in the two parts are viewed from different angles. Occasionally we have dwelt at length on supposedly well-known matters. That, because we felt that basics often are left out of sight, as in the rephrased maxim: 'The trees are not seen behind the forest'.

Literature citations in the General Part are often omitted, especially if given in the Specific Part.

# NOTES TO ITEMS OF SPECIFIC PART

## KEYS AND GROUPING

Gross morphological differences are the most convenient means for distinguishing taxa. Sometimes there is a single readily observable character that sets a species or a group of species apart from all others. These instances, however, are not encountered very often, the usual situation being that of overlapping for taxonomic indices. In such cases a group of characters has usually to be taken into consideration in the characterization of species. Thus, in composing the general key for species, we have used readily observable morphological features, adding further features requiring more detailed investigation such as pollen morphology and chromosome number. In a *Medicago* key (Lesins & Gillies, 1972) intended for workers dealing with living material, a character such as chromosome number alone was sufficient to build a dichotomous key upon. In the present key of species, using mainly characteristics of macroscopic nature, related taxa may not be grouped together. Rather, some well-distinguishable characters such as straight versus coiled pods, or violet vs. yellow corolla, have caused separation of closely related species and putting together unrelated ones.

In groupings of higher taxonomic rank (subgenera, sections and subsections) we have tried to keep in mind the degree of species similarity to be expected with the assumption of common ancestry. We realize that this objective has not been achieved in a number of groups, especially in subgenus *Orbicularia*, which comprises species with phylogenetic origin possibly as divergent as that between members of other subgenera of the genus. In such cases practical considerations for easier surveying have resulted in some more artificial arrangements, especially where only gross morphological characters were available in estimation of relationship.

In grouping according to evolutionary relationship, hybridization is the most valuable tool, the production of progeny being a satisfactory proof of a relatively close relationship between the parents. The reverse, sterility between crossing partners, as an indicator for non-relationship, has its exceptions (e.g., *M. murex*). Where similarities in a number of morphological traits exist, further testing by hybridization and other techniques is warranted. In this respect we consider that hybridization, cytological, chemical and every other available method of establishing similarities or differences among taxa should be a normal procedure in taxonomy.

9

DESCRIPTION

*Name of Taxon and Authorship.* At the heading of each species description, we have given the name which at present seems to be the correct one, although in some instances it may not be the most widely used. The author's name is written without abbreviation so that some readers, not being specialists, may not wonder what a single or few letters stand for. Spelling out was the more desirable because in some otherwise respectable college dictionaries one looks in vain for the meaning of L. or DC. The author's name in full also may emphasize that, for instance, *M. turbinata* Allioni is not the same as *M. turbinata* Willdenow. In synonyms, author names are abbreviated.

In this place, mention should be made of the name of the *Medicago* monographer Urban (1848-1931). His initials are written I, but occasionally also J. According to Loesener (1931), his baptized name was Ignatz. The German capital handwritten J is the same as I except that the former has a dash across the middle. The mixup may have arisen in that in a signature the capital German I may have been assumed to be a J.

Literature citations in connection with taxonomic names are abbreviated. We felt that readers interested in checking original descriptions would be trained botanists for whom a source listed in full would be superfluous. Synonyms of below-species rank are given on only a few occasions, since these are well covered in works by Urban (l.c.), Vassilczenko (1949, perennial species), and Heyn (1963, annual species). We have checked, however, the original descriptions of species mentioned here. In general, the photographs of plant parts: branches with inflorescences, pods, seeds, and in some instances pollen, will make it clear what we meant under a certain name. On a few occasions where a variation was found that required a changed or a new taxonomic name, we described and pictured it, and applied a Latin or a Latinized name to it. That, we think, is the tribute that may be given to Latin for easier identification of names in the written text. Talking about Latin diagnosis as required by the International Code of Botanical Nomenclature: In the past, Latin was taught in every school and Latin was the only language of communication between educated men. Nowadays, Latin generally is not taught in schools, is not used in verbal communication, and practically not in written correspondence. Moreover, botanical Latin is a special language different from the classical one, and taxonomists are warned that help from colleagues in University Classic Language departments may not be of much use to them (Stearn, 1966). A claim that diagnosis written in Latin would make it more uniform and generally understandable is an anachronism. Any living language with present translation possibilities would serve the purpose much better. Therefore, we did not follow that requirement of the Code.

*Habit and Longevity.* There is only one shrub, *M. arborea*, with perennial, woody growth habit. Twenty-one species are herbaceous perennials with winter-surviving roots, and 34 species are annual herbs dying after fruiting.

In this latter group we counted *M. lupulina*, though under greenhouse conditions the plants usually continue fruiting repeatedly. Some gradations of this nature have also been observed in other annuals, some of which die after the aboveground parts are cut off, while a few (*M. orbicularis, M. intertexta* and others of sect. *Intertextae*) regenerate, though with diminished growth. It may be noted that stripping annuals of their unripe fruits may be used to extend flowering period (important in hybridization work) and hence the lifespan.

Growth habit varies from prostrate to upright, and is greatly influenced by growing conditions. In closed plant communities the plants tend to grow upright in order to get their share of light; without competition the plants of the same variety may either lie flat with only their distal portions upright (decumbent), or remain clearly prostrate. Under growth chamber conditions, without competition, under which most of the observations reported here were made, the upright growth habit, if noted, can be considered a characteristic of a particular taxon or accession.

The length of the plant was measured on fruiting branches or stems. In annuals these are, in fact, branches, though sometimes referred to as stems: they all originate from the single stem which often ceases elongation and branches overtake it in length. In addition, internodes between basal branches may be rather short, therewith giving the plant a many-stemmed appearance. Under certain climatic conditions, annuals germinating in fall may branch in the form of a rosette (McComb, 1974). In perennial herbs we deal with multiple stems which originate from the crown, the compact transition region between perennial roots and the one-season aboveground shoots. Under greenhouse conditions the recorded plant length is usually greater than that observed in natural growing sites. Nevertheless, the potential growth as recorded under greenhouse conditions may give an estimate for comparison with other taxa. Records obtained in nature, often on plants growing under very dry conditions or in overgrazed areas, may be underestimates of the potential plant vigour.

*Hairiness of Aboveground Vegetative Parts* ('aboveground' usually omitted in text). The hairiness may vary considerably, even from plant to plant within a species, although a few taxa have dense hairiness as a specific character (*M. marina, M. lanigera, M. arborea*). Aging plant parts may lose their hairs, becoming glabrate. Upper sides of leaves as a rule are less hairy than their undersides, in a couple of species they are characteristically glabrous. The two kinds of hairs, one-celled simple and many-celled (usually glandular, i.e., with roundish sticky tips), are often found together. A more thorough scrutiny in regard to glandular hairiness is likely to be undertaken in future, since glandular hairs seem to provide resistance to certain insect pests (Shade et al., 1975). In a few species (*M. scutellata, M. disciformis*) glandular hairiness is a species characteristic, and in some other species a few of the forms have well expressed glandular hairiness.

*Inflorescence and Adjacent Parts* (Fig. 1). The inflorescence in *Medicago* is a raceme; sometimes the number of florets is reduced to 1-2. The two

11

stipules are adnate to the stalk of the leaf, the petiole. In some taxa the stipule bases are close together, in some others more apart. This feature has not been assigned much significance in taxonomy and is only occasionally mentioned in the text. Usually each of the stipules is shaped like half of an arrow. The margin, especially the outer margin, may be entire, toothed with smaller or larger teeth, or it may be deeply slit or laciniate (feathery). Variation may sometimes be found on a single plant, where the stipules may be entire at the upper nodes of a stem or branch and more or less toothed at the lower nodes. A few taxa have characteristic stipules such as laciniate (*M. laciniata* var. *laciniata*), or entire, or with only slightly toothed margins (e.g., *M. minima*, *M. doliata*). The stalk of a raceme, the peduncle, arises from the axil of a leaf. Its length is measured after the fruits have begun to develop,

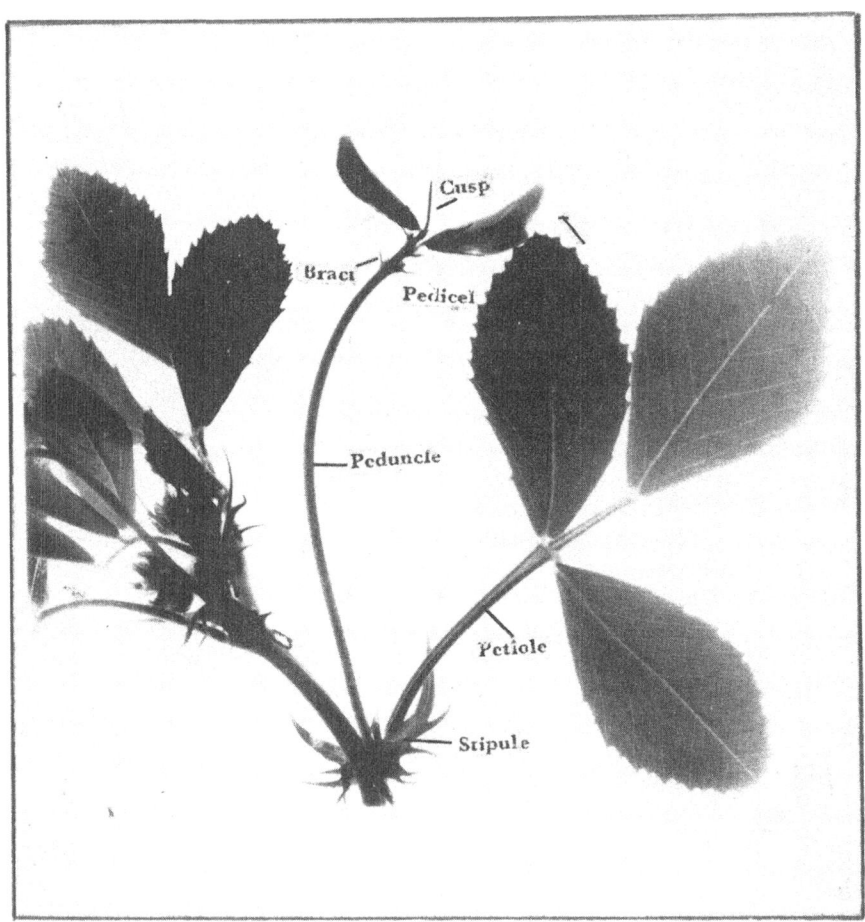

*Fig. 1.* Inflorescence and adjacent parts.

12

at which stage it ceases elongation. Its length is compared with that of the leaf petiole, since this ratio usually has taxonomic value.

Florets are borne on stalklets, or pedicels, which may vary in length. Beyond the uppermost pedicel the peduncle may extend to a cusp (awn). The length of pedicels and their inclination with respect to the peduncle may have some taxonomic value, e.g., recurved (*M. prostrata*), or upright (*M. falcata* ssp. *romanica*). As the peduncle arises from the axil of a leaf, so the pedicel arises from the axil of a rudimentary leaf, the bract. Its length, compared with that of the pedicel, has taxonomic significance in some instances. The bracts remain after the pedicels with florets or pods are shed, so that by counting the number of bracts, one may determine how many florets there have been on a particular raceme.

*Floret* (Fig. 2, -1 to -5). Parts of a floret are calyx, corolla, stamens, and a pistil. The length of the floret is measured from the base of the calyx to the tip of the corolla. The calyx consists of a bell-shaped tube with five teeth which may be wide or narrow (awl-shaped). The length of teeth is compared with that of the tube, which in turn is compared with the length of the pedicel. The corolla is typically papillionaceous, consisting of a standard (also called banner or *vexillum*), two wing petals (*alae*), and a keel (*carina*) consisting of two fused petals. The standard is longer than either the wings or the keel in all species except *M. arborea* where it may be equal to or shorter than the keel. In section *Platycarpae*, the wings are almost as long and wide as the standard. In subsection *Papillosae*, the standard is widened in its lower middle part (Fig. 36,-b). Wings may be longer or shorter than the keel. This ratio has a distinct taxonomic value. The shape of the standard (obovate, ovate, or with parallel sides) has also taxonomic value.

Stamens are ten. Nine of the filaments are fused, forming the staminal column (sheath), while the tenth, facing the standard petal, is free. Linnaeus (1754) includes *Medicago* in the group of *decandria* (= having ten stamens), *diadelphia* (= having two groups of stamens). The pistil surrounded by the staminal column consists of the ovary with one, two to many (20+) ovules, the style, and the stigma. Investigations by Armstrong & White (1935) in *M. sativa* showed that the stigma is covered with a hyaline membrane, the injury of which by striking against the standard petal, an insect's body, or other object is considered a prerequisite for fertilization. Thus, if this stigmatic surface is ruptured it becomes penetrable by germinating pollen, as was already suggested by Burkill (1894). A viscous fluid under the stigmatic membrane probably aids pollen germination.

The tripping mechanism (Fig. 2) is characteristic of the entire genus *Medicago*; it is present in some species of *Trigonella* and a few other Leguminosae (e.g., *Indigofera*). The obvious purpose of the mechanism is to ensure cross-pollination.

Before tripping, the staminal column is under tension produced by expansion-restrained cells of the column (Larkin & Graumann, 1954). It is kept in its tension-loaded position mainly by friction between the back of

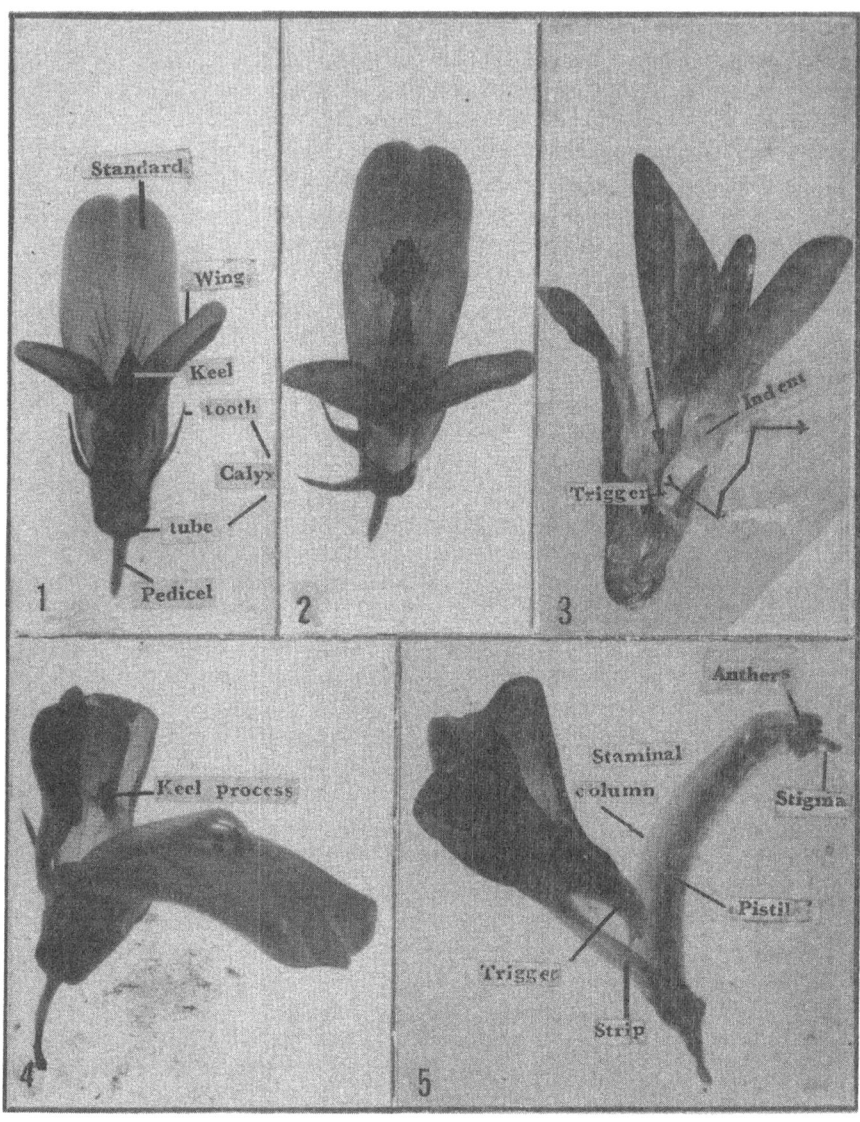

*Fig. 2.* Nos. 1-5, parts of floret and the tripping mechanism. Nos. 1 and 2, untripped and tripped floret, resp. In No. 3 straight arrow points to the gap between trigger horns. Beginning of the zig-zag arrow indicates where the spreading movement starts, its zig-zag body shows the transfer of movement to wing indent, and arrow's end indicates the pulling of wing-keel union off the staminal column. Nos. 4 and 5, floret's inside parts.

the keel petal (the strip) pressed against the backside of the staminal column. This situation is comparable to that of the driving belt (strip of the keel) on the pulley (back of the staminal column). There is needed only a slight counter-balance to keep the column in place, which is provided by the keel processes pressed against the column from the front. The release of the column, the tripping, usually is brought about by the trigger mechanism. This consists of two outgrowths (horns or triggers) in the lower part of the wing petals. The horns lie over the lower front-part of the staminal column. There is a gap between them (arrow, Fig. 2,-3). An insect trying to reach the nectaries at the base of the staminal column thrusts its proboscis into the gap, consequently spreading one or both horns. They are connected by a rigid tissue to the upper part of the wings where wing indents as outgrowths are fitted into the keel processes. Thus the movement, like pulling a trigger, is transmitted to keel processes causing them to spread. The staminal column then shoots out so that the anthers and the stigma touch the insect's body strewn with pollen from previous flower visits. Therewith cross-pollination is accomplished.

The outlined mechanism of tripping is based on observations: On a *M. sativa* plant on which the keelbase (strip) was weak, it usually broke. The result was that almost every floret was 'tripped' but without the spreading of keel that characterizes normally tripped florets. Thus, friction between the keel strip and the back of the staminal column is the main force restraining the column from tripping. As noted above, the balancing force keeping the column from bending is provided by the upper part of the keel embracing the column from behind and pressing keel processes against it from the front. The staminal column exerts only a fraction of its tension-strength against the keel processes, the main force being absorbed in the friction against the keel strip.

The wing outgrowths fitted into keel processes have no part in restraining the staminal column from tripping. This is inferred from the observation that occasionally it was possible to pull the wing outgrowths out from the keel processes without the column becoming tripped. Thus the wing outgrowths are part of the trigger mechanism. The mechanism in action may be observed by slightly bending one of the horns sideways with a needle or a similar object, which causes the staminal column to snap out. The tripping mechanism has been described by Burkill (l.c.), by Lesins (1950) and recently by Heszky (1972). The strength with which the staminal column of *M. sativa* and of related taxa shoots out is considerable. On one occasion a honey bee was found trapped to death under the staminal column of an alfalfa floret.

The tripping mechanism has evolved obviously in concordance with the available pollinators in times of the older, perennial *Medicago* origin. Most of the pollinators of perennial *Medicago* are ground-nesting bees. One of these bees, *Megachile rotundata* F., presently semi-domesticated as alfalfa pollinator, usually touches the trigger in the floret from one side, therewith activating the trigger mechanism but avoiding a direct hit by the staminal column. The honey bee, *Apis mellifica* L., has evolved probably under forested environments where *Medicago* has not been endemic; hence it generally prefers other plants as pollen source. Moreover, the colony's need for pollen is much less than that for nectar which is needed for overwintering. This is in contrast to the needs of most species of wild bees which require nectar only for their subsistence and preparation of bee-bread, but winter survival is in the form of dormant queens or pupae. Honey bees, however, soon learn to obtain nectar without tripping the floret by thrusting the proboscis under the trigger horns from the side of the floret. This manner of obtaining nectar, if pollen is not needed, is also used by generally good alfalfa pollinators, e.g., *Bombus terrestris* L. (Åkerberg & Lesins, 1949), and by drones.

Self- (i.e., automatic) tripping is a constant feature in annual *Medicago* species. In perennials self-tripping takes place to varying degrees (Lesins, 1950, 1961c; Lesins et al., 1954), and may be considered as a means of ensuring progeny during insect-deficient periods. Under what conditions, outside and within the plant, the self-tripping is brought about is not well understood. Koperzhinsky (1946) made an attempt to explain it on the basis of increased turgidity in cells of the staminal column caused by the conversion of different amounts (hence the differences in tripping rates) of available disaccharides to monosaccharides at elevated temperature with an adequate supply of water. Observations by us and other researchers on the heritability of automatic tripping have been made on several occasions, though direct studies on the number of factors involved have not been undertaken.

*Flower Color in Medicago* is a more or less intense yellow except in *M. sativa* and *M. daghestanica* which have violet florets. The yellow color is produced by flavonoids and carotenoids. Nine flavonoids (Cooper & Elliott, 1964) and 10—11 different carotenoids were found in petals of diploid *Medicago falcata*, and as many as 25 carotenoids were identified in the 22 perennial species investigated (Ignasiak & Lesins, 1972, 1975). Only a few minor carotenoids, however, were one-species characteristic, and a few more were characteristic of species groups. In violet petals of *M. sativa*, Lesins (1956a) found 3—5 anthocyanin pigments. They segregated as a unit in most of the crosses, except in one case where separation of one pigment from the rest was recorded. The observed loss of 4—5 anthocyanins by a single mutation indicated that a gene for an anthocyanin precursor, or a complementary factor was involved, or a compound anthocyanin gene behaved as a single mutational locus. In crosses of *M. sativa* with *M. glomerata*, four factors involved in production of anthocyanin color were recorded (Lesins, 1968), and a latent complementary factor was uncovered in *M. papillosa* (as *M. dzhawakhetica*) in crossing it with white-flowered *M. sativa* (Lesins, 1961a). As a genetic marker in crosses with yellow-flowered taxa the anthocyanins are a useful tool, since with yellow substances they produce green color, especially showy in petals of the first, $F_1$, generation plants.

The term, variegated flower color, applied to a population means that its different plants have different flower color. This, as a rule, is observed in offspring from hybridization, natural or artificial, of yellow with anthocyanin-flowered taxa, usually *M. falcata* x *M. sativa*. The multitude of colors, possible and encountered, may be understood considering the multitude of factors that may be involved in color production, as indicated above. Some of the possible color patterns found in the cultivar Ladak are shown in the upper part of the color plate I (facing p. 19). Variegated flower color, as applied to a single plant, means that its florets change color on aging. The usual pattern of change is from violet to green to greenish-yellow with different degrees of intensity depending on the content of pigments present; also, from pale cream to yellow in flowers where none or only a slight amount of anthocyanin pigments are present. Three racemes (a, b, c) with differently colored florets from *M.* x *tunetana* are shown on color plate I (lower part). On differently aged racemes of the same plant the color differences may be rather striking (see also Barnes 1972). In plants of hybrid origin, in which only slight amounts of yellow pigments are present, this may most easily be detected by inspecting the lower part of the keel in the throat of the floret. In the pure anthocyanin-colored florets this part usually is colorless; in yellow-flowered ones it has some yellow pigmentation which shows up in the hybrids. Color change on aging is probably due to change in flavonoid content, as the content of carotenoids was found to be about the same in young and aged florets in one case investigated (Ignasiak & Lesins, 1975). It was also found that some degradation of anthocyanins was taking place in older petals (Lesins, 1956a). In some occasions on wilting, especially after tripping, some increase in violet color was also observed.

*The Fruits* (Fig. 3,-1 to -8) are legumes, here called pods, known also as burrs or burs. In only a few species the pods are almost straight to sickle-shaped, the majority of species having spiral pods showing 3/4 of a coil to several coils (called also whorles, turns, or volutions). The two flat sides of a coil are called its faces, the narrow side its edge. Coil formation or spiralling starts at an early stage of pod development, and its pattern has some taxonomic value. In some taxa tight spiralling starts early and coils are formed while the pod is still within the calyx (Fig. 3,-1); at the other extreme, the young pod grows in loose coils out of the calyx (Fig. 3,-2), and tighter spiralling and contraction follow later. There are intermediate coiling patterns, as when tight coils are formed at an early stage of development but immediately protrude between the calyx teeth (Fig. 3,-3), or when the young pod rises somewhat from the calyx and then coils at the side of the calyx teeth (Fig. 3,-4). Actually, in all species except *M. noëana*, the coiling axis after contraction is oriented at an angle with the long axis of the calyx. Hence observations on the coiling of young pods should be made early, preferably before the petals are completely shed. It is possible that some disagreements expressed in the literature on the coiling patterns of young pods are due partly to pod observations at different stages of development, and partly to some within-species variation in coiling pattern. In *M. noëana*

17

*Color plate I.* Upper part, racemes of different plants from cultivar Ladak. Lower part, racemes (a,b,c) with differently colored florets from three plants of *M.* x *tunetana*.

the calyx remains appressed to the base of the pod as a regular star (Fig. 45,-b).

We determined coiling directions as follows: If, looking toward the apex of the pod, the tip of the apical coil points in the direction of a clock hands' movement, the coiling is clockwise (Fig. 3,-5); if it points in the opposite direction, the coiling is anticlockwise (Fig. 3,-6). In comparison with Urban's (1873) designations, clockwise refers to 'right' and anticlockwise to 'left' coiling. Kihara (1972) has argued strongly for use of the reversed right-left terms. However, as long as the meaning of terms is defined there should not be any misunderstanding. Clockwise and anticlockwise coiling directions are found in section *Pachyspirae*. In other sections the coiling direction is clockwise, except in the species *M. radiata* of section *Hymenocarpos* where on the same plant coils may turn clockwise or anticlockwise. Here, however, the term 'strongly bent' rather than 'coiled' is a more appropriate term as the pods usually do not make a full circle (Fig. 16); if they do, the tip haphazardly passes the base on one side or on the other.

The coiling direction is an excellent genetic marker in inheritance studies. Anticlockwise coiling is a dominant, one-gene determined character (Lilienfeld & Kihara, 1956); hence, from a pistillate parent with clockwise coiling, pollinated by a plant with anticlockwise coiled pods, any $F_1$ plants with anticlockwise coiled pods are known to be of hybrid origin. It should be kept in mind, however, that manifestation of any genetically controlled character in any living organism is dependent on its whole genetic constitution. On hybridization, unusual genetic backgrounds are sometimes created. An illustration of this is provided by the hybrids between *M. littoralis* (UAG No. 266, clockwise) from Czechoslovakia and *M. littoralis* (UAG No. 2157, anticlockwise) from the Canary Islands. The progenies of this hybrid included plants which had some pods with clockwise and some anticlockwise coiling in the same raceme (Lesins et al., unpubl.). Previously, we and other researchers have observed cases where pods started coiling in one direction, then changed coiling and continued in the 'correct' direction.

Pod color may vary from straw yellow to black; in most species the color is greatly affected by the environment. Moisture in the form of rain or artificial sprinkling during pod development makes them darker, as reported by McKee & Ricker (1913) for both natural and experimental conditions. Pods of some species (e.g., *M. rugosa*, *M. noëana*) are more resistant to darkening than those of other species.

Pod shape is rather variable. Apart from straight and sickle-shaped pods, coiled pods show several variations, including globular, cylindrical, disklike and other shapes. A few of these, such as the cup-shaped pods of *M. scutellata* and the nutlets of subgenus *Lupularia* are rather characteristic. Pod size is quite variable within species, although there are species with generally large pods (e.g., *M. intertexta*, *M. ciliaris*) and others with small ones (e.g., *M. coronata*). In our study the diameter of pods was measured on the middle coil (spines not included) of several pods. In figures, the actual diameter (or length of non-spiralled pods) is given at their side by an inked

19

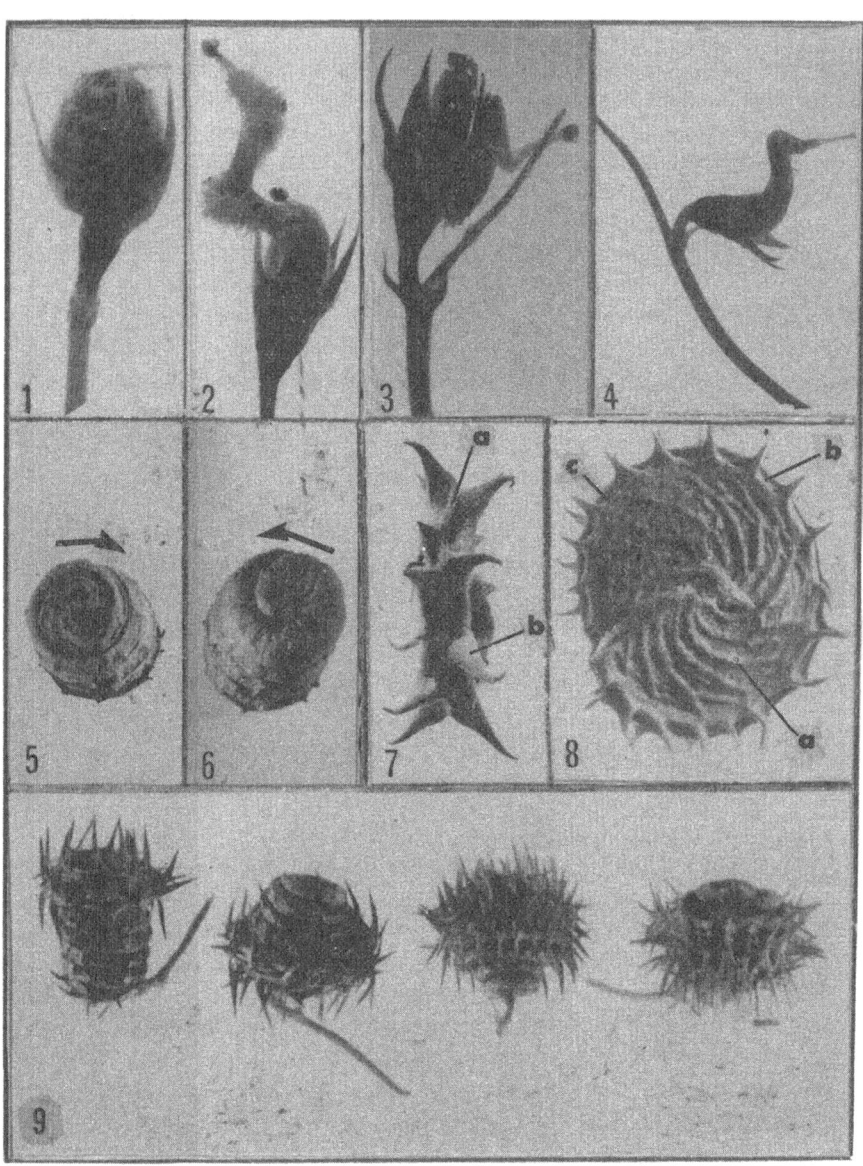

*Fig. 3.* Nos. 1-4, young pods: 1) retracted within calyx (*M. doliata*), 2) emerging from the calyx (*M. ciliaris*), 3 and 4) turning sideways through the calyx teeth (*M. arabica* and *M. laciniata*, resp.). Nos. 5 and 6, coiling direction: 5) coil tip pointing clockwise (*M. tornata* var. *striata*), 6) coil tip pointing anticlockwise (*M. littoralis*). No. 7, edge of coil with elevated dorsal suture (a), and base of spine (b) embedded in spongy tissue (*M. truncatula*). No. 8, face of a coil with radial veins (a), groove (b) between lateral vein (c) and the dorsal suture (*M. polymorpha*). No. 9, partly spineless pods in *M. intertexta*.

bar; this may provide an estimate of pod size. Pod length or height is mentioned in quotations of other authors.

Pod surface may be glabrous or covered with simple (one-celled) or with compound (many-celled) hairs, or both. Many-celled hairs usually have glandular tips, an exception being the non-glandular many-celled hairs of *M. lanigera*. Another characteristic kind of many-celled hairs is found in *M. papillosa* where the hairs are rather rough, flat, semitransparent, with their glandular tips usually broken off at pod maturity. In some species (*M. orbicularis, M. lupulina*), glandular hairiness is reported to vary with growing conditions and the stage of plant development (McKee, 1918). A study (Lesins, 1961b) on hairs in *M. sativa* x *M. papillosa* (as *M. dzhawakhetica*) has shown that probably genetic and cytoplasmic factors control their development. $F_1$ plants had pods with many-celled, but smaller, hairs than *M. papillosa*. Instead of the expected segregation in further generations in plants with glandular and plants with simple-haired pods, both simple and compound hairs were found on pods of the same plant. It was also found that the compound hairs became smaller in hybrids of advanced backcross generations in which the contribution of *M. sativa* genomes was increased. This may be another case of character expression failing to follow a simple inheritance pattern because of a thorough rearrangement of the genetic background due to hybridization. In contrast, in hybrids of the closely related taxa, *M. intertexta* x *M. ciliaris*, glandular hairiness was dominant and a simple segregation ratio (3:1) was found (Lesins et al., 1971), while in another cross, *M. suffruticosa* ssp. *suffruticosa* x *M. hybrida*, glandular hairiness was recessive, segregating in $F_2$ in a 1:3 ratio (Lesins, 1969). It should be noted that glandular hairs of *M. suffruticosa* are much smaller and few-celled compared with those of *M. ciliaris*.

Pods may be spineless or spiny, or tuberculate. There are taxa without spines, especially among perennial *Medicago*; in annuals the spininess is almost universally present. Members of section *Intertextae* are spiny without exception; in the two large sections, *Leptospirae* and *Pachyspirae*, there is only one consistently spineless species in each: *M. lanigera* in *Leptospirae* and *M. soleirolii* in *Pachyspirae*. A number of other species in the two sections have both spiny and spineless forms.

Regarding the usefulness of spines for species distribution and survival, our observations differ with views of other authors in that we consider the spines in *Medicago* to be primarily a means of anchoring the pods on the spot, rather than primarily for transportation by furred animals. In searching for pods in natural habitats, we have often found them in crevices or at the bottom of slight ground depressions, interwoven with older debris and covered with later deposits which had to be raked off to find them. This is in conformity with their favored habitat which for many annuals consists of steep, rocky hillsides, and often almost barren cliffs. Thus, if the pods happened to be blown or washed down to the foot of the hill the seedlings would be smothered by graminaceous and other more moisture-adapted plants. This applies not only to species with stocky spines, often without

21

hooks, as found on most members in section *Pachyspirae*, but also to species having pods with thin, hooked spines such as *M. minima*, *M. laciniata* and others of section *Leptospirae*. Under near-desert conditions spiny pods of *M. laciniata* anchor in wind-blown sand leeward of desert shrubs, as was observed on the outskirts of the Moroccan Sahara. True, species and varieties with hooked spines are more readily transported by animals. Species' ultimate establishment in new growing sites, however, depends on their suitability for the new sites; it may be admitted that spines provide a better foothold increasing the variety's tenacity to hold out until more favorable growing conditions occur. In section *Intertextae* where preference for heavy moist soils is prevalent, we observed the same usefulness of spines for anchoring the pods: on sloping ground after heavy rains we found the pods of *M. ciliaris* and *M. muricoleptis* on the top of small columns of clay, the spines securely interwoven with the clay particles, whereas the more loose soil around them had been washed away.

Spines are inserted in two rows, one on each side of the dorsal suture, except in *M. radiata* where they are in a single row, forming an extension to the fringe of the dorsal suture. Spines may be inserted at various angles to the face of the coil: from perpendicular (at $90°$) to the same plane as the face of the coil (at $180°$); if spines of the same coil are leaning over its dorsal suture the angle of inclination is more than $180°$. Usually the degree of inclination is somewhat variable even on the same pod, where on end coils the spines may be more bent, and on the middle coils more upright, especially if they are long. Length of spines, their number per row and angle of inclination are given in descriptions and may be helpful in identifying taxa. The bases of spines may be embedded in spongy (called also parenchymatous, loose cellular, or alveolar) tissue (Fig. 3,-7b) which may provide a further feature for characterizing taxa.

The spiny condition is usually inherited as a dominant character, hence it is useful as a marker in hybridization studies. In most cases investigated, the mode of inheritance was monohybrid (Lesins & Erac, 1968a; Lesins et al., 1970; Lesins & Singh, 1973). Considering that different accessions have different spine lengths, there was little doubt that the character was governed by more than one pair of alleles. This indeed was found in *M. rotata* x *M. bonarotiana* (Lesins et al., 1976). Some variation in spininess in pods of the same plant and even in coils of the same pod has been observed by Heyn (1963) as well as by us.

A dividing line between tubercled and spiny pods on the one hand and tubercled and spineless (smooth) pods on the other is sometimes difficult to draw, especially since the pod appearance may change with pod maturation. Thus, on one occasion in *M. daghestanica* we classified pods in a raceme as spineless when green, but had to reclassify them as spiny at maturity. The reverse may also be true when short spines are overgrown by spongy tissue at pod maturity, as in *M. turbinata*. A curiosity regarding spininess was noted in *M. intertexta*. There, on a single plant, pods with spine-deficient coils were observed; also in its progeny some similarly deficient pods were found (Fig. 3,-9). No regular inheritance pattern, however, could be established. There was no difference between progenies from seeds taken from spineless and spiny coils as to number of ovules per pod, number of mature seeds per pod, or in viability of pollen; latter, if present, would have indicated some genetic change in the reproductive cell line. It may be assumed that the cause for spineless coils was somatic mutations of a periclinal nature. The particular plant and some of its progenies probably had some genetic background permitting such somatic mutations to take place. A conclusion that may be drawn is that a single mutation can change spiny into spineless state, since two mutations taking place simultaneously is of a rather low probability.

The dorsal suture is located in the middle of the coil edge (Fig. 3,-7a), along which, in a few taxa (*M. falcata*), dehiscence of the pod takes place. The coil edge is one of the most species-characteristic morphological features of the plant. It may be wide as in *M. coronata* and *M. praecox*, where it is sometimes called the dorsal plate (region), or it may be paper-thin as in *M. orbicularis*; it may be sculptured with grooves and ridges, with the dorsal suture elevated or lying in a groove. The coil face has a number of facial (radial) veins or nerves (Fig. 3,-8a) which radiate outward over its surface from the ventral (placental) suture. Veins often have species-characteristic value with regard to their number, curving, branching, netting (anastomosing), and also their prominence. In this connection it should be mentioned that, in section *Pachyspirae*, the facial veins are often covered with spongy tissue which should be cleared off in order to see the veins, which in some types are hard to detect. Facial veins may end in a lateral (submarginal) vein (Fig. 3,-8c) lying parallel to the dorsal suture; in other forms they may run into the dorsal suture, or into the bases of the spines, or they may end in a veinless zone on the outer part of the coil face (Figs. 55,-f and 57,-e). In veinless-zone pods, the lateral vein is in the margin of coil edge.

Where a lateral vein is seen separately, its location usually has taxonomic value. It may be on the same level as the dorsal suture, or above or below it. A groove between the dorsal suture and the lateral vein (Fig. 3,-8b) may also be characteristic. A veinless zone is present in two species of section *Pachyspirae* (*M. murex* and *M. turbinata*) and in two of section *Leptospirae* (*M. tenoreana* and *M. disciformis*). A veinless zone, observed superficially, appears sometimes to be present in *M. doliata*; when cleared of spongy tissue, however, some veins can usually be seen to traverse this area and enter a lateral vein on the edge of the coil. Moreover, in species with a true veinless zone it is darker than the central part of the coil face and appears to consist of a horny substance.

The number of coils per pod is characteristic in some species. *M. radiata*, *M. arborea*, *M. rupestris* and *M. shepardii* have less than two coils. Within *M. tornata* there is one form with consistently less than two coils per pod, but some other forms have five and more coils per pod. Sometimes there may be a variation in the same plant, depending on how many ovules are developing. In predominantly cross-pollinated species this, in turn, depends on

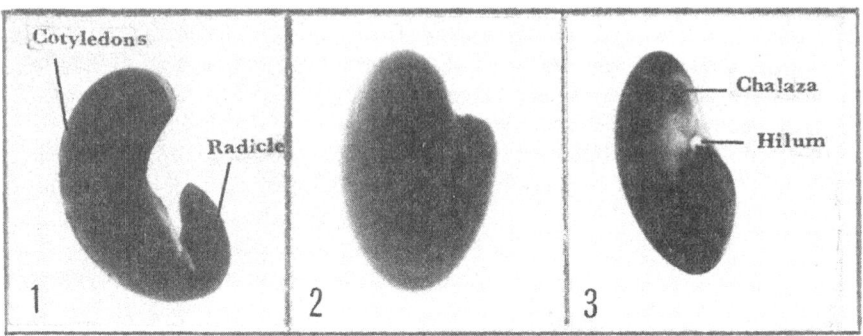

Fig. 4. Seed and its parts: 1) seedcoat removed (*M. tornata*); 2) seedcoat on, side view (*M. falcata*), 3) front view (*M. granadensis*).

whether florets are self- or cross-pollinated. Thus, in *M. sativa*, cross-pollinated florets developed more coils per pod and more seeds per coil than selfed ones (Lesins, 1950). In self-pollinated species, the number of developing seeds very likely also influences the number of coils, though conditions in nature which are influential in the number of developing ovules have not been studied. Our observations at hybridization of self-pollinated species showed that pods with fewer seeds had also fewer coils.

*Seeds* (Figs. 4, 5 and 6). Seedcoats in most species are smooth. In *M. orbicularis* and *M. ruthenica* they are verrucose (warty, Fig. 5,-30 and Fig. 6,-45), and in *M. radiata* and *M. heyniana* they are, in addition, ridged (Fig. 6,-39 and -62, resp.). In seeds freed from the seedcoat (Fig. 4,-1), cotyledons (seed leaves) and radicle (primary rootlet) are readily seen, while the epicotyl (primary stem) is hidden between the cotyledons. Outlines of cotyledons and radicle are discernible on the intact seed (Fig. 4,-2); they are useful in species identification. Of value in recognizing one of the subgenera is the orientation of long axes of radicle and seed in the pod: In *Orbicularia* the long axis is at about the right angle (in *M. carstiensis* at 45°) to the ventral suture. In this subgenus, moreover, the radicle is equal in length to the cotyledons (Fig. 5,-10,-30,-34, and Fig. 6,-39,-45,-62), except in *M. carstiensis* where it is about two-thirds the length of the cotyledons (Fig. 5,-6). In the other three subgenera the long axis of the radicle is parallel or nearly parallel to the ventral suture, and its length varies. It may be shorter than half the length of the seed (Fig. 4,-1), or longer (Fig. 4,-2), or about half the length (Fig. 4,-3). The radicle may be bent away from the cotyledons at a certain angle, or follow closely the outline of the cotyledons, which in this region may be straight or incurvate. The tip of the radicle may form a distinct outgrowth away from cotyledons (the beak, e.g., Fig. 5,-12), or it may remain appressed to them (e.g., Fig. 5,-5).

On the surface of the seed (Fig. 4,-3) two spots are discernible: the hilum (scar), where the seed has been attached to the placenta, is usually round in outline and located close to the tip of the radicle; and the chalaza,

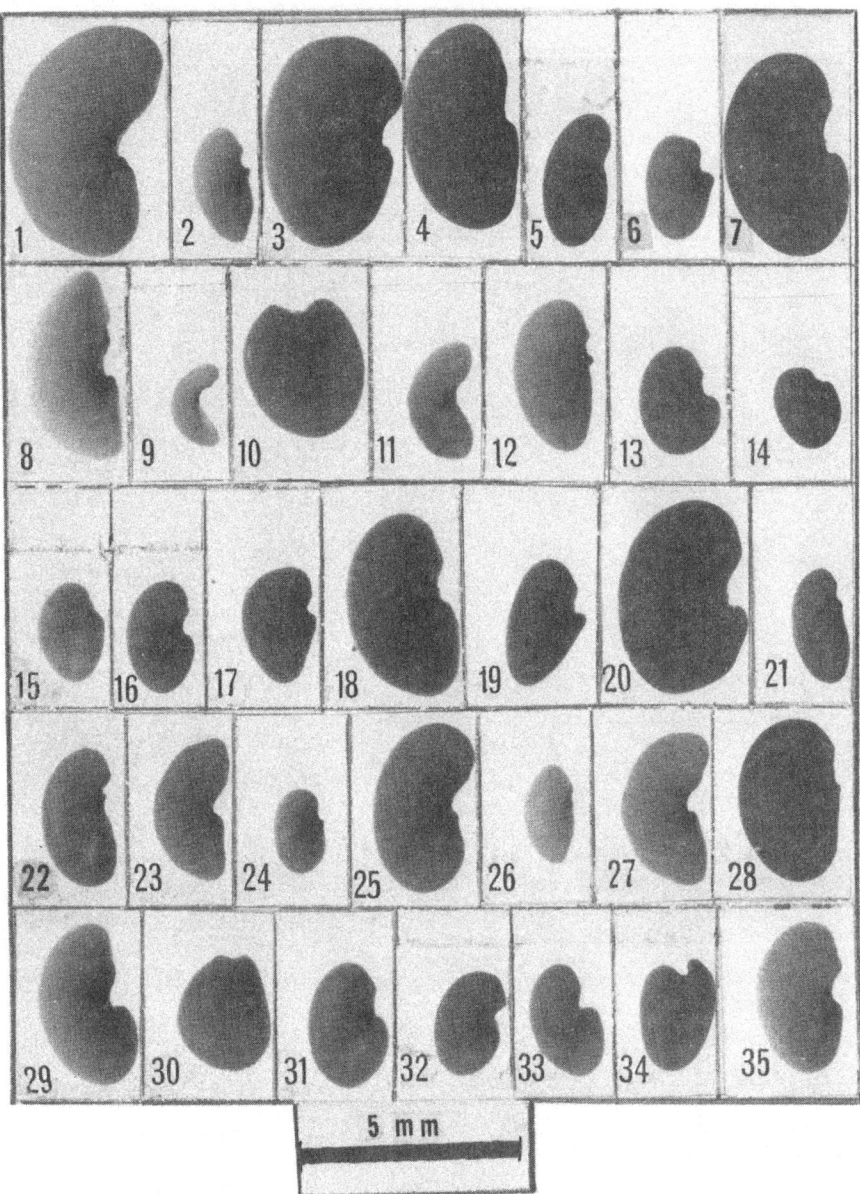

*Fig. 5.* Seeds of *Medicago* species: 1, *doliata*; 2, *arabica*; 3, *arborea*; 4, *bona-rotiana*; 5, *cancellata*; 6, *carstiensis*; 7, *ciliaris*; 8, *constricta*; 9, *coronata*; 10, *cretacea*; 11, *daghestanica*; 12, *disciformis*; 13, *dzhawakhetica*; 14, *falcata* $2n = 16$; 15, *falcata* $2n = 32$; 16, *glomerata*; 17, *glutinosa*; 18, *granadensis*; 19, *hybrida*; 20, *intertexta*; 21, *laciniata*; 22, *lanigera*; 23, *littoralis*; 24, *lupulina*; 25, *marina*; 26, *minima*; 27, *murex*; 28, *muricoleptis*; 29, *noëana*; 30, *orbicularis*; 31, *papillosa* ssp. *macrocarpa* $2n = 16$; 32, *papillosa* ssp. *microcarpa* $2n = 32$; 33, *pironae*; 34, *platycarpa*; 35, *polymorpha*.

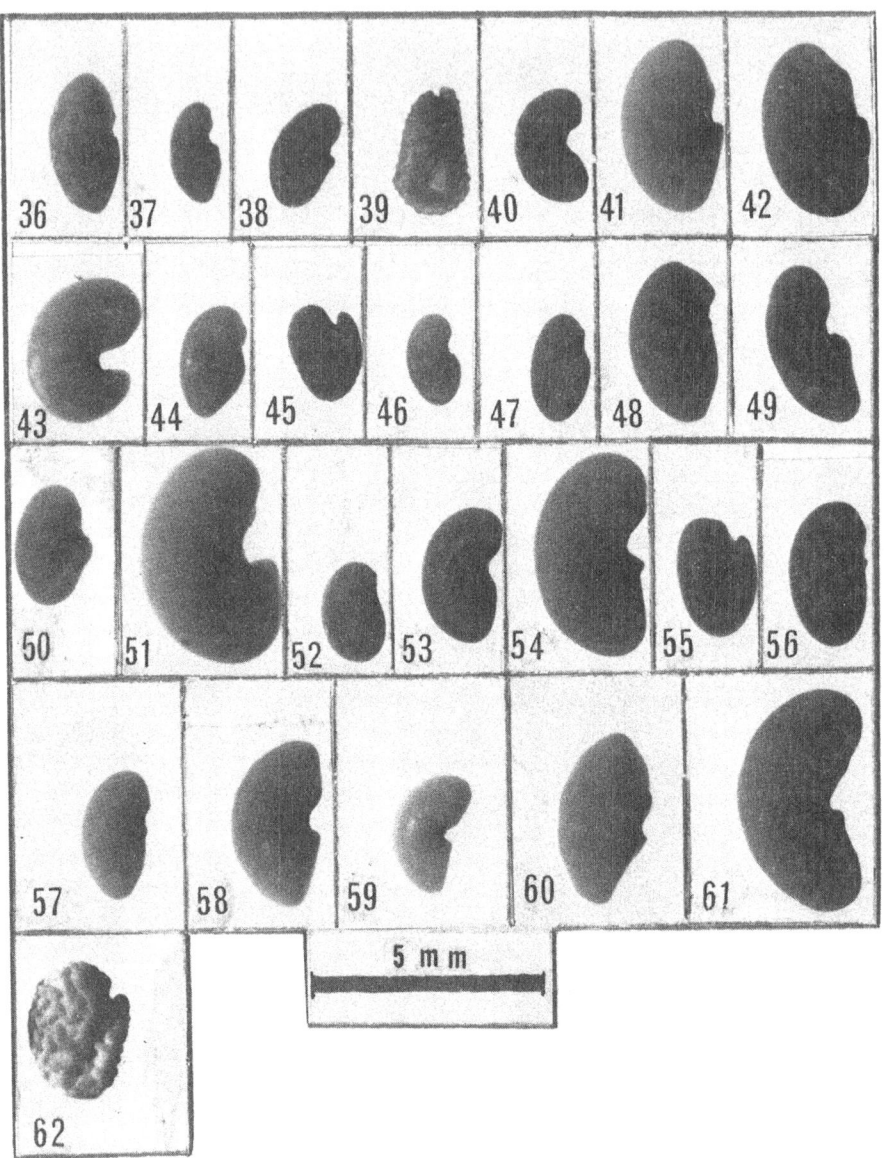

*Fig. 6.* Seeds of *Medicago* species: 36, *praecox*; 37, *prostrata* 2n = 16; 38, *prostrata* 2n = 32; 39, *radiata*; 40, *rhodopaea*; 41, *rigidula*; 42, *rotata*; 43, *rugosa*; 44, *rupestris*; 45, *ruthenica*; 46, *sativa* 2n = 16; 47, *sativa* 2n = 32; 48, *sauvagei*; 49, *saxatilis*; 50, *karatschaica*; 51, *scutellata*; 52, *secundiflora*; 53, *shepardii*; 54, *soleirolii*; 55, *suffruticosa* ssp. *suffruticosa*; 56, *suffruti-cosa* ssp. *leiocarpa*; 57, *tenoreana*; 58, *tornata* var. *tornata*; 59, *tornata* var. *striata*; 60, *truncatula*; 61, *turbinata*; 62, *heyniana*.

where the embryo is connected to the outer covering (integument). Around these spots the color may differ from the rest of the seedcoat. A differently colored stripe of characteristic shape may join the two spots. The place where imbibition of water naturally starts is the chalazal area. It should be noted that artificial scarification for the purpose of enhancing imbibition and germination usually involves sides of the seed, that is, different parts of the seed than the natural place of water intake. As a consequence, seeds tend to swell disproportionately, often breaking the radicle from the cotyledons (especially in thick-coated seeds like those found in section *Intertextae*).

Cotyledons of developing seedlings in *Medicago* are, in only a few cases, of taxonomic value (almost sessile in *M. lupulina* Fig. 9,-2, or rather long as in *M. laciniata* Fig. 9,-7), but are useful as intergeneric characters (Fig. 9,-1 to -6). Distinction between taxa is facilitated by taking into account the various details of seed morphology and appearance. Seed shape, however, may be influenced by the available space within the pod during seed development. Thus, three ovules instead of two, developing in a coil of the same size, will be pressed together, with the result that the seed shape will assume a triangular rather than a kidney-shaped form (Fig. 54,-g).

Seed size varies greatly within the same species; still several species are characteristically rather small-seeded (e.g., *M. coronata*, *M. minima* Fig. 5,-9 and -26, resp.), and others are large-seeded (e.g., *M. doliata*, *M. arborea* Fig. 5,-1 and -3, resp.). The weight per 1000 seeds given in the descriptions, together with the scale provided with illustrations of seeds (Figs. 5 and 6) give a fair idea about the size of the seeds. It may be noted that seed size may vary within the same pod, so that seeds in the apical coils may be half or one-third the size of those in the middle coils (e.g., in *M. doliata*). In species descriptions, the range of seed weights is given where two or more accessions were tested, and 'about' or 'approx.' is said if only one accession was weighed.

Seeds may be separated within a pod by a wall or partition (*septum*) of spongy (parenchymatous) tissue (Fig. 37,-e), or they may be only a thin membranous partition, or none. Between the two latter conditions there may be difficulty in assessment, since in dry pods the partition may have been broken, hence difficult to detect.

The time required for apparently ripe seeds from the same plant to become germinative may vary over an extended period: some may germinate in a week's time, others only after several months (Lesins et al., 1976). This physiological character of uneven germination is different from that caused by differences in seedcoat permeability, reported as softening of seed by McComb & Andrews (1973). In our experiments all seeds were scarified simultaneously, making their seedcoat permeable so that they imbibed water and swelled in a few hours. This imbibed-seed dormancy obviously is genetically determined as a means for enduring long periods of unfavorable conditions permitting repeated attempts of a pod to establish progeny. The earliest germinating seed will produce a rootlet, thus strengthening the pod's anchorage even though the plantlet itself may not survive. Later germinating seeds, due to their physiological unripeness or impermeable seedcoat, would have a firmer foothold and better chances of establishing themselves and reaching the fruiting stage. In herbaria and while digging up plants, occasionally a pod may be seen attached to the upper part of the root. When opened, the pod may contain some germinative seeds. The differences in seed dormancy may aid in survival on barren hillsides, or in dry sandy soil, or when flooded, to which condition especially annual *Medicago* appear to have been tailored during evolution. Pods dropped in a protected place may retain viable seeds for a very long time (see description of *M. orbicularis* and *M. heyniana*).

*Pollen* (Figs. 7, 8). The appearance of natural pollen was studied on grains mounted dry or immersed in paraffin oil. Spindle-shaped or cylindrical

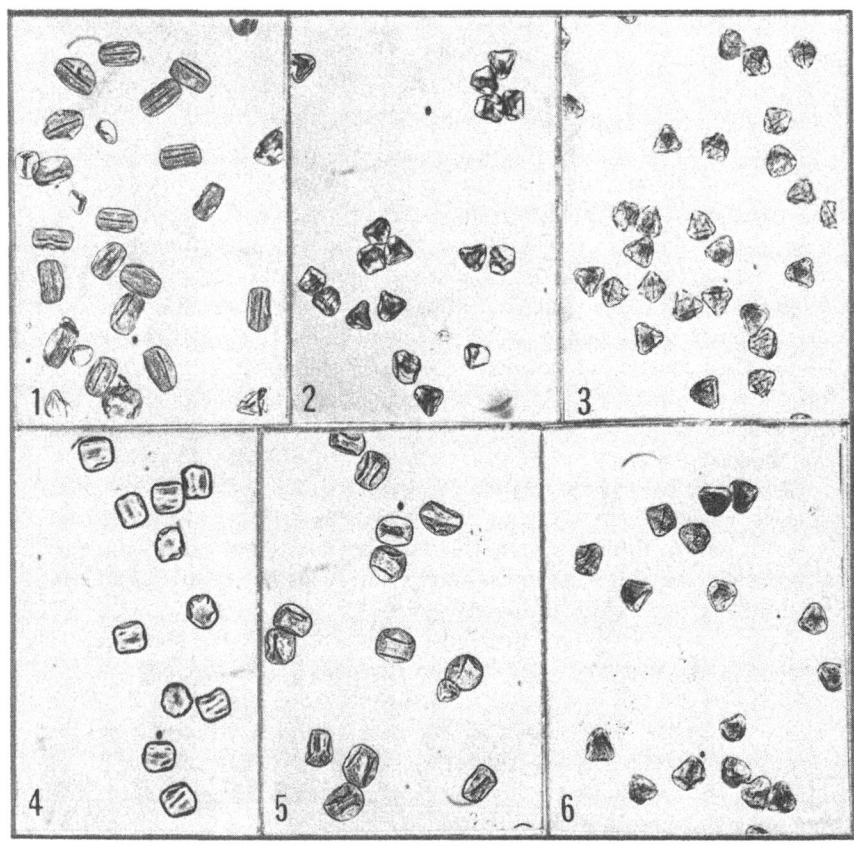

*Fig. 7.* Unexpanded (natural) pollen grains: 1) *M. turbinata*, 2) *M. granadensis*, 3) *M. suffruticosa*, 4) *M. rotata*; 5) *M. rigidula*, short cylinders; 6) *M. rigidula*, angular pyramids.

28

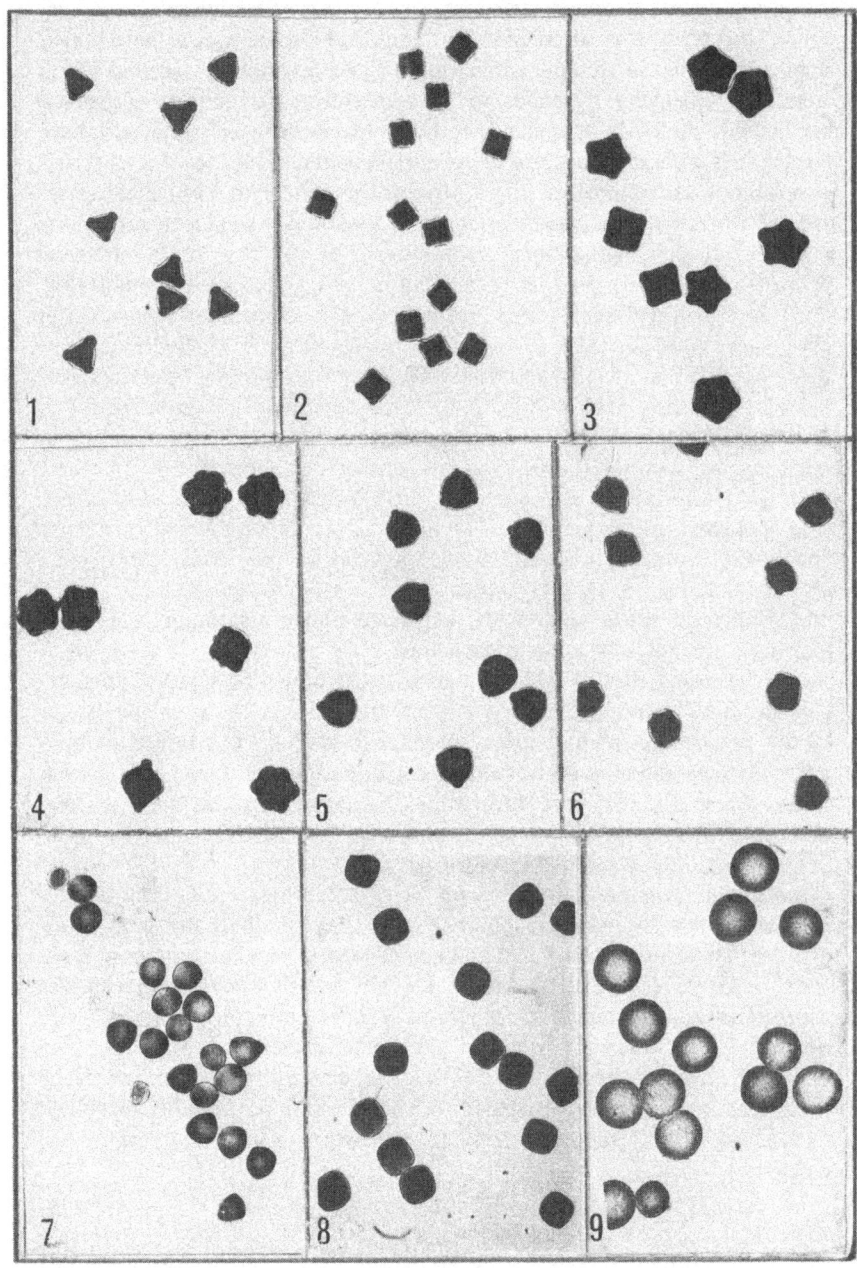

*Fig. 8.* Nos. 1-6, autoclaved pollen grains: 1) *M. rupestris*, 2) *M. granadensis*, 3) *M. bonarotiana*, 4) *M. suffruticosa*; 5) *M. rigidula*, from short-cylinder pollen; 6) *M. rigidula*, from angular pyramids. Nos. 7-9, swollen (expanded) pollen grains: 7) *M. sativa*, 8) *M. granadensis*, 9) *M. hybrida*.

grains with three germinal apertures, seen as three furrows along the long axis of the grain, were most prevalent (Fig. 7,-1); some species have pollen shaped more like a spindle, others more like a cylinder. A second group comprised triangular pyramids or an admixture of these with tetragonal bisphenoids similar in appearance to pyramids. Some species may also have bisphenoids with an admixture of cylindrical pollen. In *M. rigidula* we found accessions with cylindrical (Fig. 7,-5) and pyramidal to bisphenoid grains (Fig. 7,-6). A few species (*M. rotata, M. bonarotiana* and *M. shepardii*) have characteristically block-shaped grains (Fig. 7,-4) with four or five germinal apertures (Fig. 8,-3). In *Intertextae* the natural grains are pyramids (Fig. 7,-2) not distinguishable from other taxa with pyramidal pollen. After expansion, however, they become cube-shaped (Fig. 8,-8), different from those of all other taxa. On autoclaving, their four germinal apertures become very clearly visible (Fig. 8,-2). This technique was also helpful in disclosing intrinsic differences in *M. rigidula* between cylindrical and pyramid-shaped grains: the former type after autoclaving showed three apertures (Fig. 8,-5), the latter four to six (Fig. 8,-6). Expanded grains of *M. sativa* (Fig. 8,-7) and other members of section *Falcago* have a rounded-triangular appearance. After autoclaving, three apertures become conspicuous (e.g., *M. rupestris* Fig. 8,-1). Dry pollen in section *Suffruticosae* have a pyramid-bisphenoid shape (Fig. 7,-3); expanded pollen are round (Fig. 8,-9), but their six colpi become conspicuous after autoclaving (Fig. 8,-4). It should be noted that in watery substrates the pollen burst; also, once expanded under excessive humidity (as sometimes occurs in the greenhouse or in the preparation of herbarium specimens), they do not return to their original shape and so are not suitable for identification. Pollen taken from flower buds also may not have quite the same shape as those in fully developed flowers. Investigations in pollen of the taxa available to us up to 1963 have been published (Lesins & Lesins, 1963).

*Chromosome Number.* There is no taxonomic character as qualitatively distinct as the chromosome number, and although there are only a few different chromosome numbers in *Medicago*, they have been of great assistance in identification of the taxa. On several occasions where other species characteristics overlapped, chromosome counts have solved the problem. The somatic number, $2n$ found in meristems, usually of roots, of all taxa described here are presented in Table I. The determinations were made on accessions in our collection, except for one form in *M. lupulina* for which Tschechow (1932) reported $2n = 32$, and which we do not have in our collection.

In some species, if we suspected differences in ploidy level or in chromosome number, almost all accessions (more than 90 in *M. falcata*) were tested. In some others, one or only a few determinations were made depending on the number of accessions available. The original technique for chromosome investigation (Lesins, 1954) has been somewhat altered during recent years and is described under Procedural.

Data in Table I may require some additions as further material will be

Table I. Chromosome number (2n) in *Medicago* species

| Species epithet | 2n | Species epithet | 2n |
|---|---|---|---|
| *arabica* | 16 | *muricoleptis* | 16 |
| *arborea* | 32, 48 | *noëana* | 16 |
| *bonarotiana* | 16 | *orbicularis* | 16 |
| *cancellata* | 48 | *papillosa* | 16, 32 |
| *carstiensis* | 16 | *pironae* | 16 |
| *ciliaris* | 16 | *platycarpa* | 16 |
| *constricta* | 14 | *polymorpha* | 14 |
| *coronata* | 16 | *praecox* | 14 |
| *cretacea* | 16 | *prostrata* | 16, 32 |
| *daghestanica* | 16 | *radiata* | 16 |
| *disciformis* | 16 | *rhodopea* | 16 |
| *doliata* | 16 | *rigidula* | 14 |
| *dzhawakhetica* | 32 | *rotata* | 16 |
| *falcata* | 16, 32 | *rugosa* | 32 |
| *glomerata* | 16 | *rupestris* | 16 |
| *glutinosa* | 32 | *ruthenica* | 16 |
| *granadensis* | 16 | *sativa* | 16, 32 |
| *heyniana* | 16 | *sauvagei* | 16 |
| *hybrida* | 16 | *saxatilis* | 48 |
| *ignatzii* | 16 | *scutellata* | 32 |
| *intertexta* | 16 | *secundiflora* | 16 |
| *laciniata* | 16 | *shepardii* | 16 |
| *lanigera* | 16 | *soleirolii* | 16 |
| *littoralis* | 16 | *suffruticosa* | 16 |
| *lupulina* | 16, 32 | *tenoreana* | 16 |
| *marina* | 16 | *tornata* | 16 |
| *minima* | 16 | *truncatula* | 16 |
| *murex* | 16, 14 | *turbinata* | 16 |

investigated, and strains with different ploidy levels may turn up. Some further information on chromosome morphology may be expected, although, due to the small size of chromosomes (length 2 to $4.5\mu$), this is not likely to be applied to any considerable extent to the routine identification of species. In special studies of pachytene chromosomes, however, Gillies (1972a, b, c) has shown that in species of section *Falcago* which hybridize with *M. sativa*, chromosomes homologous to those of *M. sativa* could be identified, whereas non-hybridizing species (of section *Suffruticosae*) had morphologically different chromosome sets. Our attempts to find differences in chromosome banding pattern using Giemsa and acridine stains have not, so far, been successful (unpublished).

## HABITAT AND DISTRIBUTION

*Habitat.* The aim of our expeditions for collecting *Medicago* was to establish living plants at our home base, the University of Alberta, Edmonton. Consequently, we were after samples of seeds. This in turn determined the

31

time of expeditions, that is, the latter part of summer when pods would be shed and could be picked up from the ground. Hence notes on habitats and on some other items are somewhat different from those taken by plant collectors for herbarium specimens.

In the latter part of summer annual *Medicago* usually are dried up. Also, the weather and soil conditions in that period are dry, so not much variation in soil moisture could be noted. Regarding composition of soils, however, obvious peculiarities such as very sandy, of volcanic origin, or heavy clay soils were recorded. Regarding growing sites, rocky hillsides, underbrush, or along roadsides were noted. No notes were taken on composition of plant communities associated with particular *Medicago* accessions. This was because plants associated with *Medicago* at its active growing time often were also dried up and would be difficult to locate; moreover, we felt that although characterizations of plant communities may give some ideas on growing sites to botanists familiar with local floras, they would be of little value to workers interested primarily or exclusively in *Medicago*. Understandably, because of the late season, no notes could be taken on flowering time, an item regularly recorded by collectors for herbaria. About an accession, we usually recorded the place of collection as the distance from one populated town in the direction to another.

Collection sites often were at roadsides. Regarding roadside collections, it should be pointed out that different species and different varieties of a species were often found together. A few paces from the roadside there would not be any *Medicago* at all, though the surroundings would not have been disturbed by cultivation. The probable reasons for this are twofold. The first is the moisture regime: the roads are usually graded or paved, so that the rainwater flows off into deeper or shallower roadside ditches. Second, roadside soils are disturbed by roadbuilding, so in places deeper soil layers are exposed; in others they are mixed with the upper layers. Hence roadsides are well provided with moisture and, in addition, offer a wide diversity of soil composition that may suit the various taxa. Some species, e.g., *M. polymorpha*, thriving under higher moisture conditions than many other annual *Medicago*, is an often-found roadsider. The variability within a species along roadsides may be understood considering the long life of *Medicago* seeds and the easier spread of pods and seed along roads by animals and traffic. Once a plant has established itself and scattered seed in a place, the probability is great that another form would sooner or later join it. Self-fertilization in annual *Medicago* being the mode of reproduction assures the continuation of variability in this group of *Medicago*. In conclusion, a general assumption based on roadside collection that a particular area with its soil, temperature and precipitation indices is characteristic for a certain *Medicago* may be somewhat misleading.

Of course, environmental conditions at roadsides are not optimal, nor even tolerated by all species; for instance, *M. marina* grows only in seashore sands.

The environment of wild species changes continuously. In recent time, man-influenced changes of habitat have been rapid and drastic. Many species have almost suddenly been deprived of their natural habitats and are on the verge of extinction. The list of endangered species is growing alarmingly. In our expeditions to a number of Mediterranean countries we have seen, only too often, depleted, over-grazed areas where some centuries ago the ground had been well-covered with vegetation, often forested. On the cleared land herds of goats and sheep snap off every edible shoot so that the land becomes desert-like. A herd of goats roaming a rocky hillside, a habitat preferred by many annual *Medicago*, is shown in color plate II,-a (facing p. 35). This particular photo was taken on the Island of Karpathos. This small island in the Aegean Archipelago, also mentioned later, is the only place where a rare *Medicago* species, *M. heyniana*, is growing. The goats devour not only green shoots but search out ripe *Medicago* pods fallen on the ground. Perennial *Medicago*, too, are endangered by overgrazing, as they cannot mature seeds. We sought in vain for some plants with pods of the N. African *M. sativa*-related taxa (*M. sativa* ssp. *tunetana*, ssp. *faurei*, and *f. gaetula*). All we could find in their growing sites were some short-bitten prostrate stems. Deprived of the possibility to ripen seeds and weakened by continuous defoliation, such plants eventually succumb to diseases. Perennial though they are, their life-span is not unlimited.

Some taxa, however, may take refuge in niches unintentionally provided by man. Thus, *M. suffrutiscosa* ssp. *leiocarpa* could be collected between stones and rocks left over after roadbuilding, whereas in surrounding pastures the plants were not found or were in a rather weakened condition. Of some importance for plant survival along roadsides, in addition to previously mentioned reasons, is the present heavy motorized traffic which compels the herdsmen to keep their flocks away from roads because of casualties among the animals. Some partially protected niches, such as old graveyards, as in color plate II,-b, from Tunisia, as well as other tourist attraction sites, such as ruins of ancient settlements and temples, are quite rich in annual *Medicago* species. Probably because goats are not allowed to browse on Athens' Acropolis and the Lycabettus Hill, there are still to be found shrubs of *M. arborea*, whereas in the countryside they are only rarely to be seen.

Observing the effectiveness of small protected niches in preserving plant diversity, we feel that well-fenced areas of a few square meters, strategically placed across the countryside (close to the everyday routes of forest rangers), would serve the purpose of protecting some plant taxa from extinction. It may be noted that small enclosures may not be as objectionable to users of community pastures (whose votes are often important for elected politicians), as would be large sanctuaries, pasture rejuvenation, and reforestation projects. We realize that in a number of countries in many households goats and sheep are the most important domestic animals. However, some long-range solution between preserving mankind's natural resources, including plant diversity, and the present-day needs, have to be worked out soon, before it is too late.

*Distribution.* To indicate the areas over which natural and sometimes intro-

*Color Plate II.* Goats roaming the countryside on Island Karpathos (a), such landscape is typical for many annual *Medicago*. Carthaginian graveyard in Tunisia (b), protected places like this serve as *Medicago* refuge sites.

34

duced species are distributed, we have cited information found in the literature and have noted those countries where we collected particular taxa.
Occasionally we have pointed out exactly where a certain accession was
found, if it was an unusual one and appeared to be of interest to other
collectors. This was done because we have often received inquiries from
research workers interested in certain characteristics of a particular accession, who were planning to make additional collections.

Regarding dispersal: Plants generally are considered spotbound and less
mobile than most animals. This scarcely applies to dispersal of *Medicago*
seeds. In that respect their hard seedcoat plays an important role. We im-
merse seeds in concentrated sulfuric acid for 10 to 12 minutes to make the
seedcoat permeable, and that treatment does not damage the seed viability.
Thus, seeds eaten but not crushed by animal teeth are passed through their
digestive tract without impaired germination. Anyone who has used manure
for garden plots has found that weeds, wild Leguminosae among them,
appear in abundance for many years after application. *Medicago*, especially
annual species, are rather prolific seed producers, the first pods maturing
while top and side branches continue to grow and flower. Animals feeding
on green parts also take in ripe seeds which are excreted in animal drop-
pings and in turn searched for seeds by birds. Thus, in a chain dispersal,
some transfer of seeds may occur over long distances and over large bodies
of water. *Medicago* growing on seashores may have their pods water-trans-
ported, since in the pods some air is entrapped, making them floatable for
some time. Even such inland species as *M. lupulina* and *M. minima* are
reported waterborne for at least one day (Romell, 1954).

One has to keep in mind, however, that dispersal and establishment in a
new environment are quite different matters. Newcomers generally fail to
establish themselves in places already occupied by plants adapted to the
local conditions. It is only in disturbed and unoccupied environments that
they may have a chance to become permanently established, because in such
places they have time to accumulate genetic changes necessary to cope successfully with requirements of the new environment.

INTRASPECIFIC VARIATIONS, DISTINGUISHING FEATURES AND
RELATIONSHIPS BETWEEN SPECIES

We have listed intraspecific taxonomic ranks and their main characteristics
as reported by Urban (1873), Heyn (1963), and by some other authors. The
majority of those variations are represented in our collection, though we did
not particularly attempt to identify our accessions with the formally named
intraspecific taxa. Some variations, however, if conspicuous, or useful as
genetic markers, or important in challenging or supporting certain species
relationships, we have described and illustrated. We have also reported observed inheritance patterns, and on a few occasions have suggested changes
in rank or proposed new taxonomic names.

We became keenly aware of the need for emphasizing distinguishing characteristics for positive species identification after we received, in response to requests for *M. polymorpha* (as *M. hispida*), 47 samples from several sources. Half of the samples were wrongly identified (Lesins & Lesins, 1962). More recently (1975) we have received from an institution a *Medicago* seed collection consisting of 21 samples, 11 of which were wrongly identified. Thus, in our treatment of species here, whenever a confusion was likely, we have tried to point out the differences easily observable on herbarium specimens, on live plants, or those requiring detailed studies for their detection.

As stressed in discussing the grouping of taxa, we considered hybridization tests as the most valuable tool for assessing relationships between species. The results of such tests carried out by us or other researchers are included here.

AGRICULTURAL VALUE

The world's most important forage crop, alfalfa, belongs to the genus *Medicago*. The species cultivated in the North Temperate Zone is usually *M. x media*, known also as *M. x varia*; in the South it is *M. sativa*.

Alfalfa is continuously threatened by different pests and diseases. Its expanded cultivation in less suitable areas would be desirable, and improvements in quality, such as non-bloating, and in increased yielding capacity may be needed. All of these problems can be solved best by plant breeding.

For breeding of a crop, it is important to know first, what germ plasm donors are available and second, what characteristics of value these possess for incorporation into the crop. We have tried to answer the first question, that of possible germ plasm donors, by testing the crossability of all perennial and a number of annual *Medicago* species with *M. sativa*. Most of the perennials could be hybridized. We have put them together in section *Falcago*, and mention is made of their possible agricultural value in discussing specific taxa. None of the annuals could be hybridized with *M. sativa*, even though we usually adjusted the crossing partners to the same ploidy level. With *M. rigidula* we came closest to a successful hybridization between *M. sativa* and an annual species. *M. rigidula*, used as the maternal parent, retained pods on the plant up to 18 days. The seeds, however, had died long before that time, as only small seed scales were found on opening the pods. We were not successful in our attempts to culture the embryos of this cross. We think that reports that hybrids have been obtained between *M. lupulina* and *M. sativa* (Southworth, 1928; Schröck, 1943) are based on mistaking inbreeding-distorted *M. sativa* plants for hybrids.

The second question, what characteristics the taxa hybridizable with *M. sativa* harbor, we have not studied to any great extent. Here, however, George A. Stevenson, Canada Department of Agriculture Research Station at Brandon, Manitoba, has let us have his notes on the performance of perennial species in regard to characteristics of direct agricultural value, such as forage yields and winter survival, and some other observations. He used alfalfa cultivar 'Rambler' as the standard in assessing yielding capacity. We are thankful to him, and have included the information he provided, in discussion of the species.

The research station at Brandon is located at 99°53' W. Long. and 49°50' N. Lat. The climate there can be characterized as continental with an average precipitation of 400 mm/yearly (range 333 to 568 mm) so that the moisture is a critical factor for plant growth. One-third of the precipitation falls during June and July. The mean temperature for July is 19° C, and for January, −18.3° C. The maximum and minimum temperatures recorded in the last 20 years are 37.8° C and −45° C, respectively. The frost-free season is usually from the end of May to mid-September (105 days). The snow cover is usually sufficient for winter-survival of locally-adapted alfalfa varieties, but less cold-tolerant varieties and species suffer during winters with light snow cover.

Edmonton, where a few observations on winterkilling were recorded, is located more to the west and north, 113°18' W. Long. and 53°20' N. Lat. The climate, however, is not as continental because Edmonton is closer to the Pacific Ocean, even though separated from it by the Rocky Mts. Precipitation on the average is 475 mm, and the snow cover is usually quite sufficient for survival of locally adapted alfalfa. The mean temperature for July is 16.1° C, and for January, −16.1° C. The maximum and minimum temperatures recorded in the last 20 years are 34° and −48°. The growing season is somewhat shorter than at Brandon, with 100 frost-free days on the average. The temperature, occasionally lower during the winter than at Brandon, is moderated by a thicker (up to 150 cm) snow cover.

In several countries, notably in the U.S.A., at many institutions, investigations are pursued intensively on the agricultural value of species related to *M. sativa* with respect to insect and disease resistance. We have included such information in the species descriptions, if available.

Currently, annual species are becoming important in agriculturally advanced countries with suitable climatic conditions. In Western Australia alone, ten cultivars are used in rotation with cereals for pastures and as soil conditioners (McComb, 1974).

# ISOLATING MECHANISMS, SPECIATION, EVOLUTION IN *MEDICAGO*, AND RELATIONSHIP TO OTHER GENERA

## ISOLATING MECHANISMS

*Ploidy and Chromosome Rearrangement.* In *Medicago*, as generally in other organisms, ploidy and chromosome rearrangement constitute an effective isolating mechanism. Diploids, tetraploids and hexaploids are found in *Medicago*. Most of the diploids are of the $2n = 16$ type, with a basic chromosome number of x = 8; four species of diploids have $2n = 14$ chromosomes and one has members with $2n = 16$ and $2n = 14$ chromosomes. The origin of the $2n = 14$ types may be understood from the situation prevailing in this latter, *M. murex*, species. There, the basis has been chromosome rearrangement: the union of two chromosomes with the loss of the centromere region of one of the chromosomes involved (Lesins et al., 1970). The segments of both chromosomes can be recognized in the pachytene stage. Exactly how the transformation has occurred is not clear; a sequential two-step change has been assumed in our report, but there may have been some shortcut not readily visualized. If and when a routine cytological method for quick screening of massive numbers of plants becomes available, the process might be more clearly understood. A most surprising finding was that a complete interbreeding barrier existed between these two chromosomally distinct types of plants, which are not readily distinguishable morphologically. We do not know what the environmental conditions have been that have favored the $2n = 14$ type to the extent that the $2n = 16$ type has almost disappeared.

Three other $2n = 14$ *Medicago* species, *M. rigidula*, *M. praecox* and *M. polymorpha*, may have arisen in a similar fashion; they have one exceptionally long chromosome which may be the result of fusion of two chromosomes, as in *M. murex*. The fourth, *M. constricta*, has no exceptionally long chromosome; we have speculated that it may have developed by secondary chromosome rearrangement from $2n = 14$ *M. murex* (Lesins & Gillies, 1972). No $2n = 16$ types in the mentioned four $2n = 14$ species are found, nor are any tetraploid, $2n = 28$, forms known. All of them together with *M. murex* are annuals, three of which (*M. rigidula*, *M. constricta*, *M. murex*) belong to section *Pachyspirae* and two (*M. praecox*, *M. polymorpha*) to section *Leptospirae*. No close relationship seems to exist between these two sections that would suggest a common origin. There are two annual tetraploid, $2n = 32$, species (*M. rugosa* and *M. scutellata*). We have not found their diploid, $2n = 16$, counterparts even though we have searched through most of the accessions in our collection.

A number of perennial species are known to have both diploid and tetraploid forms. Tetraploids probably arise spontaneously from diploids. We have found tetraploid seedlings in progenies from diploid parents in *M. falcata* as well as in *M. sativa* (as *M. rivularis*). An interbreeding barrier is intrinsically established between plants of the two ploidy levels. In areas occupied by both the diploid and tetraploid forms of a species, it appears that the tetraploids are taking over; thus, diploid *M. falcata* and *M. sativa* are found only in isolated patches as relics among the prevailing tetraploids. Probably an additive effect on survival-important characters of four alleles at a locus, rather than only two as is possible in diploids, makes tetraploids more adaptable to varying environments. Since the ploidy level can be changed artificially with restoration of interfertility, such cases may be considered as steps toward, but not yet fully established, speciation.

At the hexaploid, $2n = 48$, level are found two perennial species (*M. saxatilis* and *M. cancellata*). The third (*M. arborea*) has hexaploid and tetraploid types. The first two are probably autoalloploids, with two genomes from some *M. sativa- M. falcata*-related taxa and four from *M. rhodopea* and *M. rupestris*, respectively (Lesins, 1970). The process of speciation in these instances may be thought to have been facilitated by complementary combinations of parental traits that have made them adapted to environmental conditions not fully utilizable by either of the parents. The differences in ploidy level between parents and hybrids provide an efficient barrier against disintegration of newly formed hybrids. Exactly what traits may have been supplied by the parental species, and what environmental conditions might have made complementation so important as to permit the origin of new species, we can only speculate. Some interesting hints on how hybrids may become new species are given by Proctor & Yeo (1973) and Straw (1956): In one instance in genus *Penstemon*, one species pollinated by hummingbirds, and another by large bees, are capable of producing fertile hybrids. A third species which resembles these hybrids is pollinated by wasps. It is very likely that hybrids between the former two species happened to arise in an environment where certain wasps needing nectar and pollen found the flowers of the hybrids satisfying their needs. The wasps would preferentially intercross the hybrids, therewith preventing disintegration by backcrossing to parental species, which with taxa at equal ploidy level might very well have taken place.

The third *Medicago* species, *M. arborea*, only recently has been found to have a hexaploid type; generally it is composed of tetraploids. Taking into account that *M. arborea* is a shrub, the species might be considered phylogenetically the oldest in the genus. There are no known *M. arborea* diploids, however, so the present types may have evolved from diploids as autoploids. The tetraploids and hexaploids grown by us, in regard to morphology of vegetative parts, are quite similar.

*Other Isolating Mechanisms.* Apart from isolating mechanisms based on differences in ploidy and chromosome rearrangement, there are several special cases. The first case we encountered was the noncrossability between *M. sativa* and *M. dzhawakhetica* (*M. papillosa*) at equivalent ploidy levels, where neither diploids with diploids nor tetraploids with tetraploids could be hybridized. It was only by chance that we crossed tetraploid *M. dzhawakhetica* with diploid *M. sativa* and obtained a triploid hybrid (Lesins, 1952). Later, more hybrids were obtained, all of which had the genomic ratio of two genomes from *M. dzhawakhetica* to one from *M. sativa*. Of particular interest was the finding that it did not matter whether the diploid *M. sativa* was the maternal or paternal parent (Lesins, 1961a). This was different from what we observed in another species cross mentioned later, where a certain partner had to be on the female side in order to get fully viable seeds. Under what conditions the isolating mechanism arose, and in what way it prevents hybridization is difficult to imagine. It is worthwhile noting that, in later generations of the *M. dzhawakhetica* x *M. sativa* cross, artificial hybrids of different, even reversed, genomic ratios were produced; thus the crucial phase of noncrossability is in the built-in repulsion of natural equivalent-ploidy plants.

A second special isolating mechanism consisted of a non-lethal chlorophyll deficiency (yellowish foliage) if one of the crossing partners was the pistillate parent. Though non-lethal under greenhouse conditions, it effectively reduced seed production. Thus, here was a maternally based, one-sided hybridization antagonism. This was observed in the above *M. sativa* x *M. dzhawakhetica* cross (Lesins, 1961a); it was found later in *M. tornata* (as *M. striata*) x *M. littoralis* (Lesins & Erac, 1968a), and still later in *M. laciniata* x *M. sauvagei* (Singh & Lesins, 1972). Lilienfield (1962) reported a similar phenomenon in crosses involving two *M. truncatula* strains, so that it appears that this kind of isolating mechanism is comparatively common in *Medicago*. The inheritance was analyzed in some detail in the above mentioned *M. tornata* x *M. littoralis* hybrids, and it was concluded that one cytoplasmic and three chromosomal genes were responsible for the observed segregation ratios. In the *M. laciniata* x *M. sauvagei* cross (Singh & Lesins, l.c.) it could be postulated that *M. sauvagei*, an endemic of Morocco, is the more likely species to have evolved the antagonistic mechanism, since in the few localities where it is growing it is surrounded by *M. laciniata*. Thus, without such a hybridization barrier, *M. sauvagei* would not be able to maintain its identity. Under what conditions *M. sauvagei* as a species was formed, how the isolating mechanism arose, and whether it had a role in the species origin in the first place, we have no clue.

A third kind of isolating mechanism was found between *M. prostrata* and *M. sativa* (Lesins, 1962). Here, fertility was low and seeds were poorly developed if *M. prostrata* was the female parent, whereas in the reciprocal cross, seedset was good. The results were the same whether these species were crossed at the diploid or at the tetraploid level. It is highly probable that the mechanism involved lies in the differences of the genomic constitution of endosperms. In *M. sativa* x *M. prostrata* on double fertilization, the endosperm has 2 *M. sativa* : 1 *M. prostrata* genomes and in the reciprocal cross it has 1 *M. sativa* : 2 *M. prostrata*, while in both crosses the embryo has the same genome ratio, 1 *M. sativa* : 1 *M. prostrata*. In legumes, the endosperm is the food supplier for the embryo until its supplies are used up, so that in a mature seed almost only the embryo is present. The poorly developed, shrivelled seeds suggested that the endosperm in this latter *M. prostrata* x *M. sativa* combination is not an effective provisioner of food for the embryo. Although this type of isolating mechanism is often encountered in plants, nothing is known regarding the process involved, and it is difficult even to guess under what conditions the two species have been competing to the extent that such an isolating mechanism could have arisen. As in the previous cases, it is basically a one-sided hybridization antagonism against one of the partners involved.

Other kinds of isolating mechanisms seem to be localized in individual chromosomes or segments. Their manifestation depends on the presence versus the absence of a chromosome, or the ratio of the carrier chromosomes to the rest of the chromosomes. In *M. sativa* ($2n = 32$) x *M. rhodopea* ($2n = 32$), from more than 500 florets crossed, only a single hybrid was obtained, which had $2n = 31$, and thus was an aneuploid with one chromosome missing (Lesins, 1972). Similarly, in *M. sativa* ($2n = 48$) x *M. cancellata* ($2n = 48$), from 200 florets crossed, a single hybrid with $2n = 46$ (Lesins, 1961b) and from an additional 76 crosses, another with $2n = 45$, were obtained; these hybrids therefore had two and three chromosomes missing, respectively. Finally, in a trispecies cross [*M. pironae* x *M. daghestanica* ($2n = 32$)] x *M. sativa* ($2n = 32$), two hybrids were obtained from 400 crosses, one with $2n = 30$, the other with $2n = 34$; thus two chromosomes were missing, or two extras added, therewith changing the genic ratio (Lesins, 1971).

An isolating mechanism, developed to the extend that in $F_1$ no viable gametes were produced, was found between *M. pironae* and *M. daghestanica*, both with $2n = 16$ (Lesins & Gillies, 1968). The $F_1$ hybrids could easily be produced and were of vigorous growth. Chromosome pairing at metaphase I was such that in more than half (55%) of the pollen mother cells some deviation from 8 bivalents was recorded. A common origin and relationship was indicated by formation of 8 bivalents in 45% of the cells analyzed. The resulting pollen, however, was 99-100% protoplasm-empty.

Another $F_1$, that from a cross of annual *M. turbinata* x *M. truncatula* (Lesins et al., unpubl.), was also completely sterile, producing only a few stainable pollen grains; moreover the hybrids, though flowering readily, were initially less vigorous than the parents and had the chlorophyll deficient, yellow-green foliage; with time, however, they measured in size up to their parents. Under what conditions these isolating mechanisms have arisen is difficult to imagine.

Death of hybrid seedlings due to lethal chlorophyll deficiency has been found in the crosses *M. murex* x *M. turbinata* (Lesins et al., 1970) and *M. tornata* x *M. soleirolii* (unpubl.). These pairs of species belong to section *Pachyspirae*. The possibility of obtaining hybrid seedlings indicates that the parental species in each case had a common ancestor. The nature of the isolating mechanism in these cases is difficult to get at. Raising the chloro-

41

phyll deficient $F_1$ seedlings to flowering stage, by grafting onto green seedlings of the parental species, might provide more information; our grafting attempts, however, were not successful.

Complete isolating mechanisms between species where no hybrids can be produced may usually be considered to have arisen deep in the past so that no speculations can be offered as to their nature. There are, however, a few cases in which morphological similarities between taxa are so obvious that there is no doubt about their close relationship. One is the already mentioned case of noncrossability between $2n = 16$ and $2n = 14$ *M. murex*; the other is between *M. granadensis* and the rest of the members of the section *Intertextae* (Lesins et al., 1971; and unpubl.).

At the other end of the spectrum of isolating mechanisms are cases where no isolation is experimentally demonstrable as far as fertilization and viability of progenies are concerned, but where clearly noticeable morphological differences imply a distinct isolation in nature, based on associated differences in physiological processes. Thus, diploid *M. falcata* (Lesins & Lesins, 1964), strain No. 14 (*M. falcata* ssp. *romanica*), growing in sand dunes at Lake Sivash and at the Azov Sea in southern Russia, has taproots penetrating up to 1 m deep into the soil without much branching, and has erect stems. In contrast, strain No. 28 (*M. falcata* var. *borealis*), growing in podzolic soils of northern Russia, has a shallow-spreading root system, and semi-prostrate stems. If the shallow-rooted strain was to be intercrossed with the taprooted one and the progeny grown in sands at Lake Sivash, the hybrids, being as a rule of intermediate character, would not be able either to get their roots down readily to moisture or to elevate their stems above the drifting sand. Conversely, hybrids grown in shallow podzolic soils of northern Russia would not branch profusely enough to maintain competitive growth against locally adapted vegetation, and would not be sufficiently hardy for winter survival. In experimental plots where the extreme environmental conditions described are absent, these physiological responses cannot be detected, especially if grown under generally less severe conditions. The origin of an isolation between strains, as the above, may be explained by such assumptions as: *M. falcata* population from nearby steppes was cut off from the parent population by a water strait. Given time to accumulate locally adaptive features and get rid of detrimental ones, a strain like the one at Lake Sivash may have evolved. As a matter of fact, ssp. *romanica* as *M. erecta* has been described by Kotov (1940) from the Island Biruchyi in the Azov Sea.

The importance of establishing an isolating mechanism by genetic drift, that is, by the accumulation of genetic changes due to chance gene mutations and/or chromosomal rearrangement neither favored nor selected against by natural selection, is difficult to assess. Genetic drift of some magnitude may arise in geographic isolation. On several occassions we have observed in hybrid progenies an impaired vitality of some plants. For instance, in $F_2$ progenies of normal parents from two *M. polymorpha* strains, we found a few chlorophyll deficient plants which produced chlorophyll deficient progeny. Most of the $F_2$ progeny, however, consisted of normal plants and produced normal progeny. We conclude that the recovery from the effects of genetic drift under natural conditions would be complete. We have also noted in $F_1$ from other crosses less good pollen than in the parents; we attribute that too, to the influence of genetic drift. In the *M. pironae* x *M. daghestanica* cross mentioned earlier, the hybrids were sterile on intercrossing as well as on backcrossing to parents. At first it was speculated (Lesins & Gillies, 1968) that under geographic isolation (the first species growing on foothills of the Alps, the second on foothills of the Caucasus mountains) differences in genic and chromosomal arrangement have accumulated due to genetic drift, resulting in complete isolation. This case, we thought, might provide a positive proof of speciation by genetic drift. Because in cases where neutral mutations have been accumulated, as supposedly in this case, to the extent that no viable gametes are produced by their hybrids, polyploidization, providing each chromosome with a fully homologous partner, would induce production of normal gametes and restore fertility (Newton & Pellew, 1929). After polyploidization, *M. pironae* x *M. daghestanica* hybrids indeed had 60% plasma-filled pollen grains. The interbreeding barrier, however, remained unshaken — out of more than 1200 florets, selfed or crossed, not a single seed was obtained. It therefore had to be concluded that some isolating mechanism other than genetic drift had been built up between these two species (Lesins, 1971). Regarding genetic drift, we now think that if environmental conditions do not require specific genetic changes for opening new physiological pathways or altering existing ones, then no permanent isolating mechanism would arise. On coming in contact again, such populations would have no real obstacle against merging. True, as Mayr (1970) points out, there are scarcely any two geographic locations which have the same environmental conditions, and hence would not require different physiological channels for their exploitation. This, however, is putting off as unanswerable the question whether by genetic drift alone, a complete interbreeding barrier may arise.

GENERAL FACTORS IN SPECIATION

*Environmental Stress.* Our view is, as presently held by a majority of biologists, that only under environmental stress may a species originate or become extinct. To explain what we have in mind, visualize the following conditions: In our undulating countryside, bogs or small swamps have developed in depressions. Assume that a cross-fertilized species is growing in the surrounding mineral soil. It cannot establish itself in the swampland because it is adapted to certain mineral soil resources, such as composition and concentration of mineral nutrition, pH value, moisture content, and aeration of the soil surrounding the roots. During a period of years, millions of seeds reach the swampland, but because of quite different resource conditions the developing seedlings die or plants succumb to competition. Mutations of all kinds take place, some of which may be in the direction permitting the plants to utilize more efficiently swampland resources such as less concentrated nutrient solution, more acid soils, or low aeration. Such mutation-carrying seeds would develop to somewhat more vigorous plants when growing under swamp conditions. However, mutations not drastically reducing the vitality of their carriers are usually small steps, and are also of rare occurrence. Hence, it is unlikely that some seeds would be endowed with enough beneficial characteristics to give rise to plants which might compete with the locally adapted swamp community. In addition, pollen coming from plants growing in mineral soil will fertilize and thus 'dilute' the progeny of the swampland's better-adjusted mutants. The importance of such immigration from basic stock has been well emphasized by Mayr (1970). How isolation may be established and diversification achieved may be imagined: The exclusion of the diluting influence on swamp-preadapted plants by pollen from mineral soil may be brought about by the removal of plant cover from the surrounding mineral soil by natural, artificial or combined processes such as clearing bush by burning, ploughing, and keeping the land under cultivation long enough so that the chances of intercrossing and selection within the swamp-adjusted race are increased. The different growth resources characteristic of the swampland may favor mutations channelling chemical and physiological processes in the swamp race to the extent that a population sufficiently different from the original one would arise, and, should a contact again be established, the intercross progenies would be of sufficiently lowered viability, constituting an interbreeding barrier. Swamp conditions would, in addition, favor mutations for morphological changes such as for hairy leaves or change in leaf area, adapting transpiration for growth under less concentrated nutrient solutions. Mutations for a shorter growing season would also be favored, since ground in the swamp remains frozen longer in the spring and killing frosts occur earlier in the fall. Under such circumstances features considered characteristic for a separate species would be established.

Another variation on the origin of swampland-adjusted species may be visualized. A mutation allowing plant roots to grow more satisfactorily in the swampland soil may be correlated with changes, say, in acidity of cell sap, so that conditions in stigma and style become more acid. Therewith, pollen may be effectively inhibited from growing down the style and fertilizing ovules. Differences in cell sap acidity in petals in different varieties of *Primula sinensis* and in other plants have been found by Scott-Moncrieff (1936). What mutational steps and enzymatic activities underlie the changed acidity is not known. Action of incompatibility, $S$, alleles is also based on certain incompatibility substances. It is reasonable to assume that by a mutation a substance may be produced in stigma and/or style inhibitive for self as well as foreign pollen to germinate and/or grow down the style. The crux in such a situation is that self pollen has access to stigmas in thousands upon thousands, hence a chance that among them a mutation to self-compatibility, an $Sf$ allele, will occur is immensely higher than for the comparatively few pollen grains coming from outside. A self-fertile plant may thus originate. In progenies of such a plant, self pollen will again have the priority of fertilizing the ovules. Consequently, an effective isolating barrier will be established between populations growing in swamp and mineral soils, and a self-fertile strain would originate. Following natural selection for other swampland-required traits mentioned above, a distinctive species may develop.

*Time Allowed for Evolutionary Changes.* This may be the decisive factor for reorganisation of one species into another, or branching, or extinction. For illustration we may think of a gradually lowering temperature over a long period of time. This may favor mutations which permit more effective photosynthesis under lower temperatures. Another case may be visualized: a slowly sinking seacoast area. In the adjacent plainland a certain species may be growing. If the water level in the ground is rising very slowly, the natural selection for mutants tolerating slightly higher water level, such as a shallower root system, may keep pace with the rate of the sinking seacoast; hence, a taxon with different characteristics than the original one may develop and remain as such, if the sinking of the area stops. The main point in such a situation is that if the water level rises rapidly, the time required for adjustments may be too short, and the species will become extinct. In our view, geological epochs characterized by species mass-extinction means that the environment has changed more rapidly than the rate at which organisms could get adjusted to the changed life conditions. Conversely, some marine and other organisms are said not to have changed during many millions of years. They must have had very similar living conditions during that time. Sea water may have poured off one continent and submerged another, but some masses of it must have retained unchanged conditions where particular species could continue their life habits without change.

*Previous History of Development.* This factor (or as it may also be termed the 'genetic inventory') may be decisive in whether a species becomes extinct or evolves further. We know that species of certain genera have specific organic compounds or they have pathways which may be switched on for metabolizing such compounds. Coumarins have been found in most *Melilotus* and in some *Trigonella* species, saponins in *Medicago*, and tannins have been found in some *Trifolium* species. [On screening of our *Medicago* collection for tannins, no species were found with appreciable tannin content (unpubl.).] It is also observed that coumarin protects to some extent *Melilotus* against attacks of blister beetle (Howe & Gorz, 1960), and high saponin content was positively correlated with pea aphid resistance in *Medicago* (Pedersen et al., 1976). Therefore, in the case of a destructive pest introduced or evolving, and attacking legume genera, it may be checked by one of the above substances. Obviously, a genus that possessed this substance in its organism would have more chances of survival and further development than another which did not have it.

For a species or genus, a long time is needed to adapt itself (if at all possible) to changed conditions, whereas another taxon, having a different genetic inventory, may be able to respond immediately or in a shorter time, and become adapted to the changed environment.

EVOLUTION IN MEDICAGO

The following chronological account of geological and climatological developments during the Tertiary Period according to Durham (1975), Hsü (1973, 1978) and Hsü et al. (1973, 1977) may provide a general background for understanding what might have influenced *Medicago* evolution, and may be of use as a reference to other non-geologists, like ourselves.

Epochs of the Tertiary Period:

| | | | | |
|---|---|---|---|---|
| Pliocene Epoch | 5.5 — | 1.86 million years ago | | |
| Miocene „ | 22.5 — | 5.5 „ | „ | „ |
| Oligocene „ | 36 — | 22.5 „ | „ | „ |
| Eocene „ | 53.5 — | 36 „ | „ | „ |
| Paleocene „ | 65 — | 53.5 „ | „ | „ |

In the Mediterranean area, climatic conditions during the first stages of the Tertiary up to mid-Miocene were warm. Palms and other warmth-requiring plants were still growing in Western Europe about 10-12 million years ago, although already earlier, approx. 16 m. years ago, since the beginning of the Burdigalian stage, climatic conditions had become drier and cooler. In Europe, thermophilic forests were replaced by montane coniferous floras. At the beginning of the Pliocene, the climate seems to have worsened, culminating in the ice age of the Pleistocene, the first epoch of the Quaternary Period (1.86 million — 15 thousand years ago). During the ice age, several glaciation and interglaciation intervals with warmer climates were recorded.

The following main geographic features in the Tertiary Period may be noted: During the early Tertiary the Tethys Seaway connected the Atlantic with the Indian Ocean, separating Africa and India from the northern continents. The Tethys closure at its eastern end was brought about by the northward movement of India at the end of the Oligocene. The subsequent fate of the Tethys is only vaguely understood. An arm of it extended northward to the Arctic along the east side of the Ural Mountains. A large part of eastern Europe and part of Asia was a huge lake, Paratethys. During late Miocene and Pliocene it extended from Vienna to Lake Aral. At some periods it drained its waters into the Mediterranean Sea; hence deposits found by the Deep Sea Drilling Project (Hsü, 1973) contain organisms that had lived in fresh or brackish waters. The inlet of Paratethys to the Mediterranean was broken up by earth movements. Remnants of Paratethys are considered to be the Black and Caspian Seas.

Mountain building processes involving the Alps, Apennines and Himalayas began in the Oligocene and continued during the Miocene. The Tien Shan range has been elevated since the beginning of the Oligocene; before that the area was nearly flat. The Pyrenees had their last uplift in the late Eocene, while the Atlas Mts. had their major uplift also during the late Eocene, with some upward movements continuing throughout the Tertiary Period.

*Medicago* is a predominantly Mediterranean genus. It comprises species widely varying in characteristics, so that it makes the task of arriving at

46

some inference on the evolution of the genus, and its relationship to other genera, rather difficult.

As our initial position, we adopt the generally accepted view that, in a related group of species, perennials with woody growth habit are phylogenetically oldest, followed next by herbaceous perennials, with the annuals being the youngest (Davis & Heywood, 1963; Stebbins, 1974). Although there are exceptions to this rule, we see no compelling reasons not to accept it for *Medicago*.

About two-thirds of *Medicago* species are annuals. They have the same tripping mechanism devised for cross-pollination by insects as their older, perennial relatives, but it is superfluous since they have all become self-pollinators. This provides a unique opportunity for speculation on the phylogeny of the genus. Assuming an original cross- or facultative cross-self-pollination system possessed by perennials, we may inquire under what conditions and at what times the transition to annuals with exclusive self-pollination may have taken place. Here the lead is provided by recent discoveries of the Deep Sea Drilling Project. Hsü (1973, 1978) and Hsü et al. (1973, 1977) reported, from the data obtained, that in the late Miocene Epoch (Messinian stage), about six million years ago, the Mediterranean basin was a hot desert. Due to closure of the Gibraltar connection to the Atlantic Ocean about seven million years ago, evaporation in a dry climate resulted in the drying up of the sea. The drilling data also show that repeated reopenings and closures to the Atlantic have taken place during this period. During repeated recessions, and complete withdrawal of the sea, large land areas became available for exploitation by plants. Presumably perennial *Medicago* existed before the closure of the Gibraltar gate. After its closure, however, they could not advance into the newly-opened areas because their perennial growth habit was not suitable for enduring inundation, or periods of severe drought and heat. Hence mutations favoring endurance of such conditions may have been required for the exploitation of this region. This has resulted in annual growth forms which could have followed recession of the sea because of their spurt of growth after occasional favorable conditions, their short life span, and their long-lived seeds. In this connection, the seeds of several annual *Medicago* requiring a period of high temperature to break embryo dormancy (Lesins et al., 1976) may have originated during this period. Another adaptation under such conditions was required, that of ability to set seed without cross-pollination. The insects, probably ground nesting bees, which were coadapted with perennial *Medicago* as their food source on the one hand and cross-pollinators of plants on the other, could not have followed *Medicago* in the new habitats. Under these hot, dry conditions food sources may not have been available for long periods, and during inundations their nests would have been destroyed. Furthermore, the ground temperature at some periods may have been scorching, also, for completion of bee's life cycle fresh water is a necessity. Hagerup (1932) reports conditions in the southern part of Sahara (Timbuctu), where ground temperature reaches 70-80° C and no insects, except ants, visit lowgrowing

47

plants which have, therefore, turned to self-pollination. He draws attention to plants of genus *Indigofera* which, despite having a tripping mechanism, no longer have use for it, although in milder climate they are pollinated by bees. This appears to be a close parallel with evolution of annual *Medicago*. Hagerup (1951) furhter reports conditions on the Faroe Islands, where pollinating insects are scarce because of the cold, rainy climate, and plants have likewise turned to self-pollination, while on the continent the same species are pollinated by insects, especially by bees. It seems reasonable to assume that these changes in *Medicago*, resulting in annual self-pollinators, have taken place in the Mediterranean basin during its desiccation period lasting about one and a half million years. Then another event transformed the Mediterranean region: Five and a half million years ago the Gibraltar gate opened fully (this event is counted as the beginning of the Pliocene Epoch). Drilling results indicate that this time the inrush of water was so cataclysmic that the Mediterranean Sea may have been filled up in less than 1000 years. Very likely, a number of *Medicago* species, and some taxa evolutionally connecting species and genera, became extinct. Some of the present islands are assumed to have been towering up to 3 km above the salt-layered sea-bottom during the major desiccation period. On such refuge peaks, floral remnants may have survived. One example may perhaps be *Medicago heyniana*, recently discovered by Greuter (1970) on Mt. Kolla, on the Island of Karpathos in the Aegean Archipelago.

During desiccation, the Mediterranean at times had been a series of lakes that separated plant populations in different areas and under different climatic conditions. What speciation in annual *Medicago* has taken place since the beginning of the Pliocene Epoch is not clear. After final refilling, the fluctuations in sea level have been governed by the Atlantic connection. Nevertheless, the lands around the Mediterranean where *Medicago*, especially the annuals, have survived were subject to great climatic fluctuations. Borings in connection with the Aswan High Dam construction on the Nile indicate that there have been rainy periods and some prolonged dry periods. One of these droughts, starting about 1.9 million years ago at the beginning of the Pleistocene Epoch, lasted for one million years.

No *Medicago* is native to the New World, hence the genus had its beginning after the European and North American continents had drifted widely apart. This had been accomplished largely by the beginning of the Tertiary Period; consequently *Medicago* may have come into existence some time in the Tertiary; its progenitors, as woody Leguminosae, possibly earlier.

Having the general geological and climatological events in mind, we may try to get some insight into the history of *Medicago*, concentrating on the characteristics of the older, perennial species. Apart from their perenniality and adaptation to insect pollination, which we have assumed took place during the period before Mediterranean desiccation, different individual species and groups have some rather divergent characteristics. In the following, the species are arranged in groups based on their interbreeding affinity:

   1) *M. arborea*, as noted above, may be counted as the oldest member of

48

the genus because of its shrub growth habit. Its present forms, tetraploid and hexaploid, cannot be considered progenitors of the next oldest, herbaceous perennial *Medicago* species, all of which are either diploids or close relatives of diploid species. A morphological character that sets *M. arborea* apart from other *Medicago* is its flower structure, the keel petal being as long as the standard or longer. A character in common with almost all other *Medicago* is its coiled pods, which may be a significant trait in assessing phylogenetic relationship, as will be discussed later. *M. arborea* is a warmth-requiring species; it does not survive outside the present Mediterranean climate, and it may be thought that it originated during the first part of the Tertiary before climatic conditions had become cooler. Some of our accessions of *M. arborea* may set seed in the growth chamber by self-pollination. This may indicate that some strains of the species have undergone stress, requiring self-pollination for survival, which may have taken place during the desiccation periods of the Mediterranean Sea.

2) *M. carstiensis* has the most puzzling characteristics of all the perennial species in the genus. For one, it requires winter dormancy in order to come to flower. It has also survived winters at three locations in Canada (Brandon, Edmonton and Ottawa, 49°50′, 53°20′ and 45°25′ N. Lat., resp.). The characteristic is difficult to reconcile with the hypothesis that the perennial *Medicago* originated in warm epochs. The species is endemic to the Karst plateau on the eastern coast of the Adriatic Sea. It possesses a shallow, spreading root system and is tolerant to shade — an almost unique character among *Medicago*. We wonder whether it has not developed under the influence of the cold climatic conditions reigning further east along coasts of the Paratethys; the Karst area being its refuge. A unique morphological trait of *M. carstiensis* is the position of its seed rootlet at about a 45° angle to pod's ventral suture. Thus it occupies an intermediate possition between that found in the other members of the subgenus *Orbicularia* in which seed rootlets are perpendicular to the ventral suture, and the rest of the genus in which the rootlets are almost parallel to the ventral suture. Of some importance in a phylogenetic assessment may be the finding that in *M. carstiensis* and *M. arborea* the lutein content in petals is the highest (72% and 80%, resp.) of all the perennial *Medicago* species (Ignasiak & Lesins, 1975). The characteristics of pods deserve special mention: They are coiled and have thin, flexible non-hooked spines.

3) *M. suffruticosa* and *M. hybrida* are growing in montane habitats of the most westerly part of the distribution of the genus, the Pyrenees and the Atlas Mts. of Morocco. They are adapted to altitudes with high humidity caused by the proximity of the Atlantic Ocean. There are a number of differences between them and the other perennial *Medicago*. In fact, *M. hybrida* has often been placed in the genus *Trigonella*. Their chromosome morphology was found to be different from other *Medicago* (Lesins & Gillies, 1972; Gillies, 1972c). Under what conditions the two species evolved is difficult to say. They are so closely related that their progenies behave as if the parents belonged to the same species. At that, a conspicuous morpho-

logical difference lies in the pods, which in *M. suffruticosa* are coiled, in *M. hybrida* almost straight. In some *M. hybrida* forms, seedset may originate from self-pollination, though intercrossing gives better results (Lesins, 1969).

4) *M. marina* is the only member in the perennial *Medicago* inhabiting seashore sands. It may be guessed that it is the only member of perennials which has succeeded in following, to some extent, the retreat of the Mediterranean Sea in its desiccation phase, maybe along the river valleys. No relationships tying it with other members of the genus have so far been found in hybridization tests. *M. marina* has coiled pods similar to those found in the perennial *M. pironae* and *M. daghestanica*; also their non-hooked spines are similar. Some accessions may set seed without cross-pollination under greenhouse conditions.

Turning to the section *Falcago*, in which all species are perennials and can be hybridized with *M. sativa*, and have closer hybridization affinity between certain members, the following groups may be discerned:

5) *M. daghestanica* and *M. pironae* are both submontane species with rather limited distribution. *M. daghestanica* can be readily hybridized with *M. pironae*, and their hybrids are vigorous but completely sterile. Their relationship to *M. sativa* is the most distant one encountered in section *Falcago*. None of the two species at their diploid level can be hybridized with diploid *M. sativa*. Only trispecies hybrids have been obtained. First, *M. daghestanica* and *M. pironae* were intercrossed, then their hybrids were chromosome-doubled. On crossing with tetraploid *M. sativa*, a few seeds were obtained.

Both *M. daghestanica* and *M. pironae* have coiled, spiny pods. A conspicuous difference is in their flower color: *M. daghestanica* having anthocyanin-colored, *M. pironae* yellow petals.

6) *M. papillosa* and *M. dzhawakhetica* are growing in montane to submontane habitats. The tetraploid forms may be hybridized, though some interbreeding barrier is present. They may also be hybridized with *M. sativa*, but under the peculiar conditions that their genomic ratios as parents with *M. sativa* must be 2:1. Their pods are coiled, and in the former species have rough, articulate hairs, while in the latter species they are hairless or have simple hairs.

7) *M. rhodopea* and *M. rupestris* have restricted distribution growing in montane areas. They may be hybridized among themselves and with *M. sativa*, though in the latter case some adjustment of chromosome level is required. Their pods are coiled; in *M. rhodopea* short spines may be present, in *M. rupestris* the pods are spineless.

8) *M. saxatilis* and *M. cancellata*. The former species is a mountain inhabitant, the latter grows in dry, rocky or sandy steppe soils. Both species are hexaploids, the former presumably having *M. rhodopea*, the latter *M. rupestris*, in its ancestry. They may be hybridized among themselves and with *M. sativa*, though in the latter case some interbreeding barrier is present, and ploidy levels should be adjusted, especially between *M. cancellata* and *M. sativa*.

9) *M. prostrata* is growing under submontane conditions. It has coiled, spineless pods and can be hybridized with *M. sativa*, though some inter-breeding barrier is present.

10) *M. glomerata* is a submontane species growing between low shrubs in shallow soils. It has coiled spineless pods. It can be hybridized with *M. sativa* almost as freely as if both belonged to one species. The basis for considering them separate species are the slight irregularities found in their hybrids at meiosis and at seed production.

11) *M. sativa* and *M. falcata* are often considered as subspecies of *M. sativa*. Indeed, concerning the inheritance of characters, the fertility, and the survival of progeny under experimental conditions, they behave as members of a single species, as shown by many investigators. Their diploid forms are both predominantly plainland inhabitants, the former occupying semi-deserts, the second steppes. Their morphological appearances, however, are conspicuously different. The former has coiled pods, the latter straight to sickle-shaped; the former has anthocyanins in the petals, the latter yellow flavonoids and carotenoids.

Considering the characteristics and indigenousness of the perennials, we may venture the following conclusions:

1) The straight pods as an indicator of a greater phylogenetic age as compared with coiled pods (Širjaev, 1935; Heyn, 1963) is not tenable for *Medicago*. The coiled-pod characteristic is present in *M. arborea*, *M. carstiensis*, *M. marina* and in diploid members of section *Falcago*: *M. daghestanica*, *M. pironae*, *M. rhodopea*, *M. rupestris*, *M. prostrata* and *M. glomerata*. These latter, in addition to diploidy, are of relic nature as indicated by their narrow distribution areas and low variability. Hybridization tests involving *M. suffruticosa* x *M. hybrida* and *M. sativa* x *M. falcata*, one crossing partner in each pair having coiled and the other straight pods, have shown that there are no impairments of viability in either the immediate or the later hybrid generations, and that the inheritance of this and other characters follows normal segregation ratios; hence no indication for phylogenetic time-gap can be found. Consequently, coiled pods are characteristic of the genus and incorporated during its origin.

2) Pod spininess cannot be considered as a comparatively recent characteristic in the genus, as believed by Heyn (1963), who assumed as a primitive form the spineless pods of *M. sativa*. In doing so, the phylogenetic older types with spiny pods: *M. carstiensis*, *M. daghestanica*, *M. rhodopea* and *M. marina*, were overlooked. It may be noted that the spines in these species are not hooked, whereas in most of the younger, annual species, hooks are present.

It may be speculated that there is some causative connection between kinds of spines in some perennials and certain groups of annuals. Thus, spines in *M. carstiensis* resemble those in section *Leptospirae* and *M. heyniana*, and spines of *M. daghestanica*, *M. pironae*, and *M. marina* resemble those in section *Pachyspirae*, as noted earlier. [Whether there is a causative tie between the long, rough articulate hairs on pods of the perennial *M.*

*papillosa* and that of long, cottony, articulate hairs on pods of the annual *M. lanigera* is difficult to say, except by hinting that both species are native in the eastern distribution area of the genus.]

3) The area of origin of perennials and, by implication, of the genus *Medicago* appears to be the northern coast of the Mediterranean, i.e., northwestern coast of the Tethys, rather than western and central Asia as assumed by Sinskaya (1950), Vassilczenko (1949) and some other authors. This becomes clear when the habitats and areas of diploid perennial species are mapped: *M. suffruticosa* and *M. hybrida* in the Pyrenees (the former also in the western part of the Atlas Mts. in N. Africa); *M. glomerata* in the Maritime Alps, possibly also in the mountains of N. Africa; *M. prostrata* and *M. pironae* in eastern outgrowths of the Alps, the former reaching the mountains of northern Greece; *M. carstiensis* in the Dinaric Alps; *M. rhodopea* in the Rhodope Mts.; *M. rupestris* in the Crimean Mts.; *M. daghestanica* in the Caucasian Mts. of Daghestan; *M. papillosa* in eastern and southern Transcaucasian Mts. and in their extension in eastern Anatolia of Turkey. Polyploid species, *M. saxatilis*, *M. cancellata* and *M. dzhawakhetica*, are found in the same or adjacent areas as their close relatives. *M. arborea* is found mainly in Greece and southern Italy, though its use as a decorative plant has blurred its natural range of distribution. None of the above species is found east of the Caspian Sea, either in mountains or on plains, as would be expected if the origin of the genus had been in western and central Asia.

The annuals, as a generally younger group, may not contribute much to the understanding of the area of origin of the genus. Their distribution is summarized by Heyn (1963) and, regarding individual species, is mentioned by us in the Specific Part of this book.

*Evolution of M. falcata — M. sativa.* The problem has often come to our mind: What conditions have caused the separation and morphological differentiation of *M. sativa* and *M. falcata*, which often are considered one species and very likely had a common ancestor? The problem is intriguing and puzzling the more because of the two species' conspicuous morphological differences in pod shape and flower color. These two sets of differences, however, cannot be considered either genetically deep-seated or of phylogenetic antiquity, as switching of these traits experimentally within or between the sets does not change the indices of primary importance in estimation of relationship: readiness for interbreeding, regularity of meiotic processes in hybrids, vigor and fertility of immediate and later hybrid generations. It may be thought that *M. glomerata* (or its like species), which has yellow flowers like *M. falcata* and coiled pods like *M. sativa*, and which even now shows almost no interbreeding barrier with these species, was once distributed farther to the east, and was ancestral to them. Paratethys with isthmuses during recessions and their disappearance during transgressions may be responsible for the division of the ancestral population. We may assume that during a long period of transgression, one part of the ancestral population was separated on the south coast of the connected Black-Caspian Sea, the other on the north coast. In the southern population, changes in flower pigments (loss of carotenoids, yellow flavonoids converted to pale or colorless anthocyanin precursors) would have been necessitated in order to become more conspicuous against other yellow-flowered members of the plant community in competition for attracting insect pollinators. The other, northern population, would meet thicker stands of plants including grasses, as is characteristic of steppes. Here, the smooth coiled pods, of an advantage for dispersal by rolling on open ground, would be of a disadvantage, preventing easy bursting and seed-scattering required for seed dispersal in a more closed, steppe-plant community. Mutational steps for straightening the pods for easier seed dispersal would lead to the selection of straight to sickle-shaped pods characteristic of *M. falcata*. It was found (Lesins, 1969) that inheritance of pod coiling in a similar coiled : straight pod pair, *M. suffruticosa* : *M. hybrida*, was determined by six genetic factors distributed on different chromosomes and having a cumulative effect. Hence a rather gradual evolutionary transformation may be implied.

Regarding the position of *M. falcata* — *M. sativa* on the evolutionary steps in section *Falcago*, it appears that they constitute the youngest group in it. *M. falcata*, especially, may be considered a latecomer in view that the rest of the section all have coiled pods and most of them are of relic nature, as indicated before. The same may apply to *M. hybrida* in section *Suffruticosae*.

Contrary to some authors (e.g., Sinskaya, 1950), we consider that younger species are more adapted to the more recent environmental conditions than the older ones which may have an impoverished genetic inventory, thereby becoming relics. In support of this view, we see the wide and continuous spread of the youngest, annual *Medicago*. This may also explain the wide distribution of *M. falcata*. After the retreat of glaciation, vast areas became open for revegetation so that even in areas with severe winters *M. falcata* has moved in. Its plasticity for adaptation to different environments has been increased by the development of tetraploid forms, which at present are prevalent.

The taxonomic significance of *M. sativa* has been exaggerated, probably because its tetraploid cultivated type, the alfalfa, is well known and is grown almost around the world. *M. falcata*, its closest relative, in turn has often been involved in the origin of cultivars; its cold resistance permitting alfalfa cultivation far outside the natural warm, dry area of *M. sativa*. Most important, the natural hybrids arise wherever the two species meet, giving rise to hybrid swarms which only too often have been considered as separate species (see Table II). A false impression has thus been formed that there exists a large group of *M. sativa* — *M. falcata* related species.


RELATIONSHIP TO OTHER GENERA

There have been attempts to find characters that could be safely used to distinguish the three related genera, *Medicago*, *Trigonella* and *Melilotus* [Trigonellae, Schultz (1901)]. *Medicago* especially has species that have some

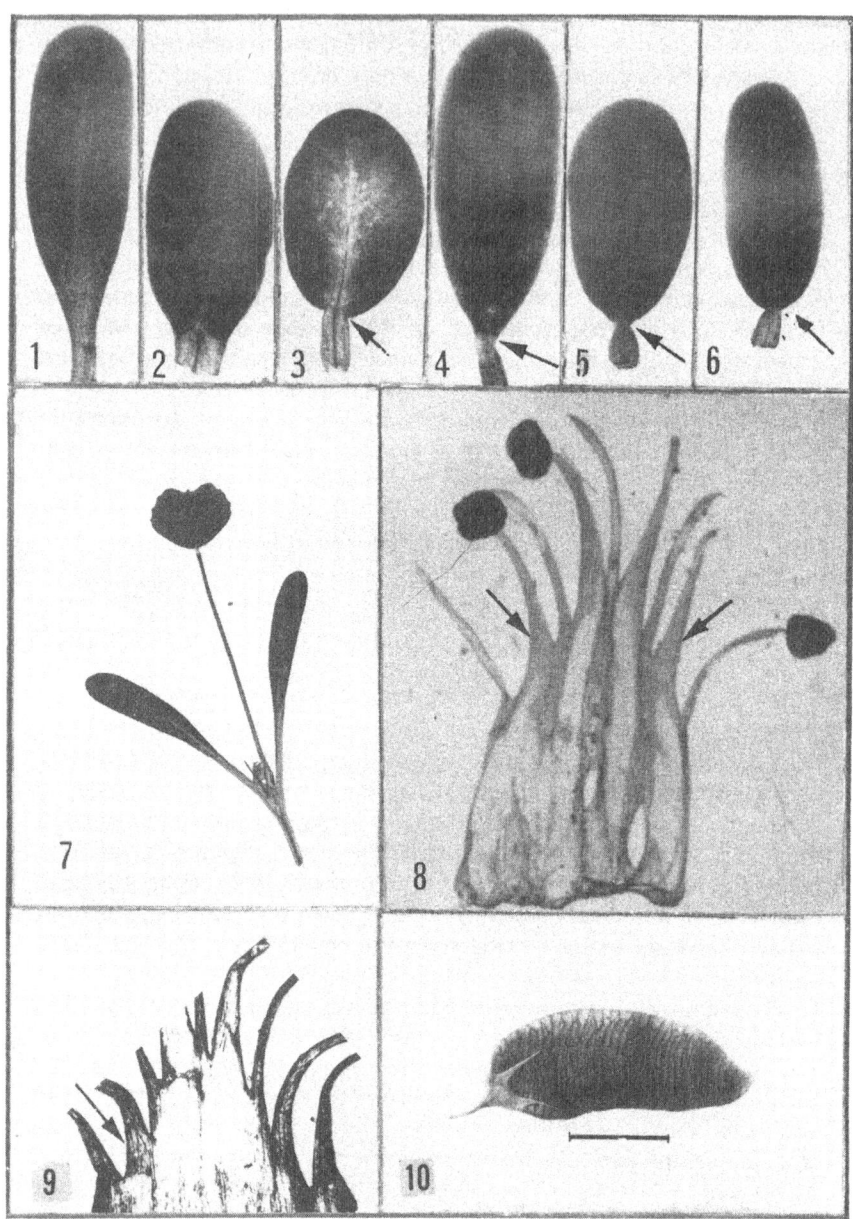

*Fig. 9.* Nos. 1-6, cotyledons of germinating seeds: 1) *Medicago tornata*, 2) *Medicago lupulina*, 3) *Trigonella coerulea*, 4) *Trigonella tibetica*, 5) *Melilotus neapolitana*, 6) *Melilotus alba.* No. 7, a seedling of *Medicago laciniata.* No. 8, anther filaments in *Medicago sativa.* No. 9, anther filaments in *Trigonella tibetica.* No. 10, a pod of *T. tibetica.* Arrows in Nos. 3, 4, 5 and 6 indicate constrictions between laminar and petiolar parts of cotyledons in *Trigonella* and *Melilotus*; arrows in Nos. 8 and 9 point to enlarged bases of filaments. Magn. in Nos. 1-6 approx. **x** 5; in No. 7, **x** 3½; in Nos. 8 and 9, **x** 14; in No. 10 the inked bar equals the natural length of pod.

characters in common with *Trigonella*. Urban (1873) thought that he had found such a clear-cut difference: In *Medicago* there is no perceptible division between the petiolar and the laminar parts of cotyledons in germinating seeds (Fig. 9,-1, -2 and -7), whereas the two parts are clearly marked in *Trigonella* (Fig. 9,-3 and -4) and *Melilotus* (Fig. 9,-5 and -6). A bulge in the petiolar part as pictured by Heyn (1963) and subsequently by Baum (1968), has not been found by McComb (1974), nor have we observed it in any *Trigonella* or *Melilotus* examined. The clearly observable boundary between petiole and lamina of cotyledons in germinating seeds has been used by Širjaev (1933) in transferring a number of species previously known as *Trigonella* to *Medicago*. Heyn (l.c.), however, indicates that, apart from cotyledon-type, most of those transfers fit into *Trigonella* in every sense. We agree with her, and in the foregoing discussion have indicated that some species treated here as *Medicago* may actually belong to a separate genus or subgenus. Baum (l.c.) thought that he had found a character that could be applied safely in discerning generic limits. This character essentially consists of the widened bases of anther filaments lying next to marginal ones. We found it in *M. sativa* (Fig. 9,-8, arrows). McComb (l.c.) could only faintly discern this character in *M. lupulina* and thought that, in general, it and other filamentous structures illustrated by Baum (l.c.) are too liable to distortion during preparation, to serve as good markers.

We studied a species which was received as *Medicago edgeworthii* Šir. (courtesy of the Curator, Inst. of System, Bot., Upsala, Sweden) in order to decide its generic position. On the evidence of enlarged filament bases it was difficult to arrive at a decision since, at least in some preparations, somewhat widened filament bases next to the marginal ones were observed (Fig. 9,-9, arrow). The cotyledons, however, showed a marked division between the petiolar and laminar parts (Fig. 9,-4).

Regarding nomenclature, the taxon cannot be *M. edgeworthii*, which supposedly is a perennial (Širjaev, 1938), whereas this one is an annual, essentially coinciding with Vassilczenko's (1953) description of *T. tibetica*. Since on the herbarium specimen which Vassilczenko had at his disposal no ripe pods were present, we include an illustration of a pod here (Fig. 9,-10). The pods are very much like those in *M. ruthenica*. The seed radicle is at right angle to the ventral suture, the florets have the tripping mechanism but are self-fertilizing, and no hybrids with *M. sativa* were obtained (Lesins, 1952, as *M. edgeworthii*). We consider it as *T. tibetica*.

The misconception that the *M. sativa* — *M. falcata* group has a comparatively ancient origin has been furthered by the similarity between *M. falcata* pods and those of *M. platycarpa*, *M. ruthenica*, *Trigonella papovii* and some other species which have been counted either with genus *Medicago*, or with *Trigonella*, or with *Melissitus*, or *Pocockia*. *M. platycarpa* and *M. ruthenica* are considered by Širjaev (1935) as being closest to the oldest section, *Ellipticae*, of the genus *Trigonella*. This is an assumption in that first, they are closely associated with the origin of both *Trigonella* and *Medicago*, and hence are of a rather early origin; and second, because of the similarity of pods, that they are closely related to *M. falcata*. Another support for such an assumption is seen in the areas of distribution: Members of *Trigonella* section *Ellipticae* and the *M. platycarpa-M. ruthenica* group, and also some other borderline species, are endemic to western and central Asia where, as part of its distribution area, *M. falcata* is also growing. However, attempts to obtain hybrids between *M. platycarpa* and *M. ruthenica* on the one hand, and *M. falcata* on the other, have been unsuccessful despite numerous attempts by us and by other investigators. The similarity, therefore, is only a superficial one. On inspection of morphological characters separating *M. falcata* from the *M. platycarpa-M. ruthenica* group, one among others is the perpendicular orientation of the radicle to pod's ventral suture of the latter group. This character, which is common also in various *Trigonella* species, is given by us as the subgeneric character for *Orbicularia*. We have not studied many of the perennial borderline species similar in type to *M. platycarpa-M. ruthenica* or the earlier mentioned annual *T. tibetica*. Our impression is that they indeed may constitute a separate genus or subgenus (*Melissitus* Medic, or subgenus *Pocockia* Grossh. of *Trigonella*). It may be noted that *M. platycarpa*, *M. ruthenica* and *M. cretacea* have the lowest petal lutein content (30%, 30.3% and 35%, resp., of all carotenoids) in the perennial *Medicago* (Ignasiak & Lesins, 1975). This may or may not be an indication of close relationship between members of this group, and analyses in the rest of the group are required to come to more definite conclusions.

The inclusion of sections *Platycarpae*, *Cretaceae* and *Hymenocarpos* in subgenus *Orbicularia* was motivated to a great extent by practical considerations, i.e., to acquaint the readers with species which often are considered as *Medicago*. The distinctness of *M. radiata* from the rest of the *Medicago* has been stressed by Heyn (1959). She pointed to the nature of spines, which in *M. radiata* corresponds to those in *Trigonella arabica*, where they also represent a continuation of the dorsal suture, and hence are not homologous to those in *Medicago* which are outgrowths of pod veins at the side of the dorsal suture.

Section *Platycarpae*, (i.e., *M. platycarpa* and *M. ruthenica*), are distributed in certain regions of Asia. Since they are often included in genus *Trigonella*, it may be of interest to draw attention to the fact that the phylogenetically older, perennial species of *Trigonella*, listed by Širjaev (1928), are all endemic to Asia: *T. elliptica* and *T. teheranica* in Persia (W. and S. W. Asia); *T. lipskii* in Bukhara (W. Asia); *T. laxiflora* in Afghanistan (S. Asia); *T. emodii* in Turkestan, the Himalayas, northern India (W. and Central Asia), and *T. gracilis* in northern India. All these *Trigonella* sect. *Ellipticae* species have short, straight or slightly bent pods, similar to those of *Medicago* sect. *Platycarpae*. A substantial difference between *Ellipticae* and *Platycarpae* lies in the flower structure: in *Ellipticae* there is no tripping mechanism (Širjaev, 1928), in *Platycarpae* it is present. Considering the intricacy of the tripping mechanism, it may be speculated that some *Trigonella* species possessing it, and having turned into annuals and self-pollinators, may be closer related to *Platycarpae* than to *Ellipticae*. One of them resembling *M. ruthenica* in pod shape also, is the earlier mentioned *T. tibetica.*

It may be thought that the perennial *Trigonella* and the *Platycarpae* groups may have originated in the Tertiary on the northeastern coast of the Tethys Sea, while the coiled-pod *Medicago* was formed on the northwestern coast. In this connection, it may be noted again that the Tethys had an arm extending to the Arctic Ocean east of the Ural Mts. isolating, to some extent, the eastern and western parts of the north coastal area. The live plant material of these groups from Asia, however, is too scanty to go on beyond guessing.

Subgenus *Lupularia* with its two species, *M. lupulina* and *M. secundiflora*, stands apart from the rest of *Medicago*. Their pods have some resemblance to those in genus *Melilotus* and to a few species in genus *Trigonella* (sect. *Capitatae*). However, though some species of *Melilotus (M. sulcata)* have concentric veins on pods as illustrated by Stevenson (1969), none of them show pod coiling, whereas *Lupularia* does, having obliquely running pod veins and twisted pod tips. In both *Lupularia* species, though the flower tripping mechanism is present, its function is utilized the least of all *Medicago* since the florets, especially in *M. secundiflora*, 'trip' before the standard petal fully opens. One may speculate that the weak tripping mechanism bridges the gap to *Melilotus* where the mechanism is absent.

In summing up the characteristics of the three related genera, several distinguishing features emerge. In *Medicago* there is one shrub species with coiled pods and many species with coiled pods bearing spines; all species have the tripping mechanism; about one-third of them are perennials. In *Melilotus* none of these features are present, the most notable difference being the lack of a tripping mechanism (Stevenson, l.c.). In *Trigonella* no woody species is known. Although about one-third of the species have a tripping mechanism as in *Medicago* (medicaginoid), this character is absent in all the perennials, as noted earlier. Neither the spininess nor the pod-coiling of the kind characteristic of *Medicago* is to be found in *Trigonella*. The possibility of extinction of progenitors and species-connecting links due to the submergence of the Mediterranean basin, introduces uncertainties into the assessment of species and generic relationships as judged on the basis of the present species. As noted earlier, common to all three genera is their comparatively late origin, i.e., after the continents had drifted apart. Further, they are to a great extent endemic to the Mediterranean region, especially the annual species.

We think that the pod-coiling may be considered as the major feature distinguishing *Medicago* from the other two genera, and that the exceptional straight pods of species *M. falcata* and *M. hybrida* are secondary derivatives, acquired during later stages of evolution under stress of certain environmental conditions.

Admittedly, much of the discussion regarding speciation in general and in the genus *Medicago* in particular, as well as relationships between genera, is hypothetical. We felt, however, that these matters had to be considered as part of a whole, and we expressed our views on them as we could see it at present. When genetical, chemical and physiological investigations are continued and further paleogeological information becomes available, the ties between genera of the Trigonelleae may become more readily surveyed and understood.

# SPECIFIC PART

GENUS MEDICAGO Linaeus, Gen. Pl. ed. 5:339 (1754).

Annual or perennial herbs or shrubs. Corolla papilionaceous with free standard and wings, the two keel petals connate; nine stamens united by their filaments forming a staminal column, the tenth stamen free; corolla and the staminal column forming a tripping mechanism for cross-pollination. Calyx bell-shaped, 5-toothed. Leaves trifoliate. Stipules adnate to the petiole. Racemes axillary, pedunculate, few-to-many-flowered. Fruit from a straight to a tightly coiled pod. Seeds 1-to-many per pod. Cotyledons at seedling stage not divided into separate petiolar and laminar parts. Basic chromosome number x = 8, some species have x = 7.

Different opinions have been expressed regarding whether the type species of *Medicago* should be considered *Medicago radiata, M. sativa* or *M. arborea* (Scofield, 1908; Britton & Brown, 1913; Grossheim, 1945, resp.). Lately *M. sativa* seems to be favored as the type species (Baum, 1970; Gunn et al., 1978).

Key to subgenera of *Medicago*:

| 1 | Pods one-seeded nondehiscent nutlets, their tips twisted in a small coil                                Subgen. *Lupularia* (Ser.) Grossh. | |
|---|---|---|
| — | Pods of other shapes, usually with more than one seed | 2 |
| 2 | Seeds with long axis at right angle to pod's ventral suture or nearly so. Radicle 2/3 of or equal to the seed in length                                Subgen. *Orbicularia* Grossh. | |
| — | Seeds with their long axis parallel to the ventral suture, or nearly so | 3 |
| 3 | Perennials. Pods straight, sickle-shaped or coiled, usually with an open centre                                Subgen. *Medicago* Tutin | |
| 4 | Annuals. Pods tightly coiled without an open centre                                Subgen. *Spirocarpos* (Ser.) Grossh. | |

| 1 | Perennials | | 2 |
|---|---|---|---|
| — | Annuals | | 25 |
| 2 | Shrubs | *M. arborea* | |
| — | Herbs | | 3 |
| 3 | Corolla yellow | | 4 |
| — | Corolla anthocyanin-colored: violet, or variegated (i.e., all shades between violet and yellow) | | 22 |
| 4 | Pods small (less than 3.5 mm long), kidney-shaped nutlets, containing one seed. Florets small (2.5-3.5 mm long) | *M. lupulina* | |
| — | Pods and florets larger | | 5 |
| 5 | Pods straight or sickle-shaped, coiled in not more than one half-circle | | 6 |
| — | Pods coiled in more than one half-circle | | 10 |
| 6 | Pods narrow, 1-3 mm in width | *M. falcata* | |
| — | Pods more than 3 mm in width | | 7 |
| 7 | Radicles and cotyledons of seeds with their long axes almost at right angle to pod's ventral suture (Fig. 18,-c) | | 8 |
| — | Radicles and cotyledons of seeds with their long axes almost parallel to the ventral suture | *M. hybrida* | |
| 8 | Pods 14-22 mm long | *M. platycarpa* | |
| — | Pods 8-12 mm long | | 9 |
| 9 | Pods wide, length less than 1.5 times width (Fig. 18,-b) | *M. cretacea* | |
| — | Pods elongated, length two or more times width | *M. ruthenica* | |
| 10(5) | Pods with spines or tubercles | | 11 |
| — | Pods without spines or tubercles | | 15 |
| 11 | Plants with rhizomes. Spines uniformly thin, 3-7 mm long. Seed with radicle at 45° angle to pod's ventral suture | *M. carstiensis* | |
| — | Plants without rhizomes. Spines thick at the base, short and rigid, or tubercles only. Seed with radicle almost parallel to the ventral suture | | 12 |
| 12 | Pods and whole plants covered thickly with felted hairs; seashore plants | *M. marina* | |
| — | Pods and plants not covered with felted hairs; not seashore plants | | 13 |
| 13 | Pods 5-7 mm in $\phi$ | | 14 |
| — | Pods less than 5 mm in $\phi$ | *M. rhodopea* | |
| 14 | Pods with glandular, articulated hairs. Standard oblong with sides parallel in the middle part (as in Fig. 21,-c). $2n = 16$ | *M. pironeae* | |
| — | Pods without glandular, articulated hairs. Standard oval (as in Fig. 20,-b). $2n = 48$ | *M. saxatilis* | |
| 15(10) | Pod surface corrugated by a net of elevated veins | | 16 |
| — | Pod surface smooth (after removal of hairs), or with only slightly elevated veins | | 17 |

| 16 | Pods with 1-3 coils, 4-6 mm in $\phi$. $2n = 48$ | *M. cancellata* |
| — | Pods with 1-1.5 coils, 3-4 mm in $\phi$. $2n = 16$ | *M. rupestris* |

16    Pods with 1-3 coils, 4-6 mm in $\phi$. $2n = 48$     *M. cancellata*

—    Pods with 1-1.5 coils, 3-4 mm in $\phi$. $2n = 16$     *M. rupestris*

17    Seeds separated within the pod by thick, spongy walls (Fig. 37,-e)     **18**

—    Seeds not separated within the pod, or separated by poorly defined walls     **19**

18    Pods covered with rough articulated hairs     *M. papillosa*

—    Pod surface without articulated hairs     *M. dzhawakhetica*

19    Leaflets nearly round, broadly ovate to obcordate even at upper nodes, (length : width, 3:2); upper side of leaflets glabrous. Unexpanded pollen grains triangular-pyramid-shaped (Fig. 7,-3)     *M. suffruticosa*

—    Leaflets oblong, narrowly obovate, at least at upper nodes, (length : width more than 2:1); upper side of leaflets always somewhat hairy. Unexpanded pollen grains spindle-cylindrical (as in Fig. 7,-1)     **20**

20    Pods small, 2-4(5) mm in $\phi$, with or without glandular hairs     *M. prostrata*

—    Pods larger, 5-9 mm in $\phi$, covered with glandular hairs     **21**

21    Pods large, 8-9 mm in $\phi$, with 1-2 loose coils, open in the center. In the keel and in the middle of the standard the yellow color is often of different intensity from the rest of the corolla (Fig. 28,-b)     *M. glutinosa*

—    Pods smaller, 5-8 mm in $\phi$, coiled in 1.5-4 tight coils with only a small opening in the center. Corolla color uniform     *M. glomerata*

22(3)    Pods with spines     *M. daghestanica*

—    Pods without spines     **23**

23    Corolla violet; standard elongated (length : width ratio 2:1), with sides parallel in the middle part (Fig. 21,-c). Pods with 1.5-5 coils, with only a small opening in the center, 3-9 mm in $\phi$     *M. sativa*

—    Corolla variegated, standard oval (length : width ratio less than 2:1)     **24**

24    Pods with 2-5 coils, with a small opening in the center, covered with glandular hairs     *M.* x *tunetana, M.* x *gaetula*

—    Pods with 0.5-2.5 loose coils, with a large opening in the center, without glandular hairs     *M.* x *varia, M.* x *media, M.* x *hemicycla*

25(1)    Pods small, one-seeded nutlets (less than 3.5 mm long), their tips twisted in a small coil     **26**

—    Pods usually larger, coiled throughout their body in at least 1¼ coils     **27**

26    Veins on face of pod running obliquely from pod's ventral suture, do not change direction before joining the dorsal suture (Fig. 10,-b). Unexpanded pollen grains spindle-cylindrical     *M. lupulina*

—    Veins on face of pod change direction before joining the dorsal suture (Fig. 11,-b). Unexpanded pollen grains triangular pyramid-shaped      *M. secundiflora*

27   Seeds with long axes almost at right angle to pod's ventral suture; seedcoats ridged or verrucose      28

—   Seeds with long axes almost parallel to the ventral suture; seedcoats smooth      30

28   Pods spiny. Seedcoats ridged      29

—   Pods spineless. Seedcoats verrucose      *M. orbicularis*

29   Coils with one row of spines; coil edge paper-thin    *M. radiata*

—   Coils with two rows of spines; coil edge wide    *M. heyniana*

30   Expanded pollen grains cube-shaped (Fig. 8,-8). Seeds black or red brown      31

—   Expanded pollen grains round-triangular (Fig. 8,-7). Seeds yellow or yellow brown      34

31   Pods large, with 8-10 coils (Figs. 69,-d; 70,-b), florets large, 8-10 mm long; seeds large, 13-17 g/1000      32

—   Pods smaller, with 5-7 coils (Figs. 71,-b; 72,-b); florets smaller, 5-7 mm long; seeds smaller, 7.5-10 g/1000      33

32   Pods and spines with glandular, articulated hairs    *M. ciliaris*

—   Pods glabrous or with a few simple hairs    *M. intertexta*

33   Pods cylindrical or disk-shaped with truncate apex and base

     *M. muricoleptis*

—   Pods more round, barrel-shaped    *M. granadensis*

34   Pods densely covered with long hairs, resembling small cotton balls    *M. lanigera*

—   Pods without long hairs      35

35   Coils imbricate like a set of bowls, with their convex parts towards apex and base, or towards base only      36

—   Coils not markedly imbricate      37

36   Coils with their convex parts towards pod base only. $2n = 32$

     *M. scutellata*

—   Coils with their convex parts towards both base and apex. $2n = 16$

     *M.* x *blancheana, M. bonarotiana*

37   Coil edges with ridges, or with wing-like elevations running obliquely or at right angle to the dorsal suture      38

—   Coil edges smooth, or spined, or with tubercles      40

38   Wing-like elevations on coil edges running at right angle to pod's dorsal suture. Pods small, 3-5 mm in $\phi$, usually with 1.5 coils

     *M. shepardii*

—   Ridges on coil edges running obliquely towards the dorsal suture. Pods more than 5 mm in $\phi$, with 2½-5 coils      39

39   Dorsal suture usually in a groove in the middle of the coil edge. Calyx appressed to the base of the pod as a regular star. $2n = 16$

     *M. noëana*

—   Dorsal suture elevated in the middle of the coil edge. Calyx appressed sideways to the base of the pod. $2n = 32$    *M. rugosa*

| 40 | Dry, unexpanded pollen grains shaped like irregular blocks (Fig. 7,-4). Pods cylindrical, spines short (0.5-2 mm), inserted in the margins of the coil edges almost at right angle to the face of the coil; apical coil concave | *M. rotata* |
| --- | --- | --- |
| — | Dry, unexpanded pollen grains of other shapes (spindle-cylinder, or triangular pyramid-shaped) | 41 |
| 41 | Pods soft-walled. Central part of each coil consisting mainly of veins with thin membranous tissue between them (coils may be pulled apart releasing the seed). Spines, if present, slender, their base with two prongs (roots) connected by a membrane, one prong inserted in the dorsal suture, the other in the lateral vein or in a veinless zone | 42 |
| — | Pods hard-walled (for release of seed, crushing of pod may be necessary). Spines, if present, stocky, their base conical, often embedded in spongy tissue. Venation on the face of the coil usually not clearly discernible | 50 |
| 42 | Face of coil with radial veins running into a veinless zone | 43 |
| — | Face of coil with radial veins running into a lateral vein | 44 |
| 43 | Pod edge grooved; spines slanted away from the apical coil; apical coil spineless | *M. disciformis* |
| — | Pod edge level; spines pointing to both apical and basal end | *M. tenoreana* |
| 44 | Pod edge level or slightly concave, completely or almost completely covering grooves between the edge and lateral veins | 45 |
| — | Pod edge grooved, grooves between dorsal suture and lateral veins observable in edge-on view | 46 |
| 45 | Peduncle usually many-flowered (up to 17 florets). $2n = 16$ | *M. coronata* |
| — | Peduncle few-flowered (1-2 florets). $2n = 14$ | *M. praecox* |
| 46 | Dorsal suture of pod lying in a groove; on pod edge alternate 4 ridges with 3 grooves | *M. arabica* |
| — | Dorsal suture of pod elevated above lateral veins | 47 |
| 47 | Plants densely hairy. Lateral veins on face of coil at 1/3-2/5 of the radius below the dorsal suture. Stipules entire or slightly toothed | *M. minima* |
| — | Lateral veins on face of coil at 1/3 or less of the radius below the dorsal suture. Stipules deeply incised | 48 |
| 48 | Florets with wings longer than the keel. $2n = 14$ | *M. polymorpha* |
| — | Florets with wings shorter then the keel. $2n = 16$ | 49 |
| 49 | Apical coil spiny; lateral veins on coil face joining as shoulders at right angle to the elevated dorsal suture (Fig. 60,-f) | *M. laciniata* |
| — | Apical coil spineless; lateral veins only slightly protruding from the coil face (Fig. 59,-c) | *M. sauvagei* |
| 50(41) | Coils spineless; coil face without lateral vein or veinless zone; coils tightly appressed | *M. soleirolii* |
| — | Coil face with lateral veins or veinless zone | 51 |

| 51 | On coil face radial veins ending in a veinless zone | 52 |
| — | On coil face radial veins ending in a lateral vein | 53 |

51   On coil face radial veins ending in a veinless zone   52

—   On coil face radial veins ending in a lateral vein   53

52   Veinless zone wide, about 1/3 of coil radius; upper side of leaves completely glabrous   *M. murex*

—   Veinless zone narrower, 1/4-1/5 of coil radius, upper side of leaves at least sparsely hairy   *M. turbinata*

53   Pods convex at both ends, subspherical or oval in shape. Young pod contracted and concealed within calyx (Fig. 3,-1) *M. doliata*

—   Pod ends truncate, pods cylindrical or subcylindrical   54

54   Coils of pod tightly appressed, with no slits between them in dry mature pods; juncture between individual coil edges not markedly depressed   55

—   A continuous or interrupted slit between coil in dry mature pods; coil edges usually sloped towards their juncture   57

55   No groove on pod edge between dorsal suture and lateral veins. Radial veins on pod face strongly curved, running almost concentric before joining the lateral vein. $2n = 14$   *M. constricta*

—   On edge of immature pods a shallow groove between dorsal suture and lateral veins, disappearing at pod maturity. Radial veins only slightly curved. $2n = 16$   56

56   Pods glabrous; dorsal suture usually not higher than margins of the edge; spines, if present, inserted at 180° to the plane of coil face or obliquely to it   *M. littoralis*

—   Pods with sparse hairs; dorsal suture usually strongly protruding in the middle of coil edge; spines inserted at 90° or obliquely to the coil face   *M. truncatula*

57   Dorsal suture in the middle of an evenly convex pod edge (puffed-up parenchymatous tissue on edge margins may need clearing off). Radial veins somewhat curved. $2n = 16$ *M. tornata*

—   Shallow groove between dorsal suture and lateral veins that disappears at pod maturity. Radial veins strongly curved. $2n = 14$

*M. rigidula*


SUBGENUS LUPULARIA (Ser.) Grossheim, in Kom. (ed.) Fl. USSR 11:134 (1945).

    Annuals, biennials or perennials. Corolla yellow. Pods one-seeded non-dehiscent nutlets, their tips twisted in a small coil. $2n = 16, 32$.

Key to species of subgenus *Lupularia:*

1.   Veins on face of pod running obliquely from the ventral suture; do not change direction before joining the dorsal suture

*M. lupulina*

—   Veins on face of pod change direction before joining the dorsal suture   *M. secundiflora*

Apart from being distinct from other *Medicago* subgenera in morphological, especially pod characters, subgen. *Lupularia* is also distinguishable in such characters as susceptibility to diseases: it is resistant to *Pseudopeziza medicaginis* f. sp. *sativae*, but susceptible to *Pseudopeziza medicaginis* f. sp. *lupulinae* (Schmiedeknecht & Lesins, 1968). Its protein composition as found in serological investigations differ from other *Medicago* species studied (Simon, 1969).

*1.* *Medicago lupulina* Linnaeus, Sp. Pl.: 779 (1753). Figs. 10\*; 5,-24.

Perennial, biennial or annual herbs; prostrate to decumbent, 20-60(80) cm long, branching from the base. Vegetative parts more or less densely covered with simple appressed or with simple and glandular upright hairs. Stipules wide, entire or irregularly toothed. Leaflets broadly ovate to obovate, 11-14 mm long, 6-11(17) mm broad; almost equally hairy on upper and lower sides, margin slightly toothed in its upper half; teeth somewhat irregular; midrib ending in a triangular tooth. Peduncle 14 to 24-flowered, longer than the corresponding petiole; florets gathered in a head-shaped cluster, with a cusp beyond the terminal floret. Florets tiny, about 2.5-3.5 mm long. Pedicel shorter than the calyx tube; bract equal to or longer than the pedicel. Calyx 1.5-2.3 mm long, covered with simple or simple and glandular hairs; teeth ± the length of the tube. Corolla yellow, about twice the length of the calyx; standard subround to broadly ovate; wings shorter than the keel. Young pod contracted within the calyx, covered with simple or glandular hairs. Mature pod an ash-grey to black nutlet with a coiled tip, spineless, about 2.5 mm long, glabrous or covered with simple or glandular hairs. On pod face 3-5 curved veins running obliquely from the center, branching somewhat in the outer part, entering the dorsal suture without change of direction. Each pod containing one seed. Single seed oval, yellowish, 1.7-2 mm long, 1-1.3 mm broad. Seed weight 1.4-1.6 g/1000. Radicle longer than half (up to 2/3) the length of the seed, its tip protruding. $2n = 16,32$.

*Habitat.* M. lupulina prefers moister soils and cooler temperatures than the expressly annual *Medicago* species. In warmer arid regions it may occur in moist meadows, or at higher altitudes where moisture and temperature conditions are more suitable for it.

*Distribution.* Europe, most of Asia, and North Africa. Growing from 33° to 60° northern latitude. We collected it at Ifrane, Morocco; at Oslo, Norway, and at localities between these latitudes. As adventitious plants we found it in Alberta (Grande Prairie, 55° N. Lat.), Canada; it is reported as an escape from cultivation in many North American localities.

*Variation Within Species.* Growing under rather different environmental conditions in widely separated sites, the species still does not show conspicuous variation. Urban (1873) discerned two varieties: 1) *typica*, usually annual, biennial, rarely perennial plants with standard not exceeding 1¾ of the calyx length, and 2) *cupaniana*, perennial plants, glandular-hairy pods

---

\*   The main figures showing vegetative parts are listed first.

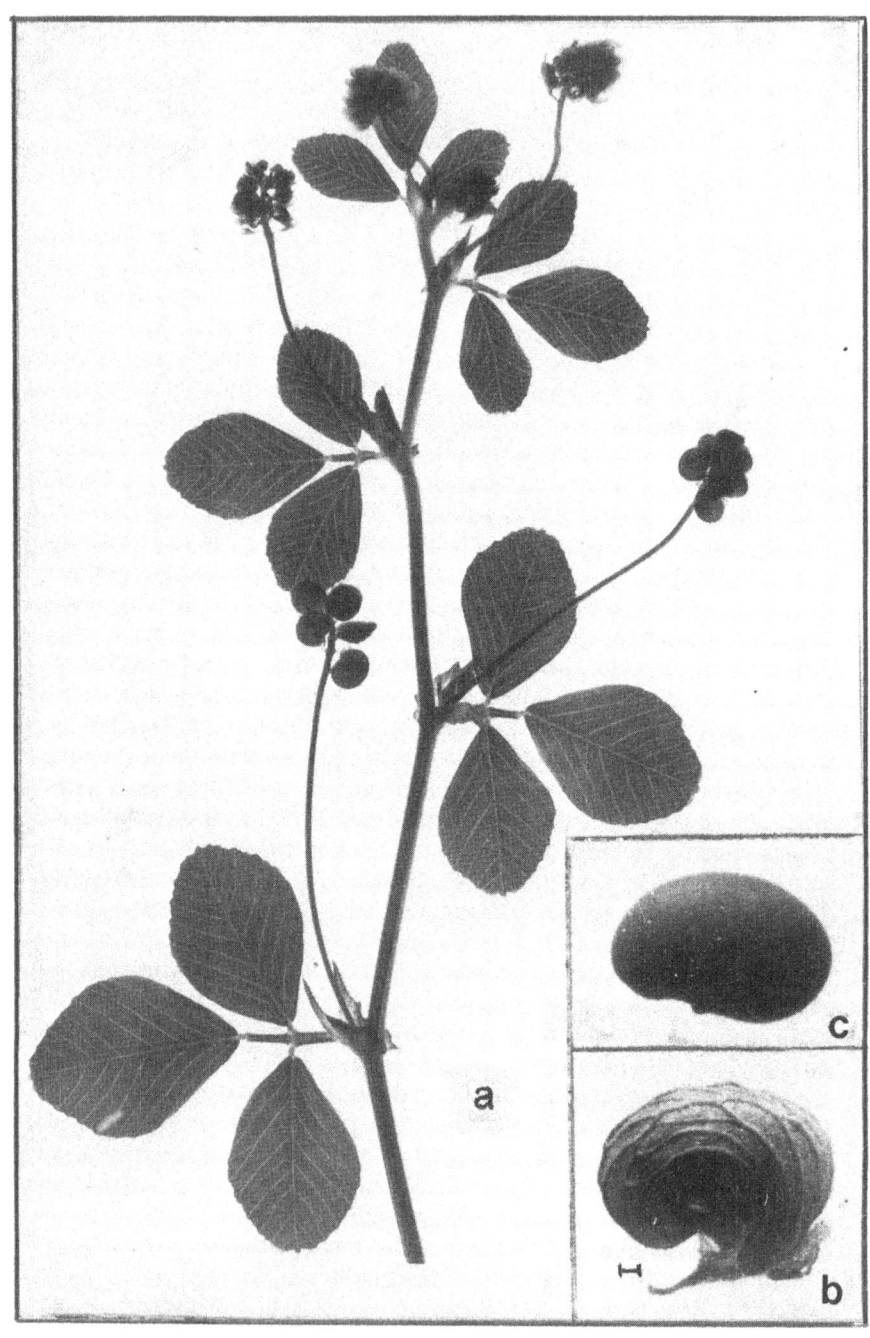

*Fig. 10. M. lupulina.* Branch (a), pod (b), and seed (c). Note: veins do not change direction on the pod face; the tip of radicle appears as a protrusion at the side of the seed.

68

and with standard more than twice the length of the calyx. However, as one form of *cupaniana* he lists *leiocarpa* with pods without hairs or with simple hairs. The var. *typica* (= var. *lupulina*) he divides into forms: *wildenowii*, with glandular-hairy pods, and *stipularis*, with broad-ovate stipules. In addition he lists some monstrous forms. We find this subdivision of *M. lupulina* acceptable, though the ratio of length of the standard against calyx, used as the basic character for distinguishing between his two varieties, is not quite satisfactory, as teeth are unequal. Calyx teeth adjoining standard petal are shorter than those adjoining the keel; thus standard may be more than twice the length of the calyx, or it may not be more than 1¾ of it, depending on which calyx side is used for measuring.

Regarding lifespan: Although according to our description *M. lupulina* is 'perennial, biennial or annual', we have never found, in the greenhouse a plant which had died after setting seed. If repeated flowering and seed setting is accepted as the basic trait characterizing a perennial plant, then all *M. lupulina* observed were perennials. One may speculate that in this species certain environmental conditions rather than inherent ability determine perennial, biennial or annual lifespan. Under more southern conditions perennial forms seem to prevail. At Ifrane, Morocco, we found plants which certainly were perennials in nature; according to Urban's classification they were var. *cupaniana* with glandular hairs on pods and on vegetative parts.

A tetraploid form with $2n = 32$ reported by Tschechow (1932) seems to be confined to central Siberia. All our accessions are $2n = 16$.

Regarding the tripping mechanism: Generally on inspection of florets we found that the staminal column had opened the keel, indicating that the column had some tension. However, the pistil sometimes does not even touch the standard, which shows that, though a tripping mechanism does exist, the tension in the staminal column is less strong than in other *Medicago* species.

*Relationship to Other Species. M. secundiflora* is undoubtedly the closest relative of *M. lupulina*. Some of our crossing attempts between the two have been unsuccessful, possibly because the crossing technique for handling such small florets has not been worked out. The nature of reported hybrids (Schröck, 1943; Southworth, 1928) between *M. sativa s.l.* and *M. lupulina* is questionable. Cytological evidence presented by Schröck (l.c.) is not convincing, since his maternal plant ($F_1$? from *M. media* x *M. lupulina*) has been cytologically unstable ($2n = 24, 32$). A pure *M. lupulina* plant reportedly found in the progeny would be the only indisputable evidence, but the possibility cannot be overlooked that a volunteer seed had not been introduced after soil sterilization, or had not survived it. Progenies of Southworth's *M. sativa* x *M. lupulina* had been examined by the late Dr. Fryer, who did not think they were of hybrid origin (personal commun.). In our experience, plants obtained from pollination of emasculated or almost self-sterile *M. sativa* plants may appear grossly different from mother plants, thereby suggesting hybrid origin. However, these usually are rare selfs which owe their distorted appearance to inbreeding. Several putative hybrids between *M.*

*sativa* and *M. platycarpa* in our crossing program turned out to be morphological aberrants. One plant from our cross between *M. sativa* and *M. lupulina* had a distinctly triangular tooth at the tip of the leaflets; it also turned out to be a *M. sativa* self, although Schröck (l.c.) assumed that the character indicated a true *M. media* x *M. lupulina* origin.

*Agricultural Value. M. lupulina* is sometimes added in seed mixtures for pastures. It reseeds readily and persists even under severe grazing conditions. Its yielding capacity, however, is low.

2. *Medicago secundiflora.* Durieu de Maisonneuve in Duchartre, Rev. Bot. 1:365 (1845). Syn *M. lupulina* var. *secundiflora* (Dur.) Fiori, Nuova Fl. Analit. Ital. 1:828 (1925). Figs. 11; 6,-52.

Annual herbs 15-25 cm long, procumbent, ascending to upright; branches round in transection, arising from near the base. Vegetative parts profusely covered with long, simple hairs giving the plant a greyish-green appearance. Stipules narrowly triangular, not joined at their base, with a few teeth in their basal part. Leaflets 5-12 mm x 4-10 mm, obovate (obcordate at lower nodes); apical 1/4 of leaflet margin toothed; midrib ending in a small, triangular tooth. Peduncle longer than the corresponding petiole with a distinct terminal cusp. Florets 3 to 10, in loose, one-sided raceme, 2-2.5 mm long. Pedicel slightly longer than the calyx tube; bract shorter than the pedicel. Calyx 1.8-2 mm long; teeth longer than the tube. Corolla yellow, only slightly longer than the calyx; standard elliptical, usually not becoming fully expanded (self-fertilization often takes place in the bud stage); wings shorter than the keel. Young pod densely covered with long white hairs; its tip (after protruding through the calyx teeth) turning in a small coil. Mature pod a greyish nutlet with dark-colored veins running obliquely from the ventral suture, changing direction at the edge of the pod. Single seed brownish, rounded on the cotyledon side, almost straight on the radicle side, 2 mm x 1.5 mm. Seed weight about 2.5 g/1000. Radicle more than half the seed length, its tip not protruding. Pollen grains bisphenoid-triangular in shape. $2n = 16$, chromosomes 3-4$\mu$ long, thus uniform in size, in contrast to those of most other annual *Medicago* (Lesins & Lesins, 1965).

*Habitat and Distribution.* Growing on calcareous, pebbly, sun-baked hill slopes in Algeria, less frequently in Morocco and Tunisia (Durieu, l.c; Nègre, 1959). Adventitiously found in southern France, Spain and Italy. After flowering and fruit setting, the plants die off and are difficult to locate, hence we could not observe them in their growing sites.

*Relationship to and Distinction from Other Species. M. secundiflora* is related to *M. lupulina* as indicated by similarities of florets and especially of pods, which in turn link them both to genus *Melilotus.* Some authors consider *M. secundiflora* a variety of *M. lupulina* (Fiori, l.c.). However, there are a number of differences between *M. lupulina* and *M. secundiflora*: Arrangement of florets in raceme; venation of pods (Fig. 11,-b vs. Fig. 10,-b); position of the tip of radicle in seed (Fig. 11,-c vs. Fig. 10,-c); shape of pollen grains (Lesins & Lesins, 1963); strictly annual (*M. secundiflora*) as

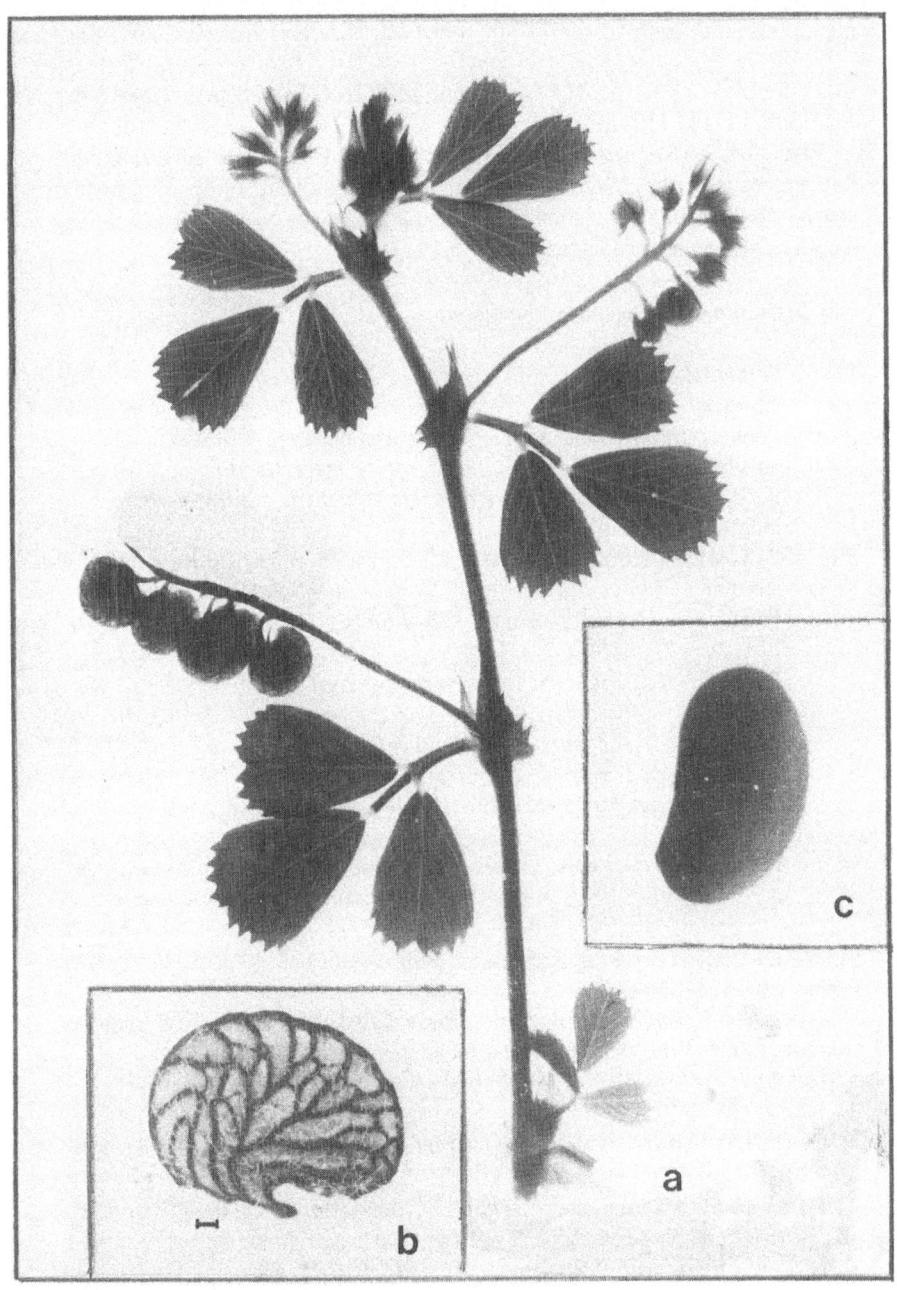

*Fig. 11, M. secundiflora*. Branch (a), pod (b), and seed (c). Note the change of direction of pod veins towards the edge of the pod.

against facultatively perennial (*M. lupulina*) lifespan; preference for hot, dry (*M. secundiflora*) vs. temperate, moist (*M. lupulina*) growing conditions.

SUBGENUS ORBICULARIA Grossheim, excl. sect. *Scutellata*; in Kom. (ed.) Fl. USSR 11:161 (1945).

Perennials or annuals. Corolla yellow. Pods usually with several seeds, flat and curved, or coiled; coiled pods usually without an opening in the centre. Radicle of seed at about a 45° to 90° angle to ventral suture, its length 2/3 or equal to cotyledons. $2n = 16$.

Key to sections of subgenus *Orbicularia*:

| 1 | Perennials | 2 |
|---|---|---|
| — | Annuals | 4 |
| 2 | Pods spiny. Radicle at 45° to ventral suture, 2/3 length of cotyledons sect. *Carstiensae* Kož. | |
| — | Pods spineless. Radicle at right angle to ventral suture, equal in length to cotyledons | 3 |
| 3 | Pods wide (width: length ratio 1:1.5 or less), dorsal suture semi-circular sect. *Cretaceae* Grossh. | |
| — | Pods less wide (width: length, 1:2 or more), somewhat bent sect. *Platycarpae* Trautv. | |
| 4 | Pods spineless. Seedcoat finely verrucose, not ridged sect. *Orbiculares* Urb. | |
| — | Pods spiny. Seedcoat verrucose and ridged | 5 |
| 5 | Pods with ¾-1½ coils, dorsal and ventral sutures thin, often fringed. Spines on dorsal suture in one row sect. *Hymenocarpos* Ser. | |
| — | Pods with several coils, dorsal suture wide. Spines in two rows sect. *Heynianae* Greuter | |

SECTION CARSTIENSAE Kožuharov. Bulg. Acad. Sci., Reports (Izvestiya) Botan. Inst. 15:132 (1965).

Perennials; spreading by rhizomes. Pods coiled, spiny. Coil edge slightly concave. Spines thin, flexible. Seeds 4-5 in one coil; seedcoat smooth.

The only representative is *M. carstiensis*.

3.  *Medicago carstiensis*. Wulfen, in Jacquin, Collect. Bot. 1:86 (1786). Figs. 12; 5,-6.

Plants 40-60 cm long, stems angular in transection, ascending to upright, arising from creeping rootstock. Vegetative parts hairless or sparsely covered with simple hairs. Stipules triangular, entire, or toothed at the lower stem nodes. Leaflets 11-22 mm x 7-14 mm, elliptical to ovate; margin toothed except for the basal part; midrib ending in a small terminal tooth. Peduncle 5 to 12-flowered, usually longer than the petiole, with a terminal cusp. Florets 8-10 mm long. Pedicel equal to or slightly longer than the calyx

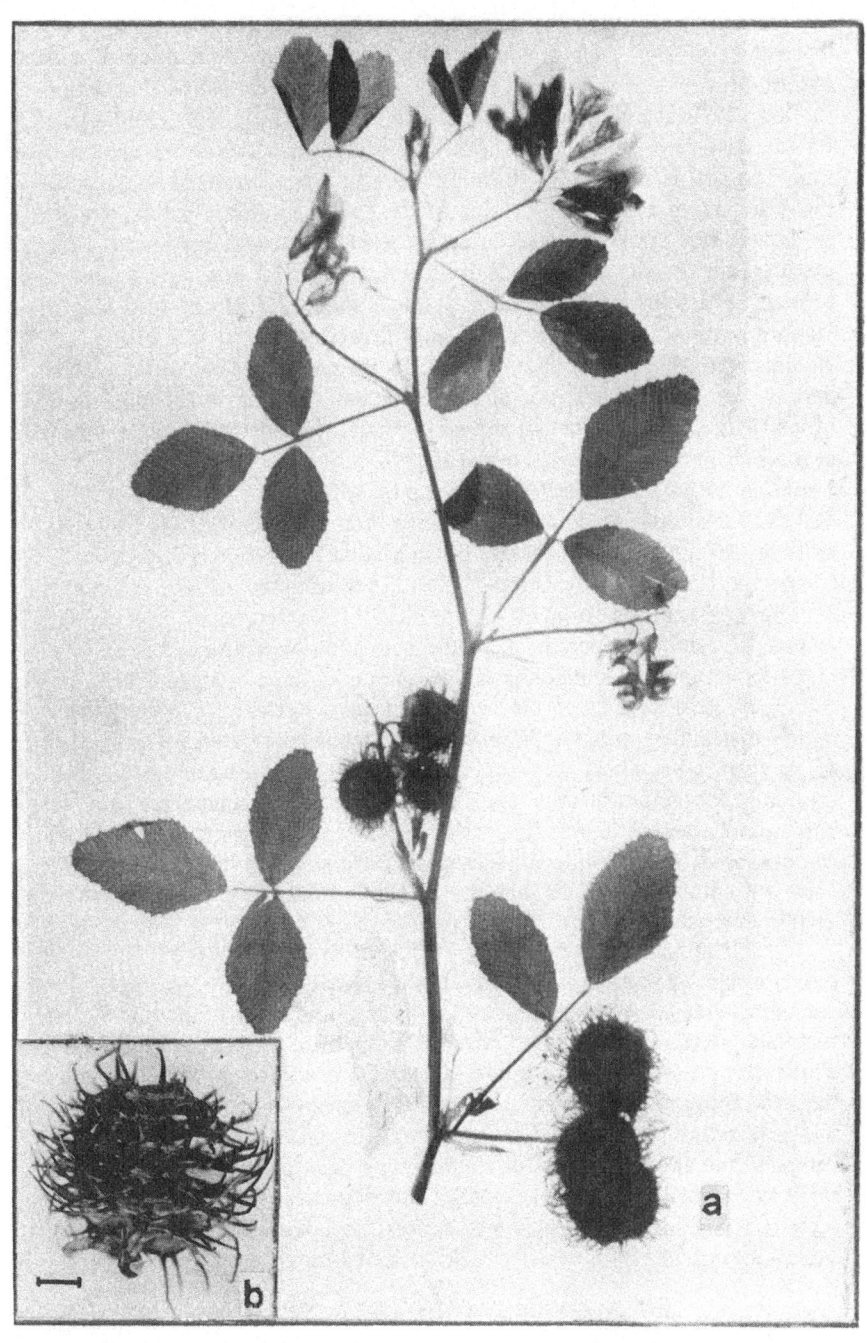

*Fig. 12,* *M. carstiensis.* Branch (a), and pod (b).

tube; bract shorter than the pedicel. Calyx 3.5-4 mm long; teeth narrowly lanceolate, shorter than the tube. Corolla yellow; standard obovate with reddish-brown veins (honey guides); wings as long as or slightly longer than the keel. Pods at maturity black, glabrous, cylindrical in their middle part, flat at basal and apical ends; coils 5-8, turning clockwise, 5-8 mm in $\phi$, spiny; on coil face 6-8 veins, forming a coarse net, veins thickening toward the edge; spines 14-18 on each side of the coil edge, thin, flexible, straight or curved, not hooked; dorsal suture in a shallow groove between the elevated margins of the edge. Seeds 1.7-2.5 mm x 1.2-1.7 mm, light yellow to brown. Seed weight 2-2.2 g/1000. Radicle about 2/3 of the seed length. Plants require winter rest for induction of flowering.

*Habitat and Distribution.* It is endemic to the coastal regions of the eastern part of the Adriatic Sea; not abundant in any growing site. Unlike most other *Medicago*, *M. carstiensis* thrives in shade. We collected it under shrubs at foothills in the vicinity of Trieste, Italy.

*Variation Within Species and Relationship to Other Taxa.* Kožuharov (1965) described a subspecies, *belasicae*, differing from the common type by silky hairs on the underside of leaves, by somewhat larger florets and pods. *M. carstiensis*, though put by Urban (1873) in the same section, *Orbiculares*, as *M. orbicularis*, differs from it in a number of characters: *M. orbicularis* is annual, *M. carstiensis* perennial, requiring in addition winter rest in order to come to bloom. This requirement is unique to *M. carstiensis* and is not found in other species of the genus *Medicago*. Further, *M. orbicularis* is widely distributed, whereas *M. carstiensis* may be considered a relic species. *M. carstiensis* has always a creeping rootstock, again a unique trait, since creeping roots found in other species (*M. falcata*, for instance) are sporadic and not characteristic for the entire species. In *M. orbicularis* pods are smooth, seeds verrucose, in *M. carstiensis* pods are spiny, seeds smooth. We agree with Kožuharov (l.c.) that *M. carstiensis* should be put into a separate section *Carstiensae*.

Hybridization attempts with *M. sativa* at both $2n = 16$ and $2n = 32$ ploidy levels were unsuccessful (unpublished results).

*Agricultural Value.* Forage production at Brandon has been about 60% that of alfalfa (Rambler). Winter survival at Edmonton was good, at Brandon winterkilling took place in winters of light snowfall. Seed production under field conditions where bee pollinators were scarce was good, indicating that self-fertilization takes place to a considerable extent. Experiments for its use for pasturing and in soil erosion control may be worthwhile because of the rapid spread of the plants by underground rhizomes. The species is moderately resistant to *Pseudopeziza medicaginis* f. sp. *sativa* (Schmiedeknecht & Lesins, 1968).


SECTION PLATYCARPAE Trautvetter, *Medicago* Division I, Acad. Imp. St. Petersb. Bull. Sci. 8:271 (1841).

Perennials. Pods spineless, flat, wide (width:length ratio 1:2.5 to 3), dorsal suture slightly curved, ventral suture almost straight. Florets with wings almost as large as the standard, both strongly bent outwards; keel very short, so that florets appear tripetalous (Fig. 14,-c). Seedcoat smooth or somewhat verrucose (Figs. 5,-34; 6,-45).

Key to main species of sect. *Platycarpae*:

1      Pods large, 15-22 mm long, leaflets broadly elliptical, seedcoat smooth         *M. platycarpa*

—     Pods smaller, 8-12 mm long, leaflets narrowly elliptical, seedcoat minutely verrucose     *M. ruthenica*

In addition to the two above mentioned species, Vassilczenko (1952) lists as closely related species under *Trigonella: T. karkarensis* Vass., *T. schischkinii* Vass., and *T. korshinskyi* Grossh. The present authors have not studied these taxa.

Members of section *Platycarpae* are considered by some authors, including Linnaeus (Sp. Pl. 1753), as belonging to the genus *Trigonella*. No doubt the section strongly deviates in many respects from the other perennial *Medicago*, which constitute the subgenus *Medicago*. Such authors as Širjaev (1934), the monographer of genus *Trigonella*, assign them to *Medicago* because of the fact that cotyledons are not divided into petiolar and laminar parts (see Fig. 9).

4.   *Medicago platycarpa* (L.) Trautvetter, Acad. Imp. St. Petersb. Bull. Sci. 8:271 (1841). Syn. *Trigonella platycarpos* L. Sp. Pl.:776 (1753). Figs. 13; 5,-34.

Plants ascending to erect, branching from the crown; stems 60-80(100) cm long, quadrangular in cross-section. Vegetative parts glabrous except for the sparsely haired underside of the leaves. Stipules triangular, not adnate at their base, with prominent teeth along the entire margin. Leaflets broadly elliptical, 20-30 mm x 20-25 mm, almost round at lower stem nodes; with nearly the whole margin serrate. Peduncle 4 to 7-flowered, longer than the corresponding petiole, with a terminal cusp. Florets 10-13 mm long. Pedicel longer than the calyx tube; bract much shorter than the pedicel. Calyx 2.5-5 mm long, sparsely covered with simple hairs; teeth triangular, shorter than the tube. Corolla yellow, tinged violet-red on the outside of the petals and along the veins (honey guides); standard with parallel sides, somewhat larger than the wings, almost twice as long as the keel. Pods glabrous, 14-22 mm x 6-8 mm, somewhat curved along the dorsal suture. Seeds brownish-yellow, 2.5-3 mm x 1.8-2 mm, 4 to 6 per pod. Seed weight about 3 g/1000. Seedcoat smooth (in contrast with the related *M. ruthenica*). Radicle as long as or only slightly shorter than the cotyledons.

*Habitat and Distribution.* The species is reported growing predominantly in moist podzolik soils, in river valleys, along edges of forests, in forest glades and meadows, between shrubs and high grasses (Balabaev, 1934; Vassilczenko, 1952). Generally it is one of the few shade tolerating *Medicago* species.

The main distribution area of *M. platycarpa* is western Siberia, especially the southwestern part. Scattered growing sites have been found as far north as 68-69° N.Lat. (Balabaev, l.c.) and at higher altitudes (of up to 3,000 m) as far south as Tien Shan and the Altai Mts. To the east, the boundaries are Lake Baikal and the River Amur. Nowhere, however, is the species found as the main component in plant communities; generally it is considered a pretty rare species.

Fig. 13, *M. platycarpa*. Branch (a), and pod (b).

*Relationship to other taxa. M. platycarpa* has been successfully hybridized with *M. ruthenica*, when used as the pistillate parent. The reciprocal combination, however, has not even stimulated ovary development (Oldemeyer, 1956), and hence the relationship between the two taxa probably is not a very close one.

We tried to hybridize *M. platycarpa* with *M. sativa*, using *M. platycarpa* at its natural chromosome level $2n = 16$, as well as at the artificially induced $2\tilde{n} = 32$. *M. sativa* also was matched to these levels by $2n = 16$ (ssp. *coerulea*) and $2n = 32$ (cultivated alfalfa). Neither at equal chromosome levels ($2n = 16$ x $2n = 16$, and $2n = 32$ x $2n = 32$) nor at unequal chromosome levels (*M. platycarpa* $2n = 32$ x *M. sativa* $2n = 16$) were any hybrids obtained.

*Agricultural Value.* At Edmonton, $2n = 16$ as well as $2n = 32$ plants came through a couple of winters unharmed. At Brandon, survival has not always been satisfactory. There, a very high variation even within a single accession was noted. Regrowth has been light, as already pointed out by Balabaev (l.c.).

5. *Medicago ruthenica* (L.) Ledebour, Fl Ross. 1:523 (1842). Syn. *Trigonella ruthenica* L., Sp. Pl.:776 (1753). Figs. 14; 6,-45.

Plants ascending to upright, branching from the crown, stems 30-50 cm long. Vegetative parts, except the upper side of the leaves, covered with short appressed hairs. Stipules triangular, entire or with a few teeth at the base. Leaflets elliptical to obovate, 7-15 mm x 2-5 mm, serrate in 4/5 of their apical part, midrib ending in a small terminal tooth. Peduncle 6 to 12-flowered, longer than the corresponding petiole, with a small terminal cusp. Florets 5-7 mm long. Pedicel longer than the calyx tube; bract inconspicuous, much shorter than the pedicel. Calyx 2.5 mm long; teeth shorter than the tube. Corolla yellow, tinged with dark purple on the outside of the petals and on the inside toward the base; standard large, usually wider in its upper part, almost twice as long as the keel; wings as long as the standard, both kinds of these petals strongly recurved. Pods slightly bent, glabrous, 8-12 mm x 3.5-4.5 mm. Seeds 1-6 per pod, greenish-brown, 1.8-2.5 mm x 1.5-1.8 mm. Seed weight 2-2.5 g/1000. Seedcoat rough, minutely verrucose. Radicle almost as long as the cotyledons.

*Habitat and Distribution. M. Ruthenica*, unlike the related *M. platycarpa*, is reported as growing in open steps, on sunny hillsides, in dry, gravelly soil (Balabaev, 1934; Vassilczenko, 1952).

Distribution area lies in the east of *M. platycarpa*, from the Transbaikal region to the Pacific Ocean. It also grows further south than *M. platycarpa*, reaching about 34° N. Lat. (Vassilczenko, l.c.).

*Variation Within Species.* Vassilczenko (l.c.) notes that there are two varieties of *M. ruthenica* deviating from the usual type; the one having much larger leaflets and pods and the other having smaller, narrower leaflets, shorter pods, and shorter, leafy stems. The first variety seems to be adapted to sandy soils, the second to rocky, gravelly hillsides.

*Relationship to Other Species.* As noted before, *M. ruthenica* has been hybridized with *M. platycarpa*, with the latter as the pistillate parent (Oldemeyer, 1956). No backcrossing to *M. ruthenica* has been successful, not even if the pollen was taken from the hybrids. Our attempts to hybridize it with *Trigonella papovii* Vass., to which it has some morphological resemblance, were not successful.

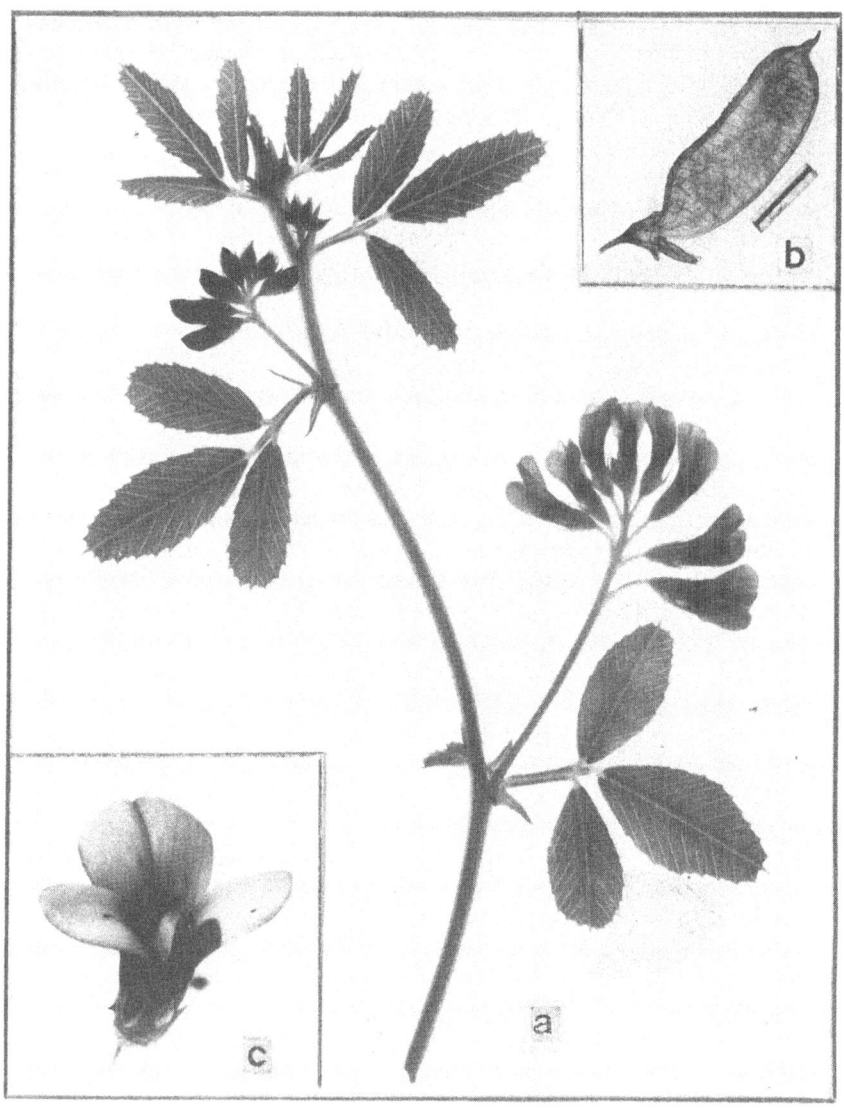

*Fig. 14, M. ruthenica.* Branch (a), pod (b), and floret (c); note the large wings and the small keel petal.

SECTION ORBICULARES Urban, Verh. bot. Ver. Brand.: 48 (1873).

Annuals. Pods spineless, with 3-7 coils. Coils with soft walls, edges pergamentaceous, paper-thin. Seeds 3-6 in one coil; seedcoat verrucose.

The section has a single representative, *M. orbicularis*. It was found to be different in serological tests from other *Medicago* species investigated by Simon (1969).

6. *Medicago orbicularis* (L.) Bartalini, Cat. Piante Siena:60 (1776). Syn, *M. polymorpha* var. *orbicularis* L. Sp. Pl.:779 (1753); *M. marginata* Willd., Enum. Hort. Reg. Bot. Berol.:802 (1809); *M. applanata* Willd. ex DC. Prodr. 2:175 (1825); *M. cuneata* Woods, Tour. Fl.:84 (1850). Figs. 15; 5,-30.

Plants 35-50(-120) cm long, glabrate; many branches start from the main stem at ground level, branching secondarily throughout their length. Stipules laciniate. Leaflets 9-18 mm x 6-14 mm, obovate, cuneate to obcordate; margin in its 1/3-2/3 apical part serrate; midrib ending in a triangular tooth. Peduncle 1 to 5-flowered, shorter than the corresponding petiole, with a terminal cusp. Florets 4-6 mm long. Pedicel longer than or equal to the calyx tube; bract shorter than or equal to the pedicel. Calyx about half the length of the floret; teeth broadly triangular, shorter than or equal to the tube. Corolla yellow, often with a violet hue on the outer side of the standard; standard obovate; wings shorter than the keel. Young pod glabrous or with sessile glandular hairs, rising in a loose spiral from the calyx, then turning sideways. Mature pod glabrate, light straw-colored or black, discoid, subspherical or cylindrical (end coils smaller), spineless; coils 3-7, usually not appressed, 9-20 mm in $\phi$, with pergamentaceous paper-thin edges, turning clockwise; on coil face 12-26 radial veins, slightly branching soon after leaving the ventral suture. Seeds 3-6 in each coil, yellowish-brown, separated by a thin wall; 2.5-3 mm x 2-2.5 mm; seedcoat verrucose. Seed weight about 5.5 g/1000. Radicle as long as the seed. On germination cotyledons show a purplish hue.

Habitat. Grows mainly in heavy soils that do not dry out readily.

*Distribution.* An omni-Mediterranean species extending to the western part of Asia. Adventitious in central Europe. Introductions in some regions for establishing it as a range plant in pastures have not been successful (e.g., in Argentina, person. commun. Dr. Burkhart).

*Variation Within Species.* As the number of synonyms indicate variations in *M. orbicularis* are abundant. Some of the pod characters which have served to demark species or subspecific ranks are: diameter and number of coils, and the direction of coil edges (spreading apart or adhering together). Heyn (1963), studying herbarium specimens, concluded that there was no discontinuity of character expression, hence no clear basis for naming separate taxa. She gives an extensive table (No. 11, l.c.) showing the variations found. We intercrossed plants: 1) pods roundish, 5.5-6 coils, 8 mm in $\phi$, coil edges appressed, with 2) pods short-cylindrical, 3 coils, 18 mm in $\phi$, coil edges not closely appressed. Unpublished $F_2$ data were consistent with the hypothesis that coil number was controlled by one gene ith two alleles, whereas pod shape and coil diameter were controlled by more than one gene. A considerable amount of recombination of phenotype for the three characters was observed, hence we conclude that the species is a highly variable one, and support Heyn's contention that further taxonomic subdivision is not justified.

Interesting variations were found in a single spot on Mt. Kolla (Island of Karpathos in Aegean Archipelago). A few seeds from a partly decayed pod

*Fig. 15, M. orbicularis.* Branch (a), and pods (b,c).

gave plants with 3 to 5-flowered peduncles and light colored pods, whereas the plants presently growing there had 1 to 3-flowered peduncles and black pods. It appears that a combination of impermeable seedcoat and slow after-ripening of seeds in dry climatic conditions may preserve different growth forms side by side for an almost unbelievably long time.

Urban (1873), as noted before, combined *M. orbicularis* with *M. carstiensis* in section *Orbiculares*, because of the similarity in size and position of the radicles and cotyledons in relation to the ventral suture in the pods. However, as pointed out earlier, the two species differ in basic morphological as well as physiological and ecological characters. Therefore, we consider the two as belonging to two different sections, though both part of the same subgenus *Orbicularia*.

*M. orbicularis* appears to be somewhat related to *M. heyniana* (Greuter, 1970). In attempted crossing between *M. orbicularis* and *M. heyniana*, however, no hybrids were obtained.

SECTION HYMENOCARPOS Seringe in de Candolle, Prodr. Syst. Natur. 2:171 (1825).

Annuals. Pods flat, with 0.5-1.5 coils, usually fringed on dorsal or ventral suture, or on both. Spines, if present, in a single row. Seeds up to 8 in one pod. Seedcoat verrucose, transversely ridged.

The section with its single representative, *M. radiata*, is often considered to belong to the genus *Trigonella*. In serological tests Simon (1969) found differences between *M. radiata* and other *Medicago* species, and similarity with *Trigonella*. Heyn (1959) indicates similarity between pods, in the spines especially, of *M. radiata* and *Trigonella arabica*. According to Heyn (1963) the section should be named *Medicago* if *M. radiata* is considered the type species of the genus.

7. *Medicago radiata Linnaeus*, Sp. Pl.:778 (1753). Syn. *Trigonella radiata* (L.) Boiss., Fl. Or. 2:90 (1872). Figs. 16; 6,-39.

Plants 15-50 cm long, ascending , branching near the base. Vegetative parts covered with lengthy, diffuse simple hairs. Stipules deeply incised. Leaflets 15-25 mm x 10-15 mm, elliptical (obtuse to obcordate in some forms); margin in its 1/3-2/3 apical part toothed or crenate; pedicel of the middle leaflet long (half or more the lenght of the lamina), lateral leaflets sessile. Peduncle 1 to 3-flowered, as long as or longer than the corresponding petiole, with a minute terminal cusp. Florets 4-6 mm long. Pedicel longer than the calyx tube; bract much shorter than the pedicel. Calyx 2.5-3 mm long, teeth as long as or longer than the tube. Corolla yellow; standard almost as wide as long; wings longer than the keel. Pods glabrous to sparsely covered with simple hairs, with ½-1½ coils, 18-28 mm in $\phi$. Dorsal, and often ventral sutures fringed, paper-thin. Pod veins starting close to the ventral suture run radially halfway to the pod edge then anastomose, some veins proceeding into the dorsal suture. Spines rather fine, short (1 mm), arising directly above the dorsal suture in a single row, some spines forked at

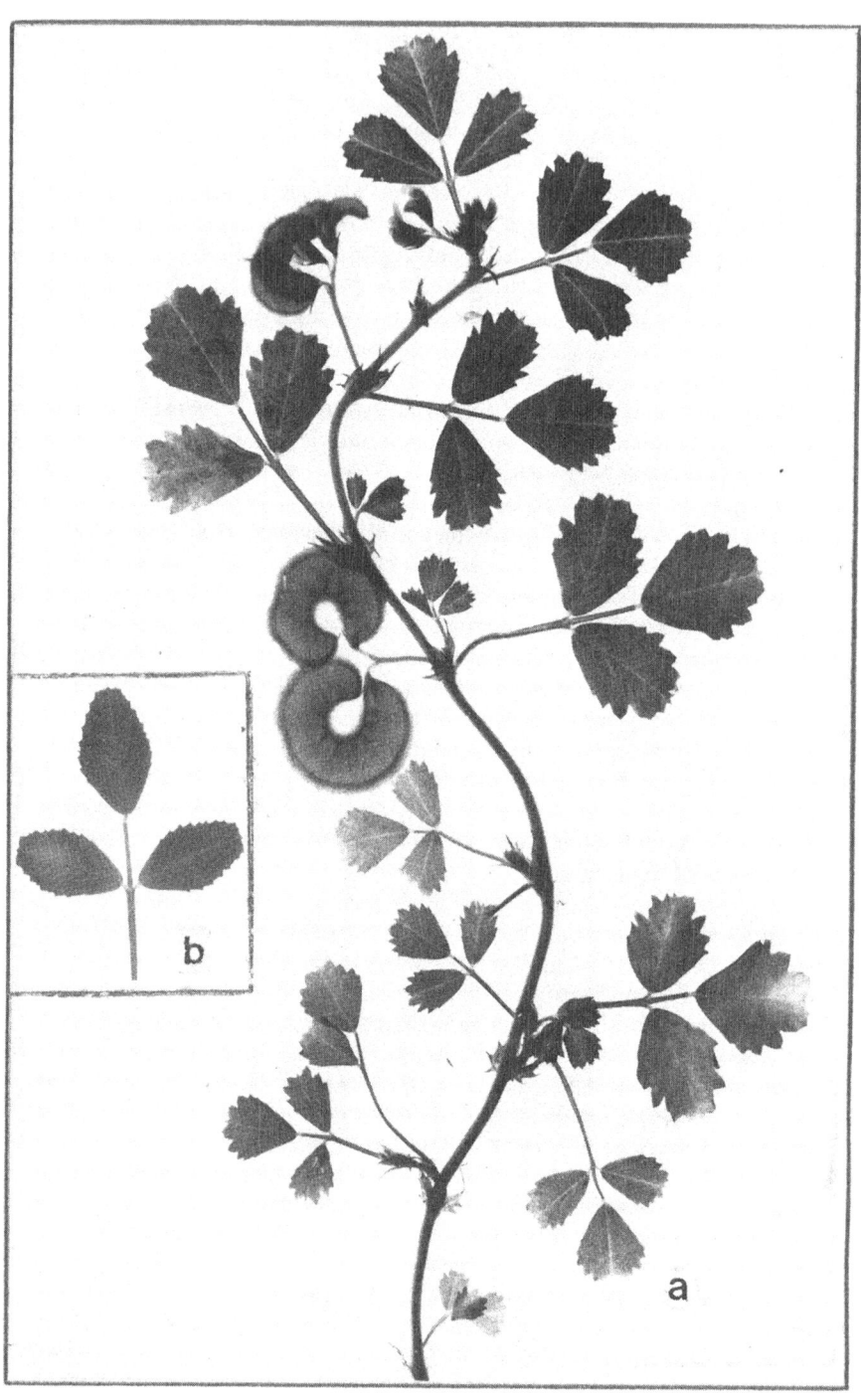

*Fig. 16, M. radiata.* Branch (a) with crenate leaves, and a serrate-margined leaf (b).

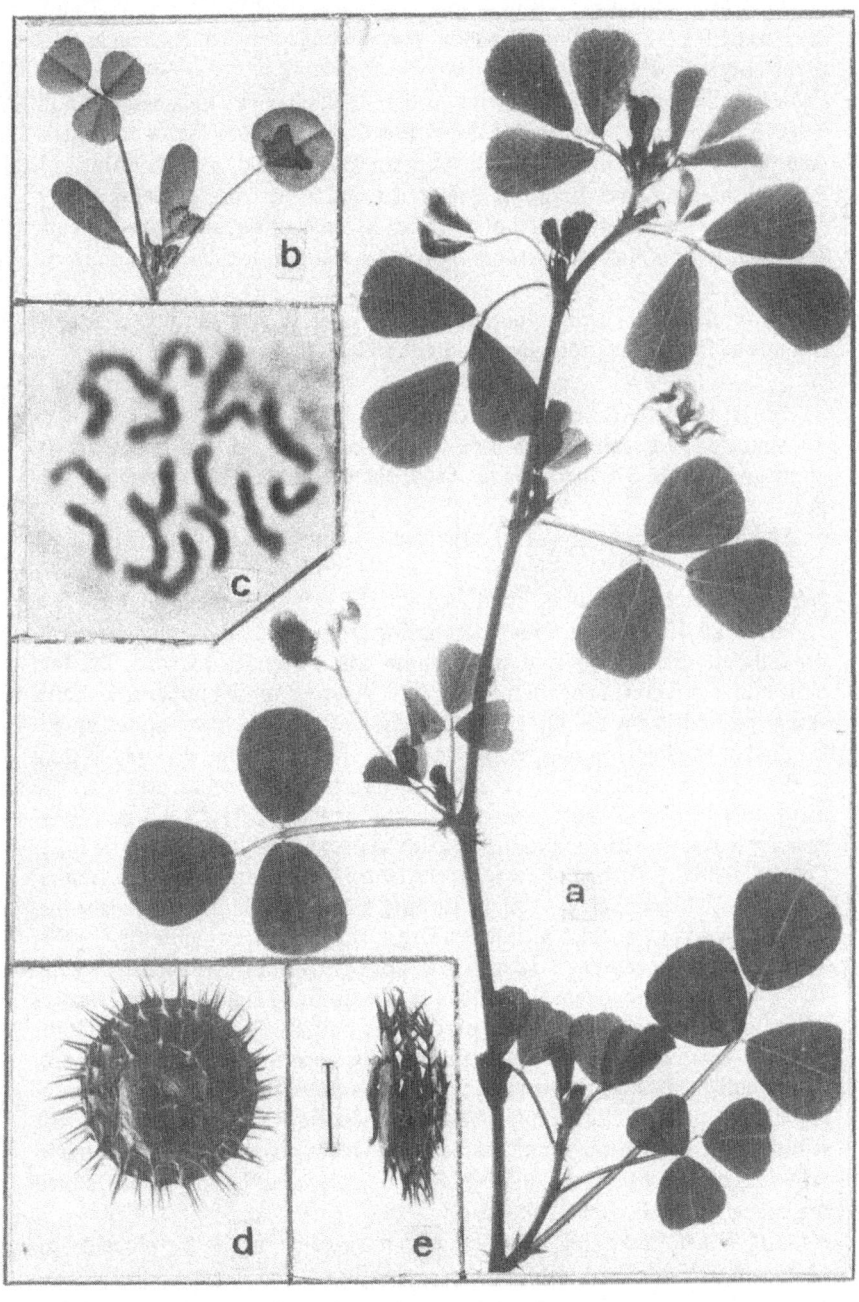

*Fig. 17, M. heyniana.* Branch (a), seedling (b); chromosome complement (c) magn. x 4,000. Pods in face (d), and edge-view (e).

the tip. Seeds brownish, 6-8 in a pod, separated, 2.7-3.2 mm x 2.3-2.7 mm.
Seed weight 1.8-2.5 g/1000. Seedcoat conspicuously ridged. Radicle as long
as the cotyledons.

*Habitat and Distribution.* Growing in dry soils, on rocky hillsides, often in
desert-like environs. Distributed from the Caucasus, Asia Minor to central
Asia. We collected it in central and southern Turkey, and in Lebanon.

*Relationship to Other Taxa.* It does not seem to be closely related to any
other *Medicago* species. As noted later, hybridization attempts with *M.
heyniana*, with which it has some morphological features in common, were
unsuccessful.

*Agricultural Value.* During copious spring rains it may develop a sizable
amount of forage, according to Kasimenko (1951).

SECTION HEYNIANAE Greuter, Candollea 25/2:189 (1970).

Annuals. Pods spiny, with 3-4.5 coils. Coils with soft walls, not tightly
appressed. Seeds 3-4 in one coil. Seedcoat verrucose and transversely ridged.

The only representative is *M. heyniana.*

8. *Medicago heyniana* Greuter, Candollea 25:190 (1970). Figs. 17; 6,-62.

Plants 20-45 cm long, branches starting from the base of the main stem,
decumbent, sparsely covered with simple hairs. Stipules toothed. Leaflets
5-10 mm x 3-8 mm, obòvate to obcordate, glabrous on the upper side, hairy
along the midrib on the underside; margin entire at the lower nodes, small-
toothed at the higher nodes. Peduncle 1 to 2-flowered, appoximately as long
as the corresponding petiole. Florets 7-9 mm long. Pedicel shorter than the
calyx tube; bract slightly shorter than the pedicel. Calyx 3-3.5 mm long,
sparsely covered with simple hairs; teeth approximately as long as the tube.
Corolla yellow; standard obovate; keel slightly longer than the wings. Young
pod arising from the calyx then turning sideways. Mature pod glabrous,
yellowish-grey, often with a violet hue, disk-shaped or short-cilindrical. Coils
3-4,5, turning clockwise, 9-12 mm in $\phi$, not tightly appressed, coil face with
10-12 veins emerging from the ventral suture, running almost concentrically,
becoming indistinct in the outer part of the coil face. Spines thin, 17-20 in
a row, 3-4 mm long, grooved at their base, not hooked, inserted at 150-180°
to the coil face with one prong in the dorsal suture, the other in the scleren-
chymatous part of the coil face. Seeds yellow-brown, 3-5 in one coil, not
separated by a partition, 3 mm x 2.5 mm; seedcoat verrucose and transversely
ridged. Seed weight about 5 g/1000. Radicle as long as the cotyledons, hence
the triangular-oval shape of the seed.

*Habitat, Distribution and Variation Within Species.* In the type locality of
*M. heyniana* (Mt. Kolla, the Island of Karpathos in the Aegean Archipelago)
at an altitude of 700 m, we collected on a dry, rocky hillside two forms:
one with one-flowered peduncles as originally described by Greuter (l.c.),
the other with two-flowered ones. This latter form originated from a partly
decayed pod, indicating that such plants had been growing there many seed

84

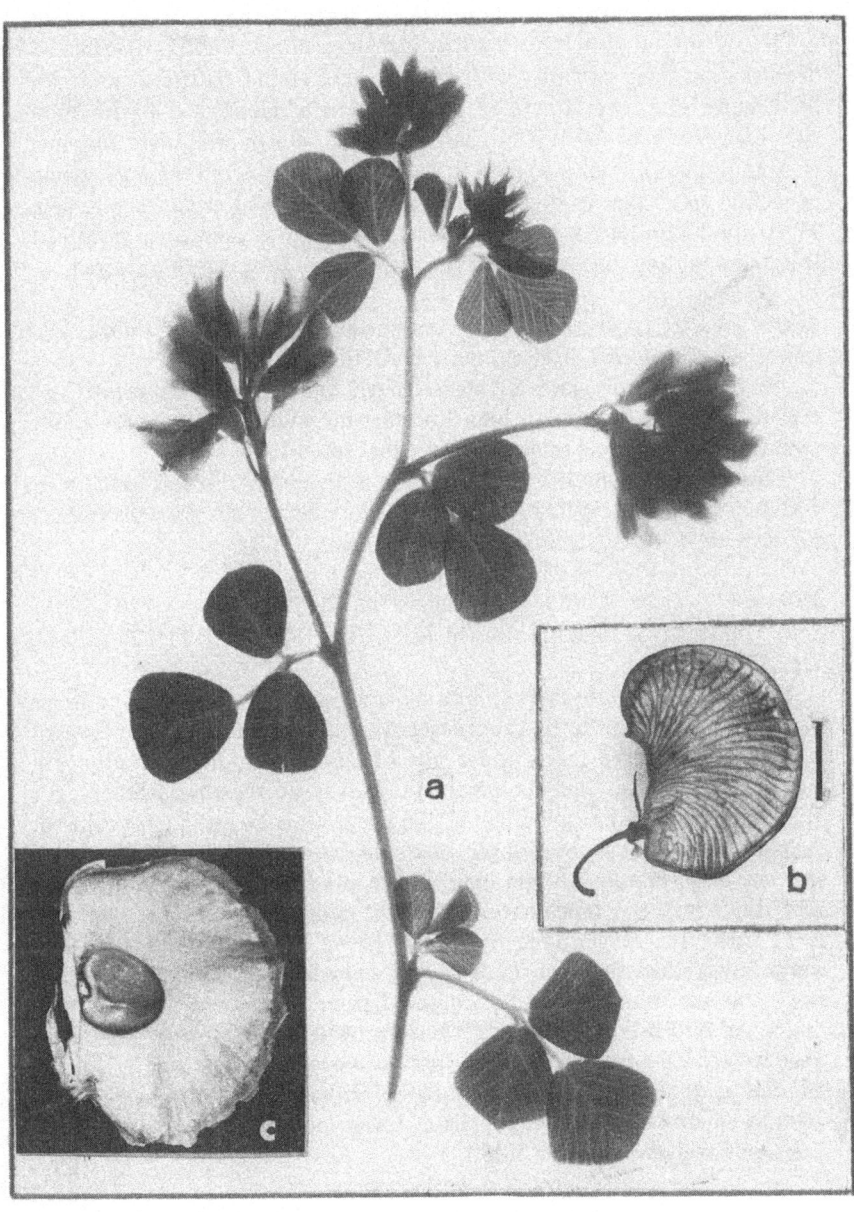

*Fig. 18, M. cretacea.* Branch (a), and pods (b,c). Note non-branching radial veins on pod face (b); a seed in a pod with its long axis perpendicular to the ventral suture (c).

generations ago. Further, we found some plants showing a darkly colored, irregularly shaped blotch on the first leaf of seedlings (Fig. 17,-b). Since *M. heyniana* has some similarity with *M. orbicularis* and *M. radiata*, as indicated by Greuter (1.c), we attempted hybridization of *M. heyniana* with these two species. From 70-100 crosses with each, no hybrids were obtained (Lesins, unpubl.). It is concluded that *M. heyniana* is indeed the only species in section *Heynianae* as described by Greuter (l.c.). The chromosomes (Fig. 17,-c) are 3-4.2$\mu$ long with centromeres at median to submedian positions, thus more uniform in size than those of annuals in subgenus *Spirocarpae*.

SECTION CRETACEAE Grossh. (Grossheim, gen. *Trigonella*, subgen. *Pocockia*, ser. *Cretaceae*), in Kom. (ed.) Fl. USSR 11:120 (1945).

Perennials; lower parts of stems tough, hardy. Pods one-seeded, not coiled, flat, wide, length less than 1.5 times the width, ventral suture almost straight, dorsal suture a semicircle. Seedcoat smooth.

The only representative of the section is *M. cretacea*. It was found to be the only species susceptible to *Pseudopeziza meliloti*, but resistant to *Pseudopeziza medicaginis* (Schmiedeknecht & Lesins, 1968).

9. *Medicago cretacea* Marshall von Bieberstein, Fl. Taur.-Cauc. 2:223 (1808). Syn. *Trigonella cretacea* Grossh. in Kom. (ed.) Fl. USSR 11:120 (1945). Figs. 18; 5,-10.

Plants ascending to erect, profusely branching from the crown; stems 15-25 cm long, tough in texture. Vegetative parts, especially stems, covered with short, appressed hairs, upper side of leaves somewhat less hairy. Stipules slender, entire. Leaflets obcordate, obovate to rhombic, leathery, 6-8 mm long and almost as wide, entire-margined; veins clearly lighter colored. Peduncle 3 to 8-flowered, longer than the corresponding petiole, with a terminal cusp. Florets 5-8 mm long. Pedicel 3-4 times longer than the calyx tube; bract leathery, much shorter than the pedicel. Calyx 3-3.5 mm long; teeth triangular, longer than the tube. Corolla yellow, standard obovate; wings longer than the keel. Pods usually one-seeded, 8-12 mm x 6-8 mm, straw-colored; veins on pod face proceed from ventral to dorsal suture almost without branching. Seeds yellow-brown, 2.5-3.5 mm x 2-2.5 mm. Seed weight 3.8-5.6 g/1000. Radicle almost the length of the seed.
*Habitat and Distribution.* According to Grossheim (l.c.), the species is growing in dry, limestone-rich soils at lower mountain zones. Endemic to Transcaucasian and Crimean USSR.

SUBGENUS MEDICAGO Tutin, in Fl. Eur. 2:154 (1968) excl. subgen. *Lupularia* et *Orbicularia*, sect. *Intertextae* et spec. *M. scutellata*.

Perennials. Corolla yellow or violet. Pods coiled, usually with an open centre, less frequently curved or almost straight. Seeds 1-2 to several per pod, their long axis parallel or close to parallel to the ventral suture. 2$n$ = 16,32,48.

Key to sections of subgenus *Medicago*:

1      Shrubs. Standard equal to or shorter than the keel. Hybridization with *M. sativa* s.l. not successful        sect. *Arboreae*
—      Herbs. Standard longer than the keel       2
2      Plants greyish with a dense feltlike pubescence; growing in seashore sands. Hybridization with *M. sativa* s.l. not successful       sect. *Marinae* Grossh.
—      Plants not with a dense feltlike pubescence; not in seashore sands       3
3      Leaflets wide (length : width ratio 3:2 and more), glabrous on their upper side. Corolla yellow. Hybridization with *M. sativa* s.l. not successful       sect. *Suffruticosae* Vass.
—      Leaflets narrower (2:1 or less), usually somewhat hairy on their upper side. Corolla yellow or violet. Hybridization with *M. sativa* successful       sect. *Falcago* Rchb.

SECTION FALCAGO Reichenbach, Fl. Germ. Excurs. 2:504 (1832).

Perennial herbs. Corolla yellow, violet or variegated. Pods with one to several seeds, straight, sickle-shaped or coiled in up to 5 coils. Florets with the standard longer than the keel; wings usually longer than the keel. $2n = 16,32,48$

Key to subsections of section *Falcago*:

1      A net of prominent veins all over the coil face or at least in its peripheral part       subsect. *Rupestres* Grossh.
—      Veins on coil face not prominent       2
2      Pods spiny       subsect. *Daghestanicae* Vass.
—      Pods spineless       3
3      Pods with a distinct, thick spongy partition between seeds       subsect. *Papillosae* Grossh.
—      Pods without partition between seeds or with a thin membranous partition       subsect. *Falcatae* Vass.

The above subsections are close affinity groups in regard to interfertility between their members, and can also be hybridized with *M. sativa*.

SUBSECTION FALCATAE Vassilczenko, Acta Inst. Bot. Komarovii Ser. 1/8:33 (1949) pro ser., emend.

Corolla yellow, violet or variegated. Pods straight, sickle-shaped or coiled; spineless; veins not prominent; partition between seeds absent or thin, membranous. $2n = 16,32$.

Key to taxa in subsection *Falcatae*:

1      Pods straight or sickle-shaped (curvature not more than a semi-circle). Corolla yellow       *M. falcata*

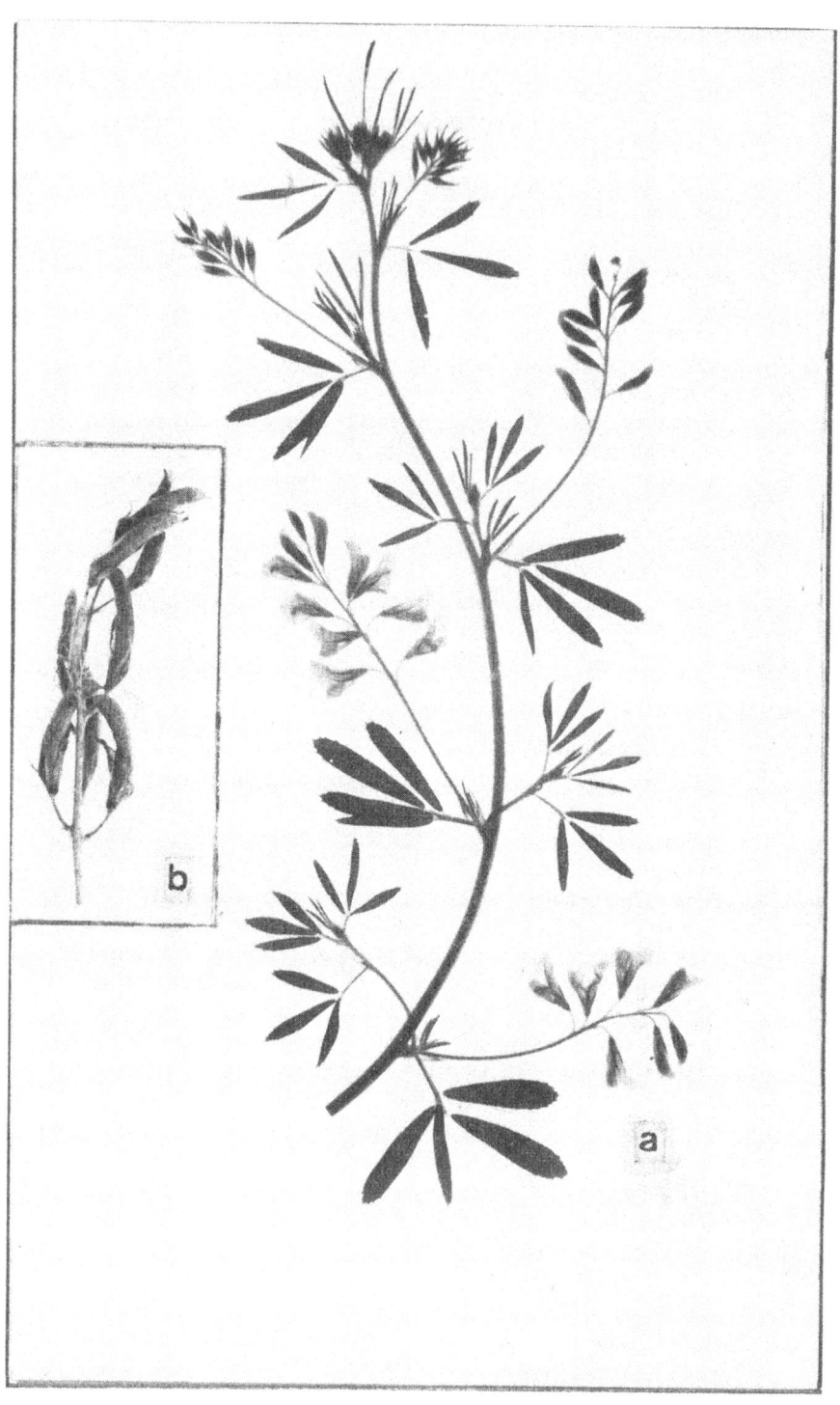

*Fig. 19, M. falcata* ssp. *romanica*. Branch (a) nat. size; raceme (b) magn.
1.5.

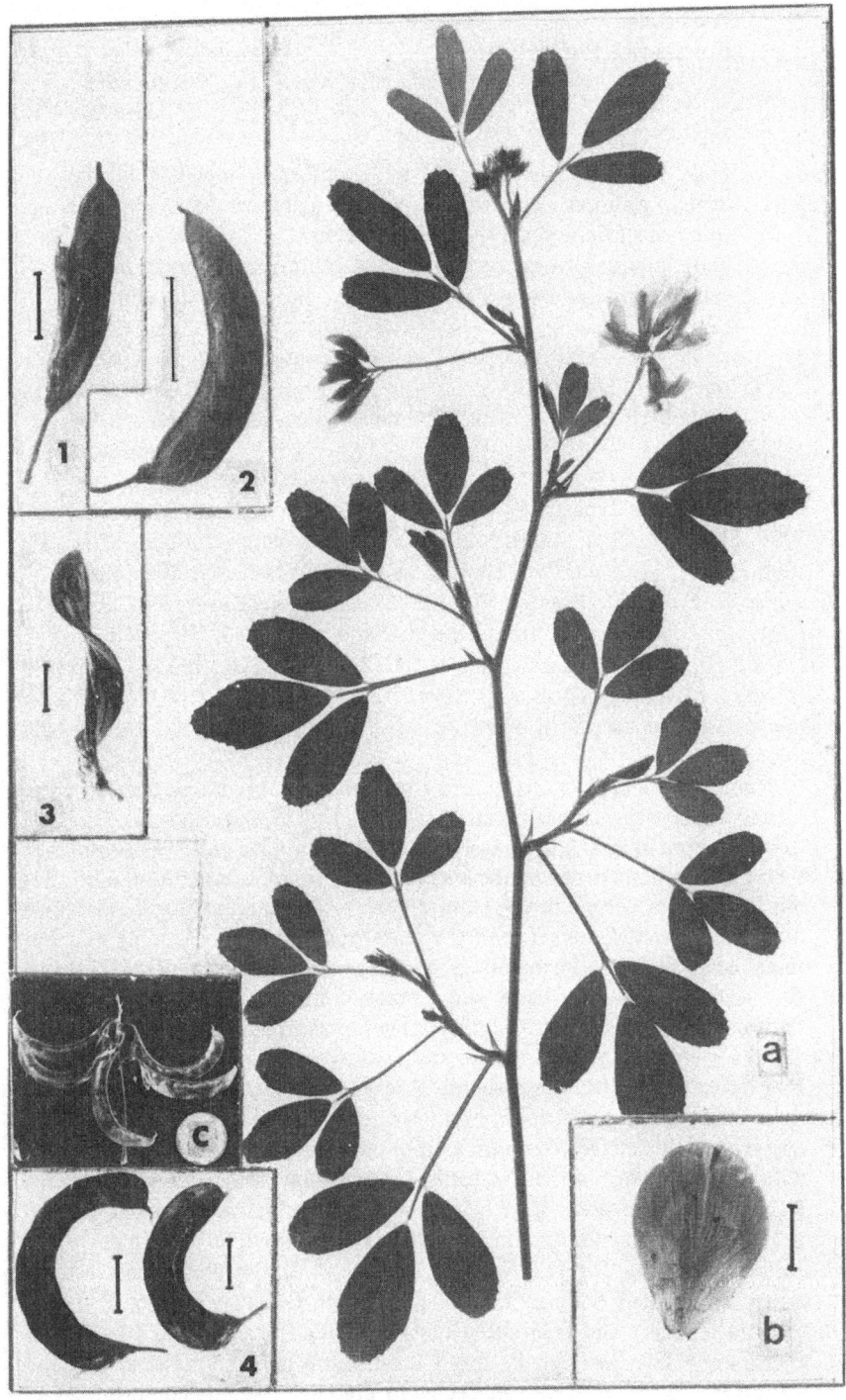

Fig. 20, *M. falcata*. Branch (a), standard petal (b), and pods: 1, ssp. *romanica*; 2, var. *altissima*; 3, var. *tenderiensis*; 4, forms with sickle-shaped pods. Raceme (c) with recurved pods (*M. borealis* Grossh.).

| | | |
|---|---|---|
| – | Pods coiled | 2 |
| 2 | Corolla violet or variegated | *M. sativa, M. × media,* |
| | | *M. × varia, M. × hemicycla, M. × tunetana,* |
| | | *M. × gaetula* |
| – | Corolla yellow | 3 |
| 3 | Pods small (2.5-4 mm in φ) with almost no opening in the centre, glabrous, with simple or with glandular hairs. Pedicels thin, long (longer than the calyx), recurved | *M. prostrata* |
| – | Pods larger, with an opening in the centre, always with many-celled glandular hairs. Pedicels sturdy, short (shorter than the calyx) | 4 |
| 4 | Pods (4)5-6 mm in φ, with 1.5-3.5 coils, with only a small opening in the centre | *M. glomerata* |
| – | Pods 8-10 mm in φ, with 1/2-2 coils, with a large opening in the centre | *M. glutinosa* |

*10. Medicago falcata* Linaeus Sp. Pl. :779 (1753). Syn. *M. procumbens* Bess., Prim. Fl. Gallic. :127 (1809); *M. sativa* L. var. *falcata* (L.) Döll, Rhein. Fl. :802 (1843); *M. glandulosa* David., Östrr. Bot. Ztschr. 52:439 (1902); *M. romanica* Prod., Fl. Rom. 1:617 (1923); *M. erecta* Kotov, Bot. J. 2:276 (1940); *M. quasifalcata* Sinsk., Compt. Rend. Acad. Sci. URSS 48 (4):281 (1945); *M. difalcata* Sinsk., Bull. Appl. Bot. Genet. Pl. Breed. 28 (1):29 (1948); *M. borealis* Grossh. in Kom. (ed.), Fl. USSR 11:391 (1945); *M. tenderiensis* Opperm. in Klokov, Bot. J. 5 (2):44 (1948). Figs. 19; 20; 4,-2; 5,-14,-15.

Plants 40-80(120) cm long; stems prostrate to upright, arising from the crown; roots well-branched, rarely taprooted, or rhizomatous. Foliage and stems covered sparsely to densely with hairs. Stipules entire or toothed at their base. Leaflets cuneate, oblanceolate to obovate, often retuse, 5-20 (24) mm × 2-10 mm, serrate in their upper third, with a terminal tooth. Peduncle 3-20 (25)-flowered, longer than the corresponding petiole, ending in a terminal cusp. Florets 7-11 mm long, in a dense headlike or slender lax raceme. Pedicel longer than the calyx tube; bract shorter than the pedicel. Calyx 3.5-6 mm long; calyx teeth slightly longer or shorter (rarely) than the tube. Corolla yellow or yellow-orange; standard ovate or obovate, its length less than twice its width; wings slightly longer than the keel. Pods straight to sickle-shaped, curvature not more than one half-circle (Fig. 20,-4), occasionally corkscrew twisted, 7-15 mm × 1.5-3 mm, glabrescent to densely covered with simple and/or glandular hairs. Seeds yellow brown or violet brown (rarely), 2-9 per pod, 1.7-2.5 mm × 1.0-1.5 mm. Seed weight 0.9-1.8 g/1000. Radicle somewhat longer than half of the seed. 2n = 16, 32.

*Habitat and Distribution.* Nests of diploid *M. falcata* are found scattered over a surprisingly wide area: from 10° (South Germany) to 85° E. Long. (Siberia, Tomsk), and from 42° (Bulgaria, Black Sea Coast) to 60° N. Lat. (northern USSR, Leningrad). In all locations except the moist area in the northwest Caucasus, *M. falcata* seems to prefer dry steppe conditions. The species is cold as well as heat tolerant, adapted to steppes with their cold winters and hot summers. The slightly bent, comparatively heavy pods of *M. falcata* are not adapted for distribution by rolling on the ground. Also, dehiscent pods scattering seeds indicate that *M. falcata* has originated under conditions where dispersal by whole pods was not suitable.

Though the tetraploid form of *M. falcata* is much more frequent than the diploid, it does not seem to spread much beyond the diploid boundaries. Exceptions are some populations at higher altitudes in mountain ranges toward its southern limits.

*Variation Within the Species.*  Except for the yellow corolla, the non-coiled pods, and the broad standard, most of the characters of *M. falcata* are rather variable. This is reflected in the many synonyms given to the species. There are well over fifty taxonomic names below the species rank if summarized from lists provided by Oakely & Garver (1917), Sumnevicz (1932), Grossheim (1945b), Vassilczenko (1949), Sinskaya (1950), and Kožuharov (1965). At the tetraploid level at which the species at present is prevalent, combinations of different characters are found. Those most commonly noted are prostrate to ascending growth habit, and a well-branched root system; these traits by some authors are considered species-characteristic.

At the diploid (2*n* = 16) level we (Lesins & Lesins, 1964) found in different accessions the following variations: roots spreading, rhizomatous, or taproots; stems prostrate to upright; racemes headlike to elongate; number of florets per raceme 4-20; pedicels bent downwards to upright; pods sickle-shaped to straight, wide to narrow, light colored to dark, without or with glandular hairs; seeds small (0.9 g/1000) to large (1.8 g/1000); leaflets narrow to wide. Moreover, there are noted differences in winterhardiness and resistance to diseases. Oakley and Garver (l.c.) not differentiating between diploids and tetraploids, reported even wider amplitudes of variation in some characters than we found in diploids.

In diploids we discern *M. falcata* ssp. *romanica* (Prod.) Hayek, Prodr. Fl. Penins. Balc. 1:835 (1927); syn. *M. romanica* Prod., *M. erecta* Kotov, Fig. 19. In distinction to most types of *M. falcata*, in this taxon stems are erect; plants have a taproot; florets are arranged in slender, lax racemes; pods are straight and narrow, borne erect on the pedicels at narrow angle to the peduncle (Figs. 19,-b; 20,-1); the whole plant is densely covered with simple hairs giving it a greyish appearance. Recently Ignasiak & Lesins (1972) found in ssp. *romanica* also some deviation from other *M. falcata* types in carotenoid content of petals. No tetraploids are known with the above characters. The taxon may be a branch evolved from the older diploid stock.

The rest of the variations encountered in diploid *M. falcata* may be considered at the rank of varieties or forms. A few accesssions with well expressed differences may be noted:

Acc. No. 137 *M. falcata* var. *altissima* Grossh. (syn. *M. quasifalcata* Sinsk.), from N. W. Caucasus. Plants of this accession are the most vigorous seen in the species. Stems are thick, semi-erect, up to 120 cm long; leaves almost glabrous; racemes elongate, with deep yellow to orange flowers and

91

slightly bent pods (Fig. 20,-2), sparsely covered with simple hairs. The northwestern Caucasus where this variety grows is one of high precipitation, which may have influenced the evolution of this vigorous, sparsely haired variety.

Acc. No. 80 *M. falcata* var. *tirnensis* Šir. (1928) (syn. *M. glandulosa* David.) from the vicinity of the town of Vratza, Bulgaria. The characteristic feature of this variety is glandular hairs on peduncles, pedicels, and especially on pods.

Acc. Nos. 556 and 1830 *M. falcata* ssp. *tenderiensis* Vass. from the vicinity of the town of Sozopol, Bulgaria. The pods of this variety are twisted corkscrewlike along their longitudinal axis (Fig. 20,-3) and are densely covered with appressed simple hairs. The number of florets per raceme, usually not more than 5, is the lowest that is found in *M. falcata*.

Acc. No. 134, from the Moscow area. This variety has been described as *M. borealis* by Grossheim (1945). Plants have semiprostrate stems, well-branched root system and underground rhizomes. The pods are large and borne on downward bent pedicels (Fig. 20,-c).

*Features Distinguishing M. falcata from Other Species.* The description of *M. falcata* as having a racemous inflorescence, crescent-shaped pods, prostrate stems and yellow flowers (Linnaeus, 1753) applies also to other species such as *Medicago platycarpa*, *M. ruthenica*, *M. cretacea*, and *M. hybrida*. These four species are not closely related to *M. falcata* as they cannot be hybridized with it. They are often considered as belonging to the genus *Trigonella*. All of them can be readily distinguished from *M. falcata* in that their pods are distinctly wider, 3.5-8 mm wide, whereas in *M. falcata* the pod width does not exceed 3 mm. In *M. platycarpa*, *M. ruthenica* and *M. cretacea* the seeds in pods are attached with their long axes vertical to the ventral suture (subgenus *Orbicularia*); in *M. falcata* the long axes of seeds are almost parallel to the ventral suture. In the three mentioned species, as in most other *Orbicularia*, the radicles of seeds are as long as the cotyledons (Figs. 5,-10,-34; 6,-45); in *M. falcata* they are not more than 2/3 the length of cotyledons (Figs. 4,-2; 5,-14,-15). In the fourth, *M. hybrida*, the pods are wider than in *M. falcata* and the leaflets are broader with the length : width of 3:2 even at the upper nodes of the stems, compared with the length:width of 2:1 or narrower in *M. falcata*. Seeds of *M. hybrida* are at least twice as heavy as those of *M. falcata* (4-4.4 g/1000 vs. 1-2 g/1000, resp.).

As a curiosity may be mentioned Grossheim's treatment of one form of *Medicago falcata* in flora USSR Vol. 11, (1945). There (Plate X, Fig. 1), a typical pod of *M. cretacea* M. B. (compare our Fig. 18,-b) is named *M. falcata* var. *cretacea* Grossh. To make matters more confusing, Grossheim includes *M. cretacea* in genus *Trigonella*.

It is not always easy to distinguish between *M. falcata* and hybrids from intercrossing it with closely related species, especially *M. sativa* with which it freely intercrosses to produce $F_1$ hybrids with no indications of meiotic disturbances or lowered progeny viability in experimental plots. However, though the first generation hybrids are easily recognizable by their greenish

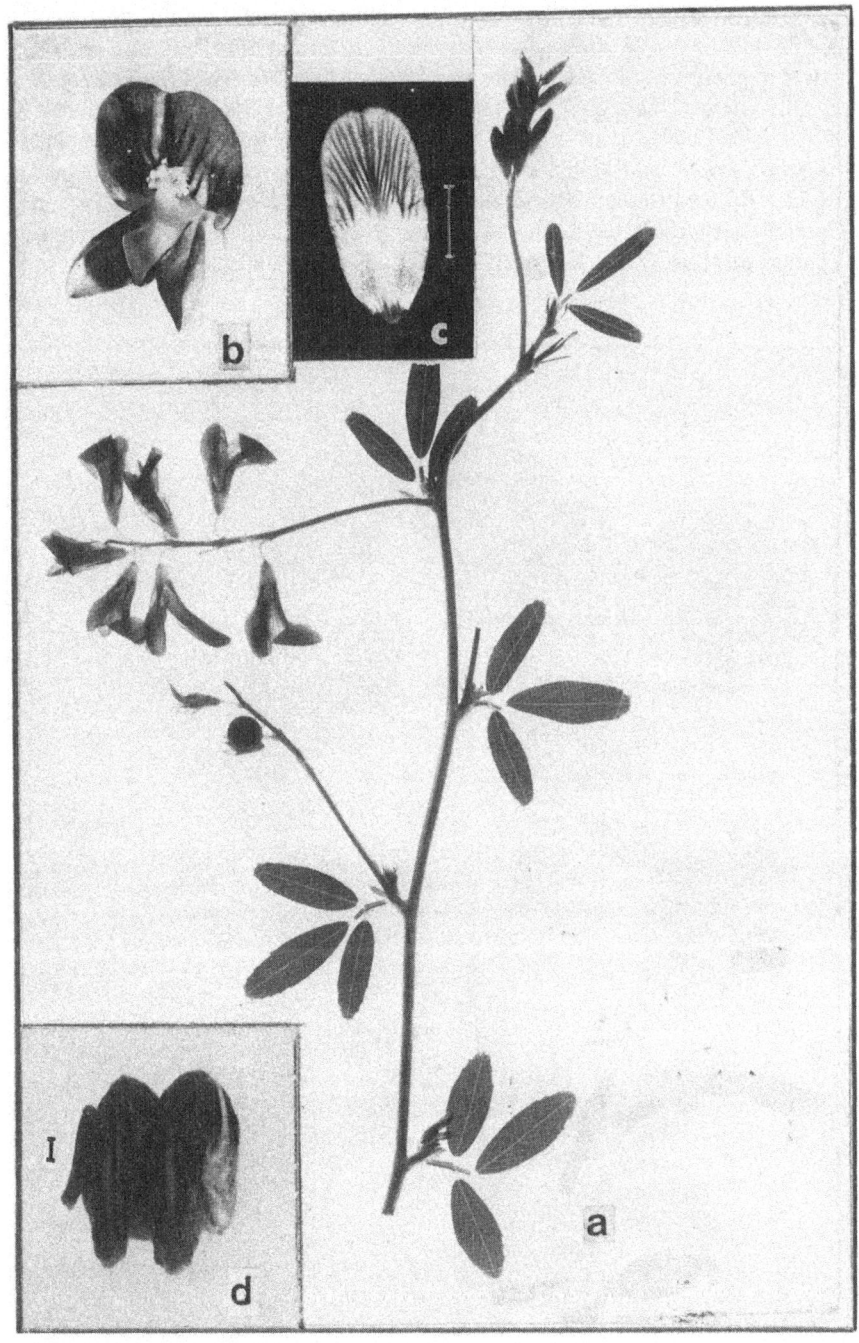

*Fig. 21, M. sativa* ssp. *coerulea*. Branch (a), tripped floret (b), standard petal (c), and pod (d). Note parallel sides of the middle part of standard.

flowers, resulting from a blending of the yellow and violet colour substances of *M. falcata* and *M. sativa*, respectively; the determination of hybrid derivatives in plants of later generations may cause difficulties. Thus the character, yellow flowers and crescent-shaped pods of *M. falcata* give with purple flowers and coiled pods of *M. sativa*, a number of recombinations, including yellow flowers and coiled pods as well as violet flowers and sickle-shaped pods. Recognition of hybrid origin is complicated by the longevity and partial self-fertility of both the species and their hybrids; thus, during a period of years, from a single hybrid may arise a population quite distinct in

*Fig. 22, M. sativa* ssp. *sativa*. Branch (a), and pod (b).

appearance from other populations. Some taxonomists are tempted to include such populations into one or another parental species, thereby obscuring the species' boundaries. Other taxonomists have given such populations distinctive varietal or even specific names. Our attempt to delineate the confused species' boundaries are given under *M. sativa*.

Hybrids of *M. falcata* with the related *M. glomerata* and *M. prostrata* may occur. We observed plants appearing to be hybrids of *M. falcata* and *M. prostrata* in the vicinity of Trieste, Italy. In hybrids involving *M. glomerata* a distinctive character would be pods coiled at more than a semicircle, with a wide opening in the centre.

Sinskaya (1950) observed in the Caucasus natural hybrids between *M. falcata* and *M. glutinosa*.

*11.*    *Medicago sativa* Linnaeus, Sp. Pl.:778 (1753). Figs. 21; 22; 6,-46,-47. Synonyms of *M. sativa* and hybrids are discussed below. Figs. 23; 24; 25.

Plants 30-80(120) cm long; stems procumbent, ascending to erect, arising from the crown. Vegetative parts covered to varying degrees with simple appressed hairs. Stipules entire or toothed in their basal part. Leaflets 8-28 mm x 3-15 mm, obovate (at lower nodes), cuneate or linear-oblanceolate (at upper nodes), serrate in their apical part; midrib ending in a terminal tooth. Peduncle 7 to 35-flowered, several times longer than the corresponding petiole, with a terminal cusp. Florets 6-12 mm long, usually in an elongate raceme. Pedicel equal to or longer than the calyx tube; bract ± the length of the pedicel. Calyx half the length of the floret; teeth ± the length of the tube. Corolla violet, lavender, rarely pink or white; standard twice or more as long as wide, with parallel sides in its middle part (Fig. 21,-c), wings longer than the keel. Young pod rises from the calyx, then bends sideways. Mature pod light to dark yellow-brown, coiled in a spiral, glabrous or with appressed simple (rarely glandular) hairs, spineless; coils 1-5, turning clockwise, 3-9 mm in $\phi$; many veins starting from the ventral suture running obliquely, branching somewhat, then anastomosing in the outer part of the coil face. Seeds yellow, or brownish, or greenish-yellow, 1.2-2.5 mm x 1-1.5 mm. Seed weight 1-2.5 g/1000. Radicle slightly over half the seed length. $2n = 16,32$.

*Habitat and Distribution.*   *M. sativa* has probably evolved under warm, dry, semi-desert conditions around the Caspian Sea. Here as described by Kisliakov (1927) for the Peninsula of Apsheron, it (as *M. coerulea*) grows under conditions where no rain falls from March to September, and very little falls in winter, so that only a few other perennial plant species (*Alhagi camelorum*, for instance) can survive. Under somewhat similar growing conditions it is found in sandy saline soils in eastern Anatolia (D.R. Cornelius, person. commun.). In such environments plant stands are not closed. The nondehiscent, roundish pods of *M. sativa* are adapted for dispersal by rolling with the wind, their spiral coils furthering an anchoring wherever loose drifting soil in encountered. It may be added that ripening of *M. sativa* pods occurs during the latter part of summer when annual desert plants have dried up

*Color Plate III.* Roller-threshing of cereals (a). Stone mills are used for grinding grains (b). In (b) standing on the right, the junior author Irma Lesins; southern Turkey, 1962.

and disappeared, leaving open ground. The strong taproot, penetrating into lower soil layers, can extract water from a considerable depth. Present plant collectors have often recorded *M. sativa* growing in stony, rocky places, which may be refuge sites resorted to by plants to escape continuous defoliation by grazing animals. Another protective characteristic against complete defoliation is the rather prostrate growth habit. Other refuge sites are roadsides and cultivated fields, where *M. sativa* may persist as a weed; owing to its vigorous root system it cannot be eradicated by field cultivation without modern machinery. Both diploid and tetraploid *M. sativa* have been collected either as field or garden weeds (Dr. Allen, Basel, person. commun.). It may be noted that pollination of *M. sativa* is accomplished by ground-nesting bees, which are native to the dry, low-precipitation regions and cannot survive in a rainy climate.

The diploid, wild *M. sativa* accessions present in our collection are: From Turkey of eastern Anatolia; from USSR of Georgian SSR, Daghestan ASSR, Stavropol Region, Volgograd Region, and Kazakh SSR; from Iran of Teheran and Kashan areas. Regarding tetraploids: Generally, tetraploids arise from diploids under natural conditions. We found both ploidy levels in one accession of *M. sativa* (as *M. lavrenkoi* Vass., seed supplied by the species author), also in two *M. falcata* accessions. Although in these cases admixture of different-ploidy seed from outside may not be excluded completely, our studies on polyembrionic seeds have shown that reduction (Lesins, 1952) as well as increase (Lesins et al., 1975) in ploidy may take place spontaneously. Also, that sporocytes with unreduced chromosome number may partake in fertilization (Lesins, 1955). The tetraploid *M. sativa*, found outside the areas of diploids in the wild, are often escapees which have acquired characteristics usually associated with wild plants, such as prostrate growth habit. Once established in a field, *M. sativa* may persist for a long time, and its pods, not too different in size and weight from cereal grains, may be carried with grain wherever it is transported.

Indeed, spreading and establishment may well have occurred at any time from the distant past to modern times. In the ancient world, communications between far-apart regions were not so very rare. Thus Israel (18th century B.C.) sent his sons to buy grain from Egypt: 'Behold, I have heard that there is corn in Egypt: get you down thither and buy for us some thence, that we may live, and not die' (Genesis, 42,2). Also, trading people such as the Phoenicians, with colonies on Mediterranean islands, in North Africa, and in southern Spain caused the spread of all kinds of seeds together with marketable goods. It may be noted that even now in some Mediterranean countries the cereals are sledge or roller-threshed (Color Plate III,-a facing p. 97). The separation of grains from chaff is accomplished by throwing the chopped material up in the breeze so that heavier grain falls closer, and the lighter material is carried farther away. The grains and seed-filled *Medicago* pods land together closer to the winnower. Moreover, milling is done on a stone base with a stone roller (Color Plate III,-b) so that the small hard-coated *M. sativa* seeds may escape crushing and later pass through the digestive tract of humans or domestic animals without losing their germinative capacity.

Generally, however, pure diploid as well as tetraploid *M. sativa* have not survived moist, cold conditions if brought outside their dry, warm climate of origin.

*Variation Within Species.* Certain quantitative differences in morphological characteristics are usually associated with ploidy levels. Tetraploids generally have a more vigorous stature than diploids and larger flowers, pods and seeds. According to Davis (Fl. Turkey, 3:488-489; 1970), the size limits between the two subspecies of different ploidy levels are:

Pods 5-9 mm in $\phi$; seeds 2.2-2.5 mm long; florets 6-12 mm long
*M. sativa* ssp. *sativa*

Pods 2.5-3.5(5) mm in $\phi$; seeds 1.6-2.1 mm long; florets 5-6 mm long
*M. sativa* ssp. *coerulea* Schmalh.

The ssp. *coerulea* Schmalh., Fl. Sred. Juž. Ross. 1:226 (1895) has been described as a separate species, *M. coerulea*, by Ledebour (Fl. Ross. 1:526; 1843), and as *M. lessingii*, Fish. and Mey. ex Kar. (in Bull. Soc. Nat. Mosc. 1:150; 1839). Fig. 21.

There are variations within each of the two subspecies so that overlap occurs in some of their morphological characters. In ssp. *coerulea* we find long, loose racemes as in Fig. 21, and rounded, compact ones as well. In some accessions the pods may be tightly coiled with almost no opening in the centre; diameter of coil 3 mm and the seed length 1 mm, whereas in other accessions these measurements may be almost doubled. Similarly, in tetraploid ssp. *sativa* (Fig. 22), we have rather variable stocks; for instance, we collected an accession (UAG 1628) in southern France that had a pod diameter of only 3.5-4 mm, and seeds about half the size of the usual tetraploid *M. sativa*.

*Relationship to Other Species.* Provided that the ploidy level is adjusted, *M. sativa* may be hybridized easily with *M. falcata, M. glutinosa* and *M. glomerata*, and with *M. prostrata*, though with the latter, some interbreeding barrier is present. Under experimental conditions the hybrid progenies with the first three species are fully viable, and under natural conditions the hybrid populations are found in ecological niches where combinations of parental characters are more advantageous for survival than those of either parent. They are also found in contact zones, in or along disturbed places such as roadsides, hedges, and ditches. Owing to their perennial growth habit and some seed production by selfing, populations of hybrid origin may persist for a long time, and may be condiered as genuine wild plants. *M. sativa* from the southern areas borders with *M. falcata* from the north, along the line from the middle of Asia to southern Russia in Europe. Along the mountain ranges of the Caucasus the two come into contact with *M. glutinosa* which is exclusively endemic there. Since almost all of the area of *M. falcata* falls within Russia, it is understandable that Russian botanists have turned their attention to it. Interest in it has been accentuated by the obviously close relationship to the cultivated crop, alfalfa.

The first taxonomist in Russia who began describing taxa of this complex as separate species was probably Grossheim who listed as species *M. virescens* (1919), *M. hemicycla* and *M. polychroa* (1925), and *M. borealis* (1945). Sumnevicz (1932) added *M. trautvetteri* and *M. schischkinii*. The most prolific in species names was Vassilczenko (see summary in Vassilczenko, 1949): *M. gunibica, M. tunetana, M. grossheimii, M. vardanis, M. grandiflora, M. tianschanica, M. tadzhicorum, M. roborovskii, M. tibetana, M. caucasica, M. kopetdaghi, M. transoxana, M. rivularis, M. agropyretorum, M. beipinensis, M. subdicycla, M. komarovii, M. alaschanica, M. alatavica, M. kultiassovii, M. lavrenkoi, M. afghanica, M. polia, M. mesopotamica, M. orientalis, M. sogdiana* and *M. ladak*. The last six species were derived from cultivated alfalfa. Sinskaya's (1950) species have the character of geographical ecospecies rather than of conventional taxonomic species. She introduced as new species *M. praesativa, M. jemenensis, M. tripolitanica, M. syriaco-palestinica, M. eusativa, M. asiatica, M. tetrahemicycla, M. quasifalcata* and *M. difalcata*. Considering her classification of *M. falcata*, similar in treatment to that of *M. sativa*, Vassilczenko (1949) observed: 'Looking over this classification it is not difficult to see that in it lacks the most important feature, − the principle of classification. ...Sinskaya's classification [is] based on a mixture of differently weighted factors of geographical, biological or ecological extraction'. In turn, the Russian botanist-agriculturist Lubenetz (1953) has raised serious criticims of Vassilczenko's and some other taxonomists' prolific creation of species names: 'Some of the species of perennial lucerne described by A. A. Grossheim and I. T. Vassilczenko are raised to the species rank on the basis of one single or a few herbarium specimens. Yet, individual herbarium specimens do not disclose the whole range of plant forms or the content of populations from different ecological conditions existing in the area of the species. Individual plants are insufficient as material to be raised to the rank of species. Therefore, I. T. Vassilczenko, as well as other botanists who have described species on the basis of one single or a few herbarium specimens, in many instances actually have raised variations or ecological geographical populations of the same taxon to species rank'.

The nomenclature situation in the *M. sativa, M. falcata, M. glutinosa* complex is such that there is hardly any taxon ranked as a species by one Russian botanist that has not been questioned by another botanist. Thus, Grossheim's species have been questioned by Troitzky (1928), Vassilczenko (1949), Sinskaya (1950) and Lubenetz (1953), and a number of Sinskaya's and Vassilczenko's species are in turn questioned by Latschaschvili (1962) and Lubenetz (1953). Lubenetz (1972) in fact rejected 14 of Vassilczenko's and 6 of Sinskaya's species as being merely ecological-geographical variations or synonyms of *M. sativa*. Thus, out of 62 species names listed by botanists as perennial *Medicago*, Lubenetz (1972) recognized at species rank only 21. Even when retaining as species *M. hemicycla, M. polychroa* and *M. tianschanica*, he admits the possibility of hybrid origin for these species. His admission of hybrid origin is based on the fact that the variability of artifi-

Table II. Our interpretation of Vassilczenko's description as separate species of taxa closely related to *M. sativa*.

| *M. sativa* according to flower color and pod coiling | Species epithet | *M. falcata* according to flower color and pod coiling | Species epithet | *M. glomerata* according to flower color, pod coiling and glandular hairiness | Species epithet | *M. glutinosa* according to flower color, pod coiling and glandular hairiness | Species epithet |
|---|---|---|---|---|---|---|---|
| | 1. *afghanica* Vass. | | 1. *glandulosa* David | | 1. *glomerata* Balb. | | 1. *gunibica* Vass. |
| | 2. *agropyretorum* Vass. | | 2. *romanica* Prod. | | | | |
| | 3. *coerulea* Less. | | | | | | |
| | 4. *kopetdaghii* Vass. | | | | | | |
| | 5. *mesopotamica* Vass. | | | | | | |
| | 6. *orientalis* Vass. | | | | | | |
| | 7. *polia* Vass. | | | | | | |
| | 8. *sogdiana* Vass. | | | | | | |
| | 9. *subdicycla* Vass. | | | | | | |
| | 10. *tadzhicorum* Vass. | | | | | | |
| | 11. *transoxana* Vass. | | | | | | |

| Hybrids between *M. sativa* and *M. falcata* according to flower color and/or pod coiling | Species epithet | Species epithet | Hybrids between *M. sativa* and *M. glomerata* according to flower color, pod coiling and glandular hairiness | Species epithet | Hybrids between *M. sativa* and *M. glutinosa* according to flower color, pod coiling and glandular hairiness | Species epithet |
|---|---|---|---|---|---|---|
| | 1. *alaschanica* Vass. | 12. *lavrenkoi* Vass. | | 1. *tunetana* Vass. | | 1. *glutinosa* M.B. |
| | 2. *alatavica* Vass. | 13. *ochroleuca* M. Kult. | | | | 2. *grossheimii* Vass. |
| | 3. *beipinensis* Vass. | 14. *roborovskii* Vass. | | | | 3. *polychroa* Grossh. |
| | 4. *caucasica* Vass. | 15. *rivularis* Vass. | | | | 4. *virescens* Grossh. |
| | 5. *falcata* L. | 16. *sativa* L. | | | | |
| | 6. *grandiflora* Vass. | 17. *schischkinii* Sumn. | | | | |
| | 7. *hemicoerulea* Sinsk. | 18. *tianschanica* Vass. | | | | |
| | 8. *hemicycla* Grossh. | 19. *tibetana* Vass. | | | | |
| | 9. *komarovii* Vass. | 20. *trautvetteri* Sumn. | | | | |
| | 10. *kultiassovii* Vass. | 21. *vardanis* Vass. | | | | |
| | 11. *ladak* Vass. | 22. *varia* Mart. | | | | |

cial hybrids between *M. coerulea* (*M. sativa* ssp. *coerulea*) and *M. falcata* almost completely corresponds with that of *M. hemicycla*; that hybrids between *M. glutinosa* and *M. sativa* repeat the polymorphism and polychromy of *M. polychroa*; and, since *M. sativa* and *M. falcata* meet in Tien Shan from where *M. tianschanica* has been described, the latter taxon may also be of hybrid origin.

We attempted to clarify the relationship between *M. sativa* and other species by studying their diploids, which are generally phylogenetically older and have less gradations in their characteristics than tetraploids. In addition, they do not interbreed with tetraploids, and are scattered among tetraploids of the same species as evolutionary relics. We had at our disposal 22 accessions of *M. sativa* ssp. *coerulea* from regions around the Caspian Sea, and 45 accessions of diploid *M. falcata* from a rather wide area covering the entire region from mid-Siberia and Leningrad in the north, to the Krasnodar region and southern Germany in the south. The two species were invariably found to differ in each of three morphological characters: 1) In *M. sativa* ssp. *coerulea* corollas were violet, while in *M. falcata* corollas were yellow; 2) in ssp. *coerulea* pods were coiled in 2-4 turns with only a small opening in the centre, while in *M. falcata* pods were straight to curved, with pod curvature never exceeding half a circle (Fig. 20,-1,-2,-3,-4); 3) in ssp. *coerulea* standards were elongated (length:width ratio at least 2:1), with parallel sides in the mid-region (Fig. 21,-c), while in *M. falcata* standards were broad (length:width less than 2:1), usually obovate with rounded sides (Fig. 20,-b). In the experimentally produced first generation hybrids, we and many other investigators have established that the flower color is invariably greenish (due to a mixture of anthocyanins, carotenoids and yellow flavonoids), while in later generations the entire color range from violet to yellow is found (color plate I facing p. 19). The first generation hybrids exhibit intermediate pod coiling with 1.5-2 loose coils with a large opening in the centre, while in later generations variability covers the range between the parental species. The first generation hybrids have an intermediate standard shape and can be distinguished from the ssp. *coerulea*; in later generations detailed measurements and pattern of inheritance have not been studied.

As a rule in taxonomic descriptions, the flower color and often the pod shape are given. Taking *M. sativa* in the central position and using these characters as yardsticks, we analyzed Vassilczenko's (1949) species descriptions. The results for *M. sativa*, *M. falcata*, *M. glomerata*, *M. glutinosa* and their hybrids are shown in Table II. It is of interest to note that both *M. sativa* L. and *M. falcata* L. as described by Linnaeus fall into the group of hybrids. This, however, might be expected since Linnaeus (1753) gives as a reference for *M. sativa* his Hortus Cliffortianus (1738) where both *M. sativa* and *M. falcata* are confused, and for *M. falcata* refers to Bauhin (1623) who writes: flowers yellow, green, and other colors. For further references see Oakley & Garver (1917).

Assuming that yellow, of different intensity, corollas, pods covered with glandular hairs and coiled in ½-2 loose coils, with a large opening in the centre, are characteristic of *M. glutinosa*, then *M. gunibica* Vass. may be considered as pure *M. glutinosa*, but four other species including the original Bieberstein's *M. glutinosa* with mixed traits appear to be of hybrid origin (*M. glutinosa* x *M. sativa*, Table II). The possibility cannot be excluded that the whole complex including *M. gunibica* are hybrids. Because of the 2n = 32 ploidy level, the group should be a secondary product in evolution as discussed under *M. glutinosa*.

*Fig. 23, M.* **x** *media*. Branch (a), and raceme with pods (b).

Finally, the diploid *M. glomerata* with uniformly yellow corollas, glandular haired pods coiled in 1½-3½ coils with only a small opening in the centre (Lesins & Lesins, 1966), was very likely one of the progenitors of *M.* x *tunetana* (Table II). Thus, the 42 species given in Table I, in our opinion, are variations on the basic three diploid and one tetraploid species, the latter of not quite certain origin.

Some other taxa of hybrid origin should be mentioned. Most often *M. varia* Martyn (1792) and *M. media* Persoon (1807), Fig. 23, are considered as separate species. This latter taxon has been shown by Urban (1877) to correspond to his artificial *M. sativa* x *M. falcata* hybrids, and, as we indi-

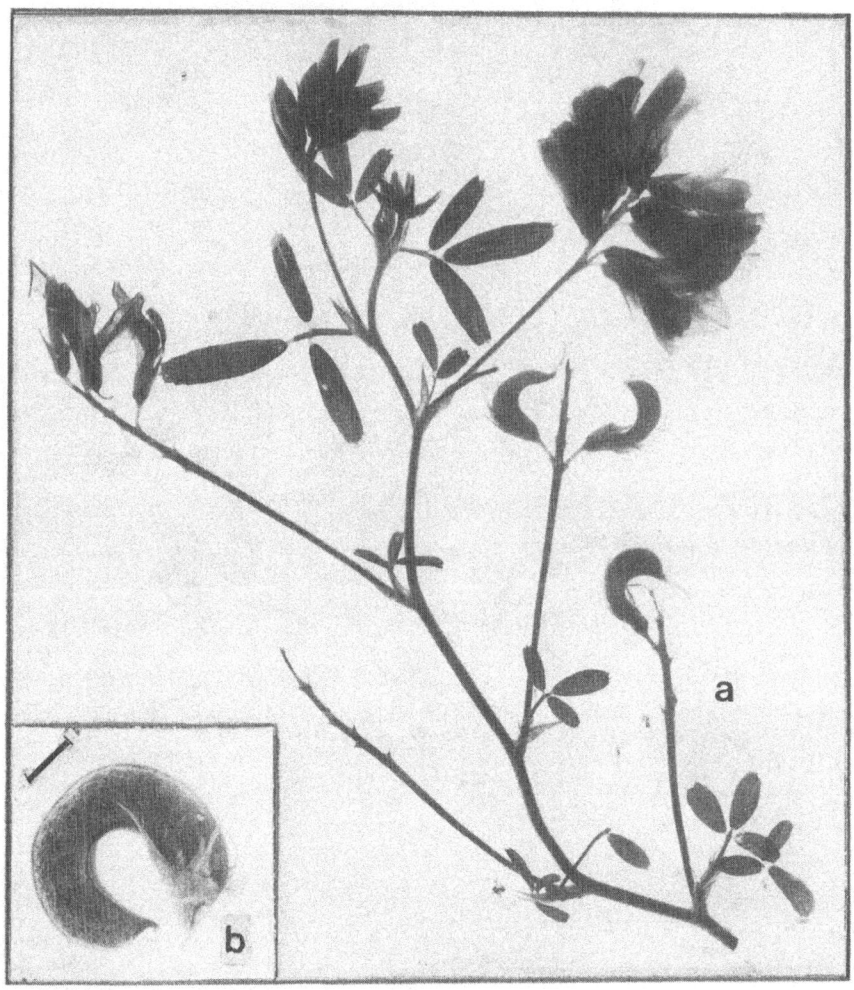

*Fig. 24,   M.* x *hemicycla.* Branch (a), and pod (b).

cated in connection with diploid *M. falcata* (Lesins, 1964), we consider also *M. hemicycla* Grossh. (Fig. 24) to be of a similar hybrid origin. Regarding Sumnevicz's species *M. trautvetteri* and *M. schischkinni* (as described from herbarium material at Tomsk University), there is some uncertainty regarding the exact nature of the material he described. Thus, although Sinskaya

*Fig. 25, M. × karatschaica.* Branch (a), and pod (b).

and Maleeva (1959) have indicated that *M. trautvetteri* is a diploid species, four samples obtained (courtesy of Dr. Vassilczenko) from its original growing site have all been found by the present authors to be tetraploids with a flower color indicative of hybrid origin. Regarding *M. schischkinii*, we have learned from botanists of the Central Siberian Botanical Garden, Novosibirsk, that, '*M. schischkinii* is an artefact; we do not have this species in Siberia' (person. commun.). Hence, it appears doubtful whether a diploid *M. trautvetteri* will be found again and whether the taxon *M. schischkinii* exists as a separate taxon. Some other described species such as *M. karatchaika* Latschaschvili (1959), we have had little opportunity to study because of a scarcity of material; a few progenies of *M. karatchaica* (Fig. 25), grown in our greenhouse, segregated for pod shape and flower color.

Much confusion in species delineation has been caused by some easygoing botanists. Thus Trabut (1917), in his aspiration of finding the origin of alfalfa, treated as species *M. gaetula*, attributing its species rank to Urban (1873), though Urban himself described it as *M. sativa*, ssp. *macrocarpa* var. *vulgaris* f. *gaetula* Urb. and indicated only in parentheses that it may be a separate species. Trabut also listed *M. coerulea* Less. as synonym of *M. gaetula* and of *M. tunetana*. This latter taxon has been described as a species only 30 years later by Vassilczenko (1949), and *M. sativa* f. *gaetula* has never been raised to species rank. Some authors have later repeated these misconstructions (Atwood & Grun, 1951; Sprague, 1959).

During our expedition to North Africa, we looked for some clues which might explain the peculiar morphological appearance reported for several indigenous *M. sativa*-related *Medicago*. These taxa are: *M. sativa* ssp. *tunetana* (Murbeck, 1897), *M. sativa* ssp. *faurei* (Maire, 1933), and *M. sativa* f. *gaetula* (Urban, 1873). All of them have coiled pods with 1.5-5.5 coils, 5-10 mm in $\phi$, covered with glandular hairs, and with only a small opening in the centre. In addition for ssp. *tunetana*, it is noted that the pods are borne on recurved pedicels. The range of color of the corollas of these taxa is recorded: yellow (ssp. *faurei*), greenish (ssp. *tunetana*) and violet (f. *gaetula*). Their habitat and distribution area is mountainous terrains of Tunisia and Algeria. We found plants most closely corresponding to *M.* x *tunetana*, at an altitude of 1600 m at Chellâla (district of Batna), Algeria. Plants collected in an area of a few square feet were of three differently colored corollas: light yellow with pinkish tinge on buds, lavender greenish, and blue green. Pods on all were twisted in 2-4 coils, and covered with glandular hairs. No doubt *M. sativa* had been one of the progenitors because of the anthocyanin pigment present in corollas. Had *M. falcata* been involved as the other parent contributing its yellow pigments to corollas, then pods would have been loosely coiled with a large opening in the centre, as is always observed on some plants in progenies of artificial hybrids involving *M. falcata*. No such plants from several hundred progenies from each type of originally collected plants were found. We consider that *M. glomerata* with its tightly-twisted pods, covered with glandular hairs, usually borne on recurved pedicels, has most likely been the other progenitor. The collected N. African plants, however, were tetraploids, whereas *M. glomerata*, as we know it from the Maritime Alps, is diploid. It may be that the tetraploids have originated from naturally tetraploidized hybrids from diploid *M. glomerata* x *M. sativa* ssp. *coerulea*, or *M. glomerata* had undergone natural tetraploidization before crossing with *M. sativa* ssp. *sativa*. It is likely that *M. glomerata* (as *M. sativa* ssp. *faurei*?) will be found in future explorations of the mountainous regions of N. Africa. *M. sativa* probably had been brought to N. Africa in prehistoric times. The occurrence of *M. sativa*-related taxa in N. Africa under natural conditions seems to be one of the cases where amalgamation of two species has resulted in new taxons adapted better to certain conditions than either of the parental species.

*History of Cultivation and Agricultural Importance.* *M. sativa* has been the first plant species to be cultivated as a forage crop. The origin of *M. sativa* cultivation (Lesins, 1976) is probably tied with the growing importance of horses in the ancient world. The first part of the second millennium B.C. was marked by a revolution in the nature of warfare brought about by a new weapon, the light horse-drawn chariot. In southern boundaries of diploid wild *M. sativa* (the present Iran), in prehistoric times from the 5th millennium B.C., lived Asiatic tribes who besides hunting and fishing pursued some agricultural activities. A kind of plough from the 4th millennium B.C., and bones of domesticated oxen and sheep have been excavated (Ghirshman, 1954). The semi-sedentary way of life furthering preoccupation with agriculture was probably imposed upon inhabitants by the nature of the region: oases of the Iranian plateau and valleys in Zagros and the Elburz Mts. The valleys 30-60 miles long and 6-12 miles wide enclosed by mountain chains restricted easy transmigration. The northern border of the wild growing *M. sativa*, the Eurasian plains, was inhabited by Indo-Europeans, nomadic horsemen. In the beginning of the 2nd millennium B.C. they moved east and south. A branch of these warrior horsemen crossed the Caucasus and moved along the folds of the Zagros Mountains and subdued the native people. The invaders, however, were assimilated by the natives. Then and there the components for raising *M. sativa* to the status of a cultivated crop were brought together: 1) the plant, 2) the horse, whose value in warfare had been discovered, and 3) the skills in agriculture of the native people. To a tribe of mixed origin, the Kassites, the horse was a divine symbol, their pottery often showing winged horses (Ghirshman, l.c.). In the middle of the 18th century B.C. the Kassites conquered Babylonia and ruled it for more than five centuries. The land of the Kassites was later known as a centre of horse breeding. It is likely that intentional seeding of *M. sativa* or alfalfa [aspo-asti, according to Hendry (1923), meaning in old Iranian horse fodder] was started to provide horses with fodder. The early agriculturists may have recognized the value of growing the more vigorous tetraploid *M. sativa* or, more likely, the spontaneously arisen tetraploids under cared-for growing conditions (irrigation of oases had been practiced before that time) naturally crowded out and replaced the less vigorous diploids. Another centre of horse breeding was around Lake Van in the northern part of Zagros, also an area of wild-growing *M. sativa*. That region was also inhabited by an amalgamation of native Hurrians and invaded Indo-Europeans. From their Mitannian Kingdom written tablets from the 14th century B.C. have been excavated that deal with chariot horse breeding, training, diet and veterinary care (Drower, 1969). Horses retained their importance for more than two following millennia. In the 9th century B.C. it is recorded (Kings Book 2, VII, 6-7) that a Syrian army, hearing the noise of horses and chariots, said

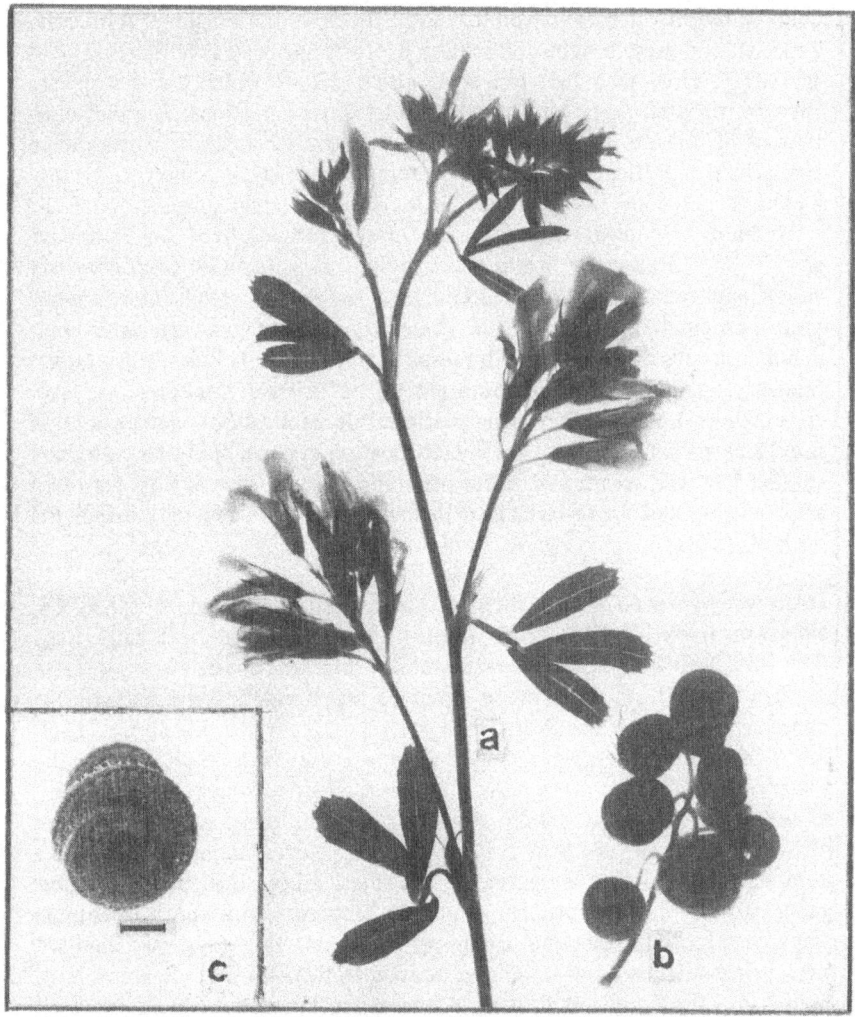

*Fig. 26, M. glomerata.* Branch (a), raceme (b), and pod (c). Note recurved pedicels on raceme.

one to another: 'Lo, the Kings of Israel hath hired against us the Kings of the Hittites and the Kings of the Egyptians, ...wherefore they arose and fled in the twilight'. Later in the 7th century B.C. the prophet Jeremiah (4, 13) forewarned of a Scythian invasion: '...his chariots will be as a whirlwind: his horses are swifter than eagles'. In the 5th century B.C. during Persian-Greek wars, *M. sativa* was introduced with the Persian army into Greece. From there it spread further to southern Italy and Europe.

*M. sativa* owes its rise to the status of the world's most important forage crop to interbreeding with the north-endemic, hardier *M. falcata*, whereupon as hybrids it became suitable for growing in the northern Temperate Zone. Of the present about 33 million hectares under cultivation (Bolton et al., 1972), more than half is seeded with stocks to which the two species have contributed their germ plasm in different proportions. The participation of *M. falcata* in different strains of cultivated alfalfa is borne out in agricultural classifications. Thus, discerning non-hardy, common, and hardy strains is based on the fact that an increasingly larger proportion of the winterhardy *M. falcata* has partaken in the formation of the respective strains. Classification by flower color: violet, variegated with predominantly violet, and variegated with predominantly other than violet color, implies more participation of *M. falcata*. Faster regrowth of some strains as compared to strains of slow regrowth means higher *M. falcata* input in the latter. Similarly, gradation in the autumn growth habit: erect, semi-erect, or prostrate (rosette), reflects increasing participation of *M. falcata* germ plasm. It should be remembered that these classification systems apply to cultivated strains. The wild-growing *M. sativa* has some features, such as slow regrowth after cutting and a prostrate growth habit, which are generally associated with *M. falcata*.

12. *Medicago glomerata* Balbis, Elencho Piant. :93 (1801). Syn. *M. sativa* ssp. *glomerata* Rouy, Fl. France 5:14 (1899); *M. sativa* ssp. *faurei*(?) Maire, Bull. Soc. Hist. Nat. Afrique Nord 24:208 (1933). Figs. 26; 5,-16.

Plants 30-50 cm long; stems erect or ascending, covered with simple appressed hairs. Stipules entire or slightly toothed in their lower part. Leaflets 7-14 mm x 2-5 mm, glabrous on the upper side, with simple hairs on the lower; at upper stem nodes narrowly obovate (often truncate at the apex), at lower ones obovate; margin roughly serrate in its apical part. Peduncle with 4 to 17 closely set florets, longer than the corresponding petiole with a terminal cusp. Florets 9-10 mm long. Pedicel longer than the calyx tube; bract shorter than the pedicel. Calyx 3-5 mm long, with appressed simple hairs, shorter than or equal to the tube. Corolla bright yellow; standard oval, wings slightly longer than or as long as the keel. Pods 4-6 mm in $\phi$, coiled in 1.5-3.5 coils, with a small opening in the center, thickly covered with many-celled glandular hairs. Pod veins anastomose shortly after emerging from the ventral suture. Seeds yellow-green to brownish, oval; 2-2.5 mm x 1.5 mm. Seed weight 2.3-2.7 g/1000. Radicle ± half as long as the seed. $2n = 16$, (32?).

*Relationship to Other Taxa.* Until we found populations uniform morphologically and with diploid chromosome number (Lesins & Lesins, 1966) it was not certain that *M. glomerata* was indeed a distinct taxon. Until then it was questionable whether it was not a segregant of *M. sativa* x *M. falcata*, or a form of *M. glutinosa*, as assumed by Grossheim (1919), or a part of *M. sativa* var. *glandulosa* (Urban, 1873). Lesins (1968), based on hybridization and cytological observations, concluded that *M. glomerata* is somewhat

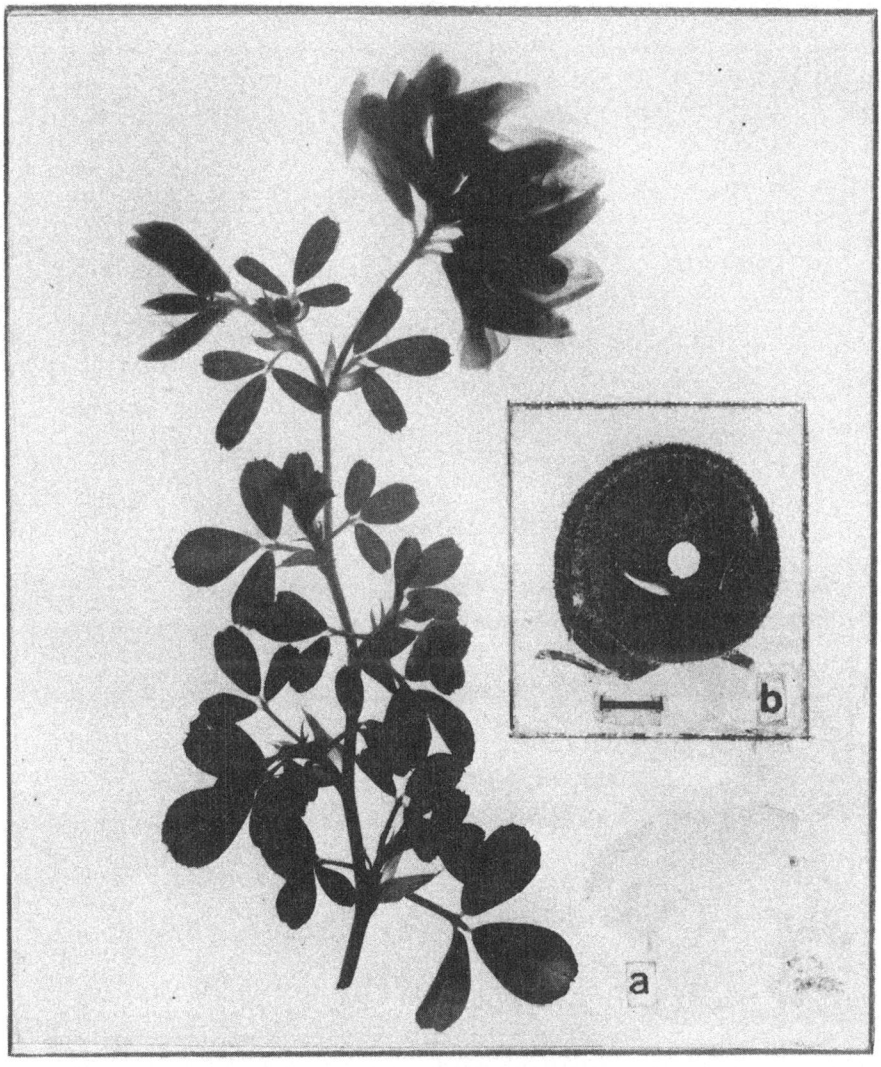

*Fig. 27,* common type of *M. glutinosa*. Branch (a), and pod (b).

109

more distantly related to *M. sativa* and *M. falcata* than these latter are to each other. Further, that its relationship to *M. prostrata* is somewhat closer than that of the latter to *M. sativa* and *M. falcata*.

It may be that *M. sativa* ssp. *faurei* Maire (l.c.), reported from N. Africa, is a tetraploid *M. glomerata*. In such a case, as noted before, existence there of such taxa as *M. sativa* ssp. *tunetana* Murbeck (often referred to as *M. tunetana*), and *M. sativa* ssp. *macrocarpa* var. *vulgaris* f. *gaetula* Urb. (often referred to as *M. gaetula*) becomes understandable. Plants of these tetraploid taxa have tightly coiled pods as *M. glomerata*, not loosely as they would

*Fig. 28, M. glutinosa (M. gunibica Vass.).* Branch (a), and floret (b); note color difference in the middle of the standard petal (arrow).

110

have been if *M. falcata* had been involved in their parentage; they have glandular hairs on pods, and pod pedicels recurved at fruiting stage, which are also characters of *M. glomerata* (Fig. 26,-b); flowers with different shades of anthocyanin indicate involvement of violet flower pigments. Thus it appears very likely that the mentioned taxa are derivatives of natural intercrossing of *M. sativa* with *M. glomerata*, though so far we know the latter only as a diploid. In Tunisia we found a cultivated alfalfa field a few miles away from wildgrowing *M. tunetana* at about the same altitude.

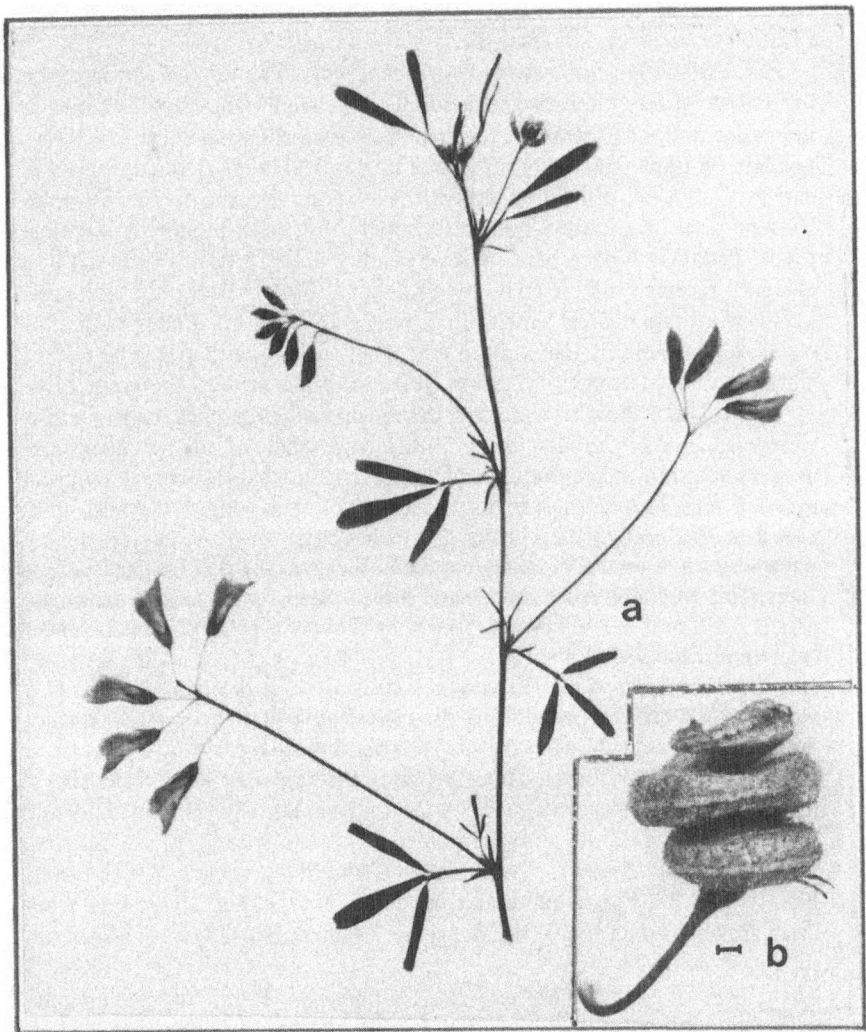

*Fig. 29,   M. prostrata.* Branch (a), and pod (b).

Our knowledge of *M. glomerata* comes from populations we collected at an altitude of about 760 m in the Maritime Alps, at Mt. Madonna della Neve. Some herbaria specimens seen by us from other localities usually had less tightly coiled pods and might be of hybrid origin. Diploids and tetraploids may be disclosed on further search in other montane habitats.

*Agricultural Value.* At Brandon, yield of forage was about 35% of alfalfa (Rambler). Winterkilling was observed during winters of light snowfall. The species may be of importance as a genetic donor for breeding of alfalfa.

13. *Medicago glutinosa* Marschall von Bieberstein, Fl. Taur. - Cauc. 2:224 (1808). Syn. *M. gunibica* Vass., Not. Syst. Herb. Inst. Bot. Acad. Sci. USSR, 11:100 (1949). Figs. 27; 28; 5,-17.

Plants 30-50 cm long; stems decumbent, ascending to erect, arising from the crown, sparsely covered with simple appressed hairs. Stipules entire at the upper nodes of the stem, toothed in their basal part at the lower nodes. Leaflets on upper side glabrous, 13-25 mm x 5-11 mm, at the upper nodes narrowly obovate, often truncate, at the lower nodes broadly obovate; margin somewhat irregularly serrate in its upper third, ending in a small terminal tooth. Peduncle longer than petiole, bearing 6 to 12 florets gathered in a compact raceme, with a small terminal cusp. Florets comparatively large, 12-15 mm long. Pedicel shorter than the calyx tube; bract shorter than or equal to the pedicel. Calyx 6-6.5 mm long, covered with glandular and/or simple hairs; teeth longer than the tube. Corolla yellow, the color often cream or light yellow in buds and freshly opened florets, changing to fully yellow after a few hours; often a different shade of yellow (lighter or deeper) is found in the middle of the floret (Fig. 28,-b); standard about 14 mm x 9 mm, ovate; wings slightly longer than or as long as the keel. Pods coiled at ½-2 coils, 8-10 mm in $\phi$, with an opening in the center; thickly covered with many-celled glandular hairs. Veins on the pod face anastamose soon after emerging from the ventral suture. Seeds oval, yellow, brownish-green, 2.4-3 mm x 1.5-1.8 mm. Seed weight about 3 g/1000. Radicle longer than half of the seed. $2n = 32$.

*Habitat and Distribution.* The taxon is considered to be adapted to comparatively moist growing conditions (precipitation 500 mm or more, mainly during spring and summer); growing at subalpine altitudes (600-2000 m a. s. 1.) on subalpine meadows, among brush, in forest glades, along river valleys, in subclay and rocky soils (Vassilczenko, 1949; Sinskaya, 1950). Endemic to Caucasia.

*Variation Within Species.* The variation found in our collection is shown in Figs. 27 and 28; Fig. 27 represents the common type, Fig. 28 the type from drier growing conditions, considered by Vassilczenko (1949) as a separate species *M. gunibica*.

*The Origin of M. glutinosa and Its Relationship.* Plants and populations with some anthocyanin color in petals we consider as hybrids, involving *M. sativa* in their parentage (see analysis of Vassilczenko's species, Table II).

Some difficulty in tracing the ancestry of *M. glutinosa* is its ploidy level

$(2n = 4x = 32)$, an indication of its secondary descent from some diploid progenitors. There is a possibility that diploid progenitors of the tetraploid *M. glutinosa* are not yet discovered or are extinct. Another possibility is that *M. glomerata* and *M. falcata* may be the progenitors of the yellow-flowered *M. glutinosa* type arisen by hybridization. The former species has yellow flowers, tightly coiled pods and articulate glandular hairs; the latter species, especially its glandular-hairy type (*M. glandulosa*, David.), has sickle-shaped pods and may have contributed to the loose coiling of *M. glutinosa* pods. In the general part of this work we advanced an hypothesis that *M. glomerata* (or its like species) may have been growing much farther east than it does now.

The third possibility may be that the yellow flowered plants, assumed here as *M. glutinosa*, are anthocyanin-free segregants of *M. falcata* and *M. sativa* crosses. There is less credibility for such an interpretation, mainly because the pods of *M. glutinosa* are densely covered with many-celled glandular hairs. Though there are forms in *M. sativa* (*M. sativa* subsp. *microcarpa* var. *pilifera* Urb.) which have glandular hairs, the thick cover and strong development of glandular pod hairs would be somewhat unexpected for *M. falcata* x *M. sativa* segregants.

The observed change in petal shade of color in early flowering stage (pale yellow) and later (deeper yellow) is probably caused by change in flavonoid pigments, as the composition and quantity of carotenoids was about the same in both flowering stages (Ignasiak & Lesins, 1975). This and the differences of yellowness in standard petals (Fig. 28.-b) are similar to the phenomenon found in accessions from N. Africa which we also consider to be of hybrid origin.

The close relationship of *M. glutinosa* with *M. sativa* and *M. falcata* is indicated by the ease of their interbreeding. Hybrids between *M. sativa* s.l. (*M.* x *media*) and *M. glutinosa* have been studied by Armstrong & Gibson (1941), and no interbreeding barrier has been observed. The naturally occurring hybrids with *M. sativa*, often considered as separate species, are noted in Table II. The occurrence of natural hybridization with *M. falcata* is pointed out by Sinskaya (1950).

*Agricultural Value.* At Brandon no winterkilling took place during the few years of testing. The forage production was about 50% of that of the cultivar Rambler. The regrowth after cutting was fair. The taxon, being endemic to the mountains of Caucasus with high precipitation during the vegetation period, may harbor valuable characters for transfer to the cultivated alfalfa.

14.  *Medicago prostrata* Jacquin, Hort. Vindob. 1:39 (1770), Figs. 29; 6,-37,-38.
     Plants 25-60 cm long; stems ascending, sparsely covered with simple or glandular hairs, or glabrate (upper side of leaves). Stipules acuminate, entire or with a few irregular teeth near their base. Leaflets thickish, 4-15 mm x 1-5 mm, narrowly wedge-shaped, narrowly obovate or obcordate; margin slightly serrate in its 1/4-1/3 apical part, usually with two prominent teeth adjacent to the apical one, hence apex appearing tridentate. Peduncle 1 to 8-flowered, longer than the corresponding petiole, with a terminal cusp. Florets

7-8 mm long. Pedicel thin, distinctly longer than the calyx, recurved; bract minute. Calyx shorter than or equal to half the length of the floret; teeth as long as the tube. Corolla yellow; standard broadly oval; wings longer than the keel. Young pod turning sideways through the calyx teeth. Mature pod light yellow to dark brown, coiled, almost without any opening in the centre, cylindrical, covered with glandular or simple hairs or glabrous, spineless.

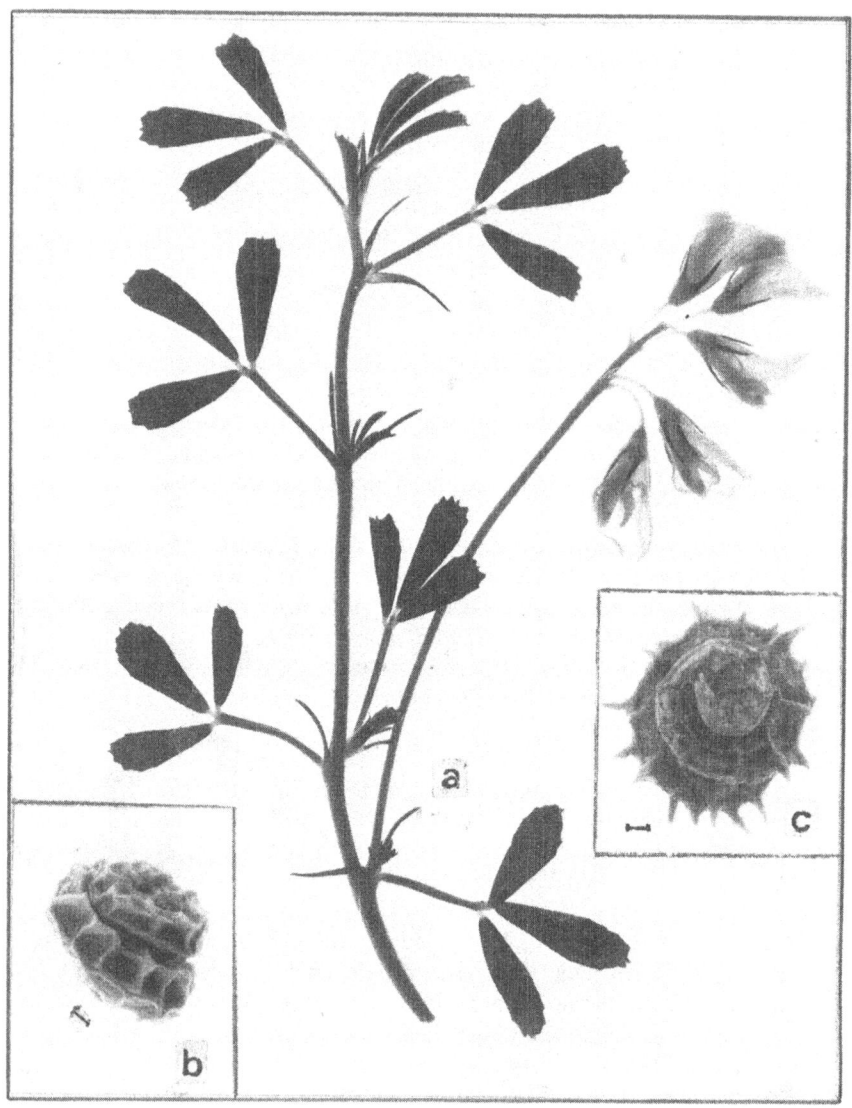

*Fig. 30, M. rhodopea.* Branch (a), and pods: spineless (b), spiny (c).

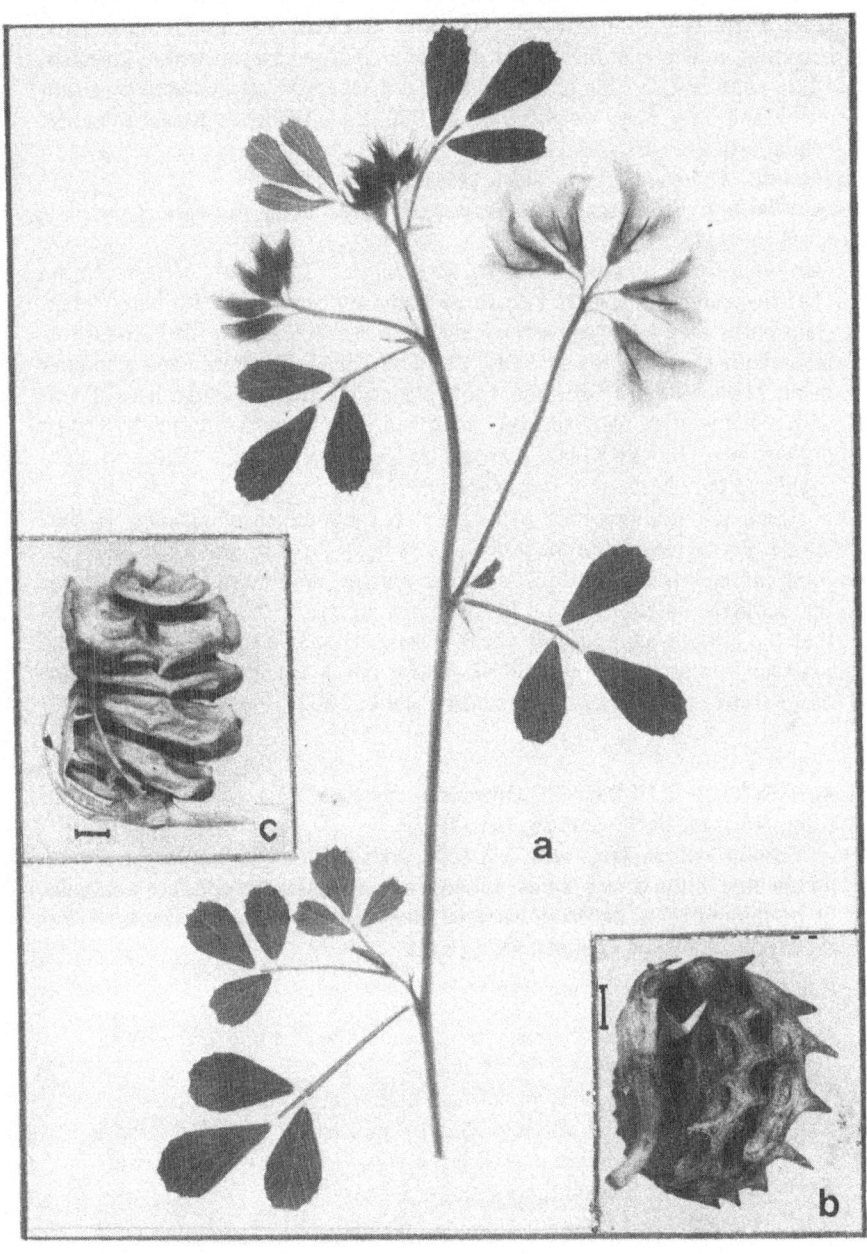

*Fig. 31*, *M. saxatilis*. Branch (a), and pods: spined pod with fewer coils (b), tubercled pod with more coils (c).

Coils 2-4.5, turning clockwise, 2.5-4(5) mm in $\phi$. On coil face 5-8 slender veins originating from the ventral suture, anastomosing in the outer part, becoming indistinct at the edge of the coil. Seeds yellow, brownish, greenish, 1-2 in each coil, 2-2.3 mm x 1-1.3 mm, not separated or separated by a thin membrane only. Seed weight 1.6-2 g/1000. Radicle slightly longer than half of the seed. $2n = 16,32$.

*Habitat.* On dry, rocky hillsides, generally in dry soil.

*Distribution.* From eastern Austria and Italy, along the eastern Adriatic coast to Greece.

*Variation in M. prostrata and Its Relationship to Other Species.* Urban (1873) divided the species into three forms: 1) *glabra* Urb., the whole plant glabrous or only younger parts somewhat hairy; 2) *declinata* Urb., the whole plant with short appressed hairs, the pod often with simple and glandular hairs; 3) *glandulifera* Urb., the whole plant with erect glandular hairs. Plants with thick whitish hairs all over, described as *M. pseudorupestris* by Hayek (1927), may be the fourth form of *M. prostrata.* Of the latter, we have examined only herbarium specimens.

There are sibling types with $2n = 16$ and $32$ chromosomes (Lesins, 1960). On crossing with *M. sativa* s.1. at both $2n = 16$ and $2n = 32$ levels, some interspecific hybridization barrier was observed when *M. prostrata* was the pistillate parent. At the diploid level crosses with *M. sativa* indicated that the anthocyanin color in petals was inherited in a simple monofactorial pattern. In crosses with diploid *M. falcata* it was found that pod shape was determined by more than two factors (Lesins, 1962).

SUBSECTION RUPESTRES Grossheim, pro ser. (incl. ser. *Cancellatae*) in Kom. (ed.). Fl. USSR 11:136, 137 (1945).

Corolla yellow. Pods with 1-5 coils; coils tightly twisted leaving almost no opening in the centre, spiny, tubercled or spineless; on coil face a reticule of prominent veins; partition between seeds present or absent. Plants of dry, rocky, or dry-steppe habitats. $2n = 16, 48$.

Key to species in subsection *Rupestres:*

| 1 | Pods spiny, or tubercled; veins on coil face ending in the lateral vein | | 2 |
|---|---|---|---|
| — | Pods spineless, veins on coil face ending in the dorsal suture | | 3 |
| 2 | Pods small, 2.5-4 mm in $\phi$ with 1-2.5 coils | *M. rhodopea* | |
| — | Pods larger, 6-8 mm in $\phi$ with 2-5 coils | *M. saxatilis* | |
| 3 | Pods small, 3-4 mm in $\phi$ with 1-1.5 coils | *M. rupestris* | |
| — | Pods larger, 4-6 mm in $\phi$ with 1.5-3 coils | *M. cancellata* | |

It is likely that the smaller, diploid *M. rhodopea* and *M. rupestris* have been involved in the evolution of the larger, hexaploids *M. saxatilis* and *M. cancellata*, respectively; thus all four species constitute a natural species group.

*15. Medicago rhodopea* Velenowsky, Sitzungsber. Böhm. Ges. Wiss. 37:21 (1893). Figs. 30; 6,-40.

Plants 10-30 cm long, more or less densely covered with simple, appressed hairs; stems arising from the crown, numerous, procumbent to ascending. Stipules usually entire, awl-shaped. Leaflets small, 8-9 mm x 2-5 mm, narrowly obovate or obcordate; margin with a few teeth in the apical part; midrib ending in a small tooth. Peduncle 4 to 10-flowered, longer than the corresponding petiole, with a terminal cusp. Florets 5-7 mm long. Pedicel longer than the calyx tube; bract shorter than the pedicel. Calyx shorter than half the length of the floret; teeth ± the length of the tube. Corolla yellow; standard broadly ovate, retuse at the tip; wings ± the length of the keel. Young pod emerges straight, then turns sideways from the calyx. Mature pod light yellow to ash grey, tubercled or spiny, lens-shaped or short-cylindrical. Coils 1-2.5, turning clockwise, 2.5-4 mm in $\phi$; on coil face some finely-marked veins run from the centre into a prominent lateral vein; from it elevated ridges, often extending to short spines, enter the dorsal suture, therewith dividing the wide groove between the lateral vein and the dorsal suture into quadrangular portions. Seeds brownish-green, 2.3 mm x 1-1.3 mm, kidney-shaped, 1-2 in each coil, separated by spongy wall. Seed weight about 1.6 g/1000. Radicle half as long as the seed or slightly longer. $2n = 16$.

*Habitat and Distribution.* Calcareous rocky sites, in lower mountain zones. Endemic to mountain ranges of Bulgaria, especially in the Rhodope Mts.

*Relationship to Other Taxa, and Agricultural Value. M. rhodopea* can be hybridized with *M. rupestris*, though evaluation of relationship by studying hybrids has not been carried out. Some hybrids have also been obtained with *M. sativa* (Lesins, 1972). Although *M. rhodopea* by itself cannot be considered of much agricultural value because of its low productivity, it may harbor some valuable characters that may be transferred to *M. sativa* by hybridization. Some plants put out in the field at Brandon were almost all winterkilled.

*16. Medicago saxatilis* Marshall von Bieberstein, Fl. Taur.-Cauc. 2:225 (1808). Figs. 31; 6,-49.

Plants up to 50 cm long; stems numerous, arising from the crown, decumbent, tough, well-foliated. Plants covered almost all over, though sparsely, with simple, appressed hairs; upper side of leaves glabrous. Stipules lanceolate, usually entire. Leaflets obovate, 9-13 mm x 4-7 mm; margin serrate in 1/3 of its apical part; midrib ending in a small tooth. Peduncle 4 to 8-flowered, much longer than the corresponding petiole, with a small terminal cusp. Florets about 12 mm long. Pedicel longer than the calyx tube; bract much shorter than the pedicel. Calyx half the length of the floret; teeth narrow, as long as or slightly longer than the tube. Corolla bright yellow; standard oval; wings slightly longer than or equal to the keel. Young pod rises straight from the calyx, then turns sideways. Mature pod straw-yellow to dark brown, cylindrical, spiny or tubercled. Coils 2-5, turning clockwise, 6-8 mm in $\phi$; on coil face 8-10 oblique veins, anastomosing after reaching 2/3 of the radius and entering a well-developed lateral vein; between it and the dorsal suture a wide groove traversed by protrusions of veins or roots of spines. Spines or tubercles, 12-14 in each row, 1-2 mm long, inserted obliquely to the face of the coil. Seeds brownish-green, 3.5-3.7 mm x 2 mm, 1-2 in each coil, separated by a spongy wall. Seed

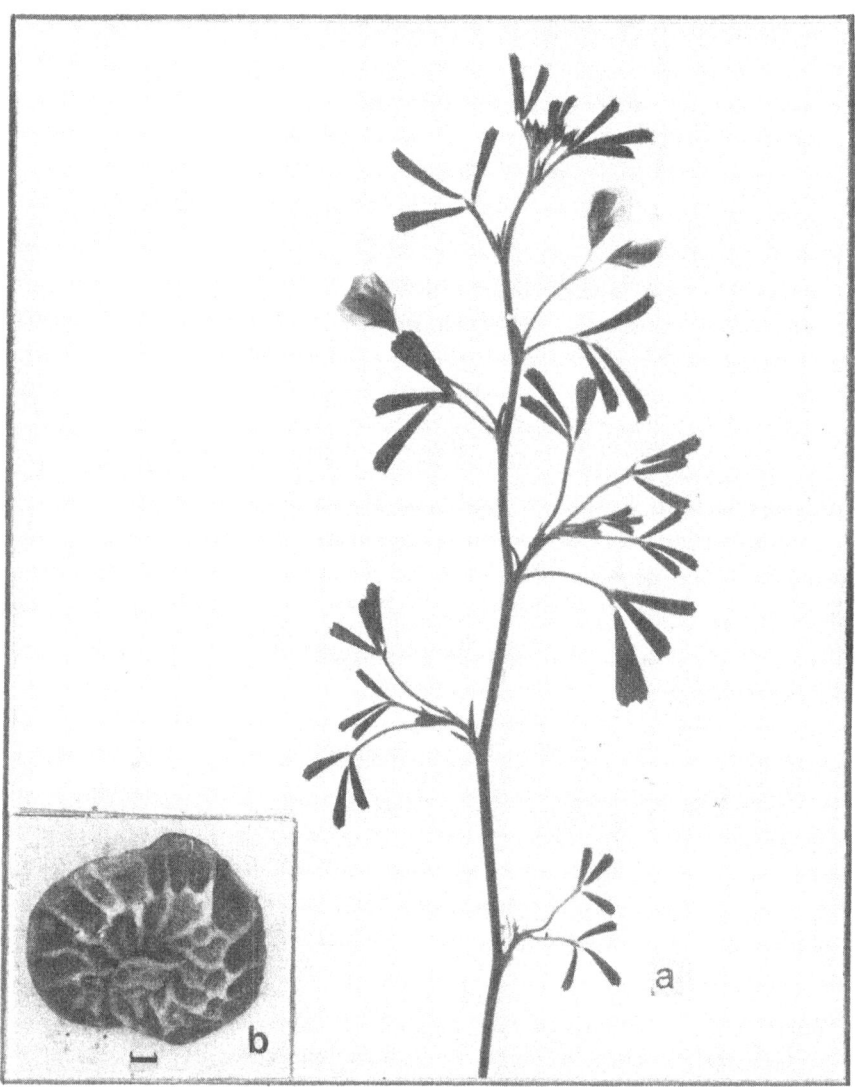

*Fig. 32, M. rupestris.* Branch (a), and pod (b).

weight about 3.6 g/1000. Radicle slightly longer than half of the seed. $2n = 48$ (Lesins & Lesins, 1963c).

*Habitat and Distribution.* Calcareous rocky sites in mid-mountain zone. Endemic to the Crimean mountains.

*Variation Within Species.* Vassilczenko (1949) cites the collector Dzevanovsky's note that in *M. saxatilis* there are two types of pods: 1) spiny, with up to 2 coils, clearly net-veined, hairy, with an opening in the centre, and 2) pods with 3-4 loose coils, indistinctly veined, inflated. We observed spiny, few-coiled, and tubercled many-coiled pods; however, on both types the hairs were sparse; even in the spiny pods the pod centre was closed.

*Relationship to Other Taxa.* *M. saxatilis* can be hybridized with *M. cancellata* and with *M. sativa*; also trispecies hybrids with *M. rhodopea* [ *(M. sativa* x *M. rhodopea)* x *M. saxatilis*] have been obtained (Lesins, 1970). Morphologically, *M. saxatilis* appears to be closest to *M. rhodopea*, especially in pod appearance.

*Agricultural Value.* The species by itself seems to be of little agricultural value. However, due to its crossability with *M. sativa*, it may be of value as a germ plasm donor to the cultivated alfalfa, provided some desirable characters are found in *M. saxatilis*. At Brandon some plants came through the winter. Regrowth was poor.

17. *Medicago rupestris* Marshall von Bieberstein, Fl. Taur.-Cauc. 2:225 (1808). Figs. 32; 6,-44; 8,-1.

Plants 10-20 cm long; stems very numerous, short, thin, leafy, arising from the crown. Vegetative parts covered with simple, appressed hairs. Stipules awl-shaped, usually entire. Leaflets small, 5-10 mm x 1.5-3 mm, narrowly wedge-shaped, truncate, with a few wide teeth in the apical part. Peduncle 2 to 4-flowered, longer than or equal to the corresponding petiole, without a terminal cusp or with a minute one. Florets 4-6 mm long. Pedicel longer than the calyx tube, recurved after flowering; bract tiny, almost undetectable. Calyx ± half the length of the floret; teeth equal to or slightly longer than the tube. Corolla yellow; standard widely oval; wings slightly longer than the keel. Young pod emerges straight, then turns sideways from the calyx. Mature pod light yellow, lens-shaped. Coils 1-1.5, turning clockwise, 3-4 mm in $\phi$, spineless, covered with simple appressed hairs. On coil face 8-11 prominent veins forming large cells before entering the thickened dorsal suture. Seeds brownish-green, 1.7-2 mm x 1-1.5 mm, 1-2(3) per pod, not separated. Seed weight about 1.5 g/1000. Radicle equal to or slightly longer than half of the seed. $2n = 16$ (Lesins & Lesins, 1966).

*Habitat.* Calcareous rocky sites, in lower mountain zones.

*Distribution.* Endemic to the Crimean mountain ranges.

*Relationship to Other Taxa.* It may be hybridized with *M. rhodopea*. Probably it may also be hybridized with *M. sativa*, as was possible with *M. rhodopea* (Lesins, 1972). It differs from *M. rhodopea* in not having well-defined lateral veins on the pod face, lacking a wall between seeds, and having fewer coils (up to 1.5). It is probably one of the parental species of *M. cancellata* (Lesins, 1970).

18. *Medicago cancellata* Marschall von Bieberstein, Fl. Taur.-Cauc. 2:226 (1808). Syn. *M. ciscaucasica* Fedtsch., Bot. Inst. Acad. Sci. USSR, Mater. Herb. 8:176 (1940). Figs. 33; 5,-5.

Plants 50-80 cm long; stems ascending to erect, arising from the crown, covered with simple appressed hairs. Stipules awl to lanceolate-shaped, entire to minutely dentate. Leaflets small, thickish, glabrous on their upper side, 8-10 mm x 4-5 mm, obovate, obcordate to narrowly wedge-shaped; margin in its apical third slightly serrate, with a terminal tooth, lying between two prominent lateral teeth. Peduncle 5 to 12-flowered, much longer than the corresponding petiole, with a distinct terminal cusp. Florets about 10 mm long. Pedicel distinctly longer than the calyx tube, bent downward at pod

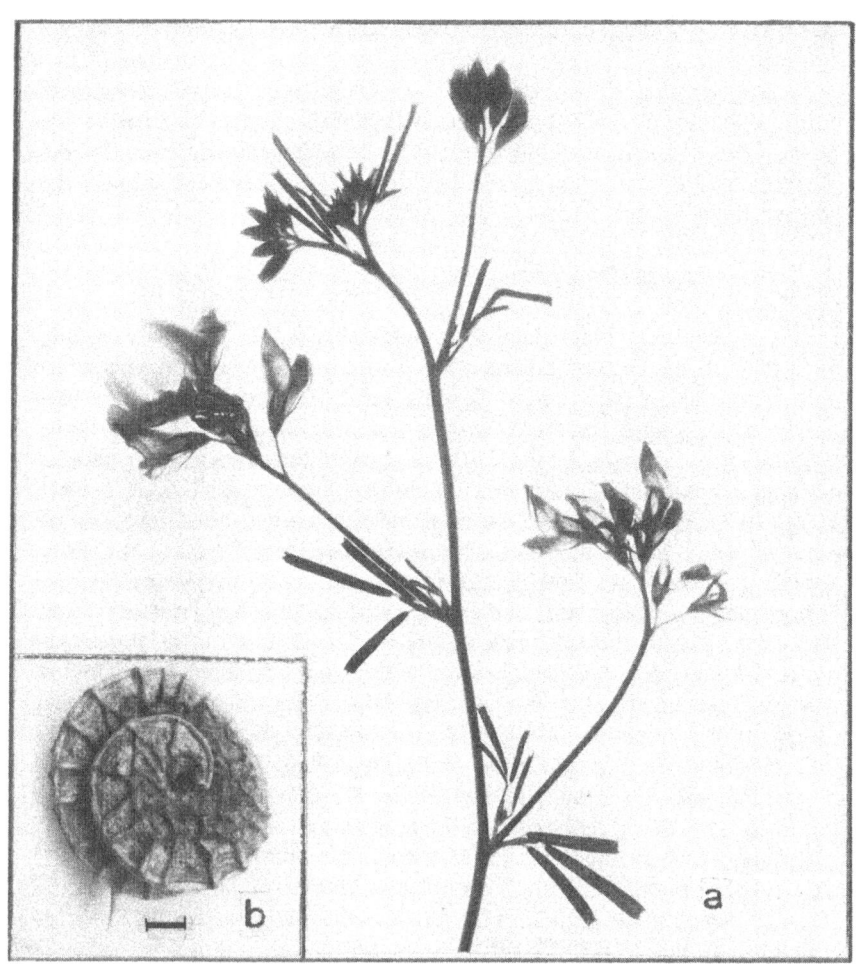

*Fig. 33, M. cancellata.* Branch (a), and pod (b).

maturity; bract much shorter than the pedicel. Calyx 4 mm long; teeth slightly longer than the tube. Corolla bright yellow; standard wide, obovate; wings longer than the keel. Young pod emerges straight from the calyx, then protrudes sideways. Mature pod light to dark-brown, lens-shaped, spineless, with 1.5-3 coils. Coils not tightly appressed, the middle one 4-6 mm in $\phi$, turning clockwise; on face of coil 9-10 veins running radially from the ventral suture, becoming thicker, forming a net of elevated veins, then entering into a thick dorsal suture. Seeds light brownish-yellow, kidney-shaped, 3mm x 1.3 mm, 1-2 in a coil, not separated. Seed weight about 2.6 g/1000. Radicle longer than half of the seed. $2n = 48$ (Lesins, 1959).

*Habitat and Distribution.* Vassilczenko (1949) reported *M. cancellata* growing in poor soils composed of sandstone and its erosion products. It seems to be confined to the southeastern European USSR, north of the Caucasus. Our accession originated from the district of Stavropol.

*Variations Within Species.* Vassilczenko (l.c.) mentions an incomplete herbarium specimen with small pods, 1.2-2.5 mm in $\phi$, denoted *M. prostrata* var. *tatarica* Less., which might belong to *M. cancellata*.

*Relationship to Other Species.* In hybridization experiments it could be crossed with *M. sativa* (Lesins, 1961), and *M. saxatilis* (Lesins, 1970). We consider it an alloautoploid composed of genomes derived from a taxon closely related to *M. sativa* s.l. (two genomes), and *M. rupestris* (four genomes). This latter species has pods very similar to *M. cancellata*, though smaller.

*Agricultural Value.* Hybridization of *M. cancellata*, thriving in poor, dry soils, with *M. sativa* would, perhaps, widen the latter's adaptability for poorer soils. *M. cancellata* survived two winters at Edmonton under field conditions. At Brandon its forage yields were about 30% of cultivated alfalfa (Rambler), seed production was poor, winter hardiness was comparatively good.

It may be a source of resistance to *Stemphylium* leaf spot (Borges et al., 1976).

SUBSECTION DAGHESTANICAE Vassilczenko, pro ser. in Acta Inst. Bot. Komarovii Ser. 1/8:79 (1949).

Corolla yellow or violet. Stems short, wiry. Pods spiny; coils (1.5)2-6, not tightly appressed; partition between seeds present or absent. Growing in sub-mountain zones of the Caucasus Mountains (*M. daghestanica*), and in the eastern Alps (*M. pironae*). $2n = 16$.

Key to species of subsection *Daghestanicae:*

| | | |
|---|---|---|
| 1 | Corolla violet | *M. daghestanica* |
| — | Corolla yellow | *M. pironae* |

*19.* *Medicago daghestanica* Ruprecht ex Boiss., Fl. Or. 2:95 (1872). Figs. 34; 5,-11.
   Plants 20-45(60) cm long; stems numerous, wiry, decumbent, arising

*Fig. 34, M. daghestanica.* Parts of herb (a), and pod (b).

from the crown, well-foliated, covered with simple appressed hairs, Stipules lanceolate, entire or with 1-2 teeth at their base. Leaflets glabrous on the upper side, small, 4-6 mm x 3-6 mm, bluish-green, obovate, obcordate; margin entire at least at the lower stem nodes; midrib ending in a small, thick tooth. Peduncle 4 to 11-flowered, longer than the corresponding petiole, with a distinct terminal cusp. Florets 6-7 mm long. Pedicel longer than the calyx tube, bract distinctly shorter than the pedicel. Calyx 4 mm long, covered with simple hairs, teeth lanceolate, ± the length of the tube. Corolla of different intensities of violet; standard broadly ovate; wings equal to or slightly longer than the keel. Young pod sparsely covered with simple hairs, protruding sideways through the calyx teeth. Mature pod greenish to dark

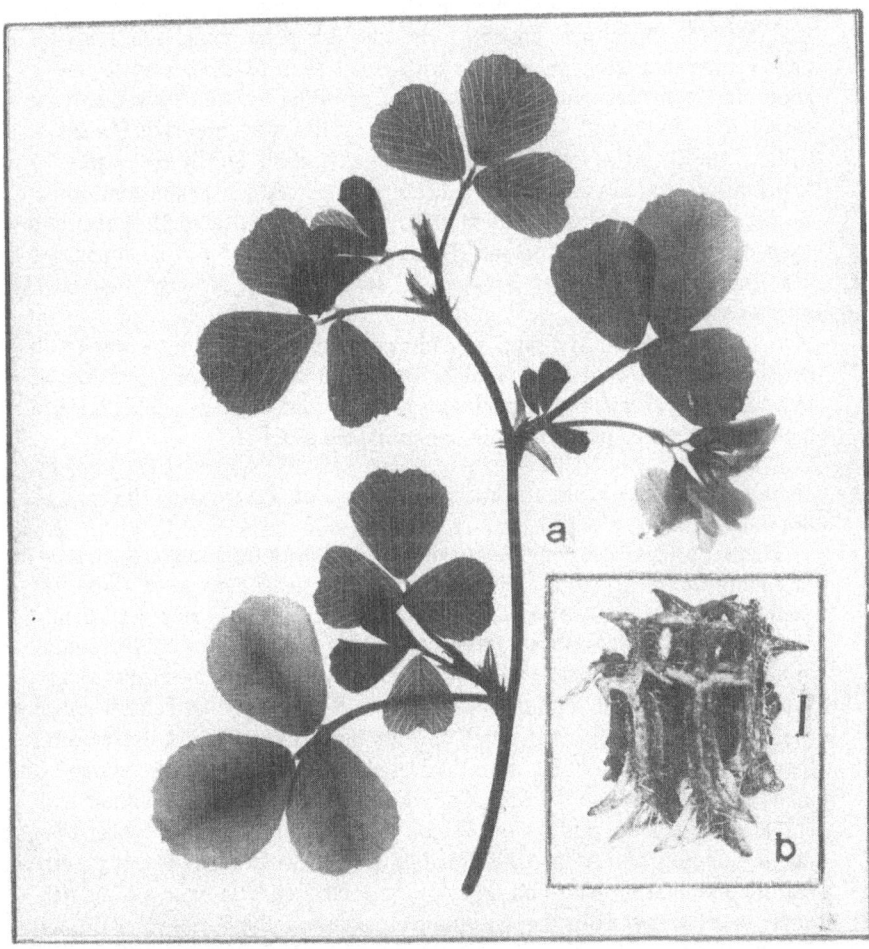

*Fig. 35, M. pironae.* Branch (a) and pod (b).

brown, cylindrical, spiny (rarely only tubercled), glabrescent. Coils 3-6, not tightly appressed, 3.5-4.5 mm in $\phi$, turning clockwise; on coil face 5-6 slightly curved veins, branching in its outer 1/3 before entering the lateral vein; between lateral vein and the dorsal suture a groove. Spines 6-10 in each row, 1-2 mm long, grooved (due to their two-rooted insertion), somewhat bent over the facial plane. Seeds greenish-yellow to brown, 2.5 mm x 1.2 mm, 1-2 in each coil, separated. Seed weight about 2.4 g/1000. Radicle half the seed length or less. $2n = 16$ (Lesins & Lesins, 1961).

*Habitat and Distribution.* Growing on calcareous, weathered rock substrate in the mid-mountain zone of Daghestan ASSR, at altitudes of 500-1500 m.

*Variations within species.* Flowers vary in color: from almost white to light violet, to fully violet.

*Relationship to Other Taxa.* The chromosome complement of *M. daghestanica* was similar to that of *M. pironae* (Lesins & Lesins, 1961) and hybridization between them was possible. The greenish petal color, a mixture of yellow pigments from *M. pironae* with violet from *M. daghestanica*, was a good marker for recognizing hybrids. The resulting hybrids, however, were completely sterile and some meiotic irregularities were observed (Lesins & Gillies, 1968). After chromosome doubling, hybrid fertility, contrary to expectation, was not improved, indicating a genic rather than chromosomal fertility barrier. On crossing the doubled hybrids to cultivated *M. sativa*, two trispecies hybrids were obtained (Lesins, 1971); hence it may be concluded that there is some affinity between *M. sativa* and the *M. daghestanica-M. pironae* group.

*Agricultural Value.* At Brandon plants came through the winter when well protected by snow or straw mulch. Due to success in obtaining trispecies hybrids with *M. sativa*, the species may be considered as a possible donor of hereditary material in the breeding of cultivated alfalfa.

20. *Medicago pironae* de Visiani, Ind. Sem. Hort. Bot. Patav.:365 (1885). Figs. 35; 5,-33.

Plants up to 40 cm long; numerous stems arising from the crown, wiry, ascending to erect, glabrate. Stipules small, acuminate, entire or toothed in their basal part. Leaflets small, almost as wide as long, 4-11 mm x 4-10 mm, obcordate to broadly obovate; margin serrate in its apical part; midrib ending in a small tooth. Peduncle 4 to 10-flowered, ± the length of the corresponding petiole, with a small terminal cusp. Florets 6-8.5 mm long. Pedicel longer than the calyx tube; bract much shorter than the pedicel. Calyx shorter than half the length of the floret, covered sparsely with simple, appressed hairs; teeth longer than the tube. Corolla bright yellow; standard oblong, with parallel sides in the mid-region, length: width ratio 2:1; wings longer than the keel. Young pod rising, then protruding sideways through the calyx teeth. Mature pod dark brown to black, cylindrical, covered with many-celled glandular hairs, spiny. Coils 1.5-4, turning clockwise, 5.5-6.5 mm in $\phi$; on coil face 9-10 radial veins, branching on its outer part before entering the lateral vein; between the lateral vein and the dorsal suture a wide, shallow furrow

124

observable in facial view as well as from the edge. Spines 8-10 in each row, ± 1 mm long, grooved from base to more than half of their length, inserted at almost 90° to the coil face. Seeds yellow, 1-2 in each coil, about 2.5 mm x 1.2 mm, not separated. Seed weight 2-2.6 g/1000. Radicle slightly longer than half of the seed. $2n = 16$ (Lesins & Lesins, 1961).

*Habitat and Distribution.* Growing on sub-mountain rocky hillsides, between low brush. Endemic to the eastern Alps in the districts of Friuli and Gorizia in northeast Italy.

*Relationship to Other Taxa.* See *M. daghestanica.*

*Agricultural Value.* At Brandon light to heavy winterkilling was observed. Forage production was about 30% of Rambler. Regrowth after cutting was poor. Should the taxon harbour some valuable characters, these may possibly be transferred to *M. sativa* (see *M. daghestanica*).


SUBSECTION PAPILLOSAE Grossheim, pro. ser., in Kom. (ed.) Fl. USSR 11:158 (1945).

Corolla yellow; stems prostrate, ascending or erect; pods spineless, turned in 2-4(5) coils, covered with simple hairs to almost glabrous, or covered with articulate, flat, semitransparent, glandular hairs. Seeds separated by a thick, parenchymatous partition. Habitat in mid-high altitudes (1200-3000 m a.s.l.) of the Pontus and Caucasian Mts. $2n = 16,32$.


Key to species of *Papillosae:*

1      Pods almost glabrous or covered with simple hairs

                                                      *M. dzhawakhetica*

—      Pods covered with articulate, semitransparent, as if membranous, glandular hairs            *M. ignatzii, M. papillosa*


The species of sect. *Papillosae* may be hybridized with *M. sativa* at the genomic ratio of 2(*Papillosae*):1(*M. sativa* s.l.).

21.    *Medicago dzhawakhetica* Bordzilovsky, Prot. Zased. Kievsk. Obschch. Estestv.:24 (1909). Syn. *M. dzhawakhetica* var. *timofeewii* Troitz., Report (Vestnik) Tiflis Bot. Gard. New Ser. 1:90 (1923). Figs. 36; 5,-13.

Plants up to 70 cm long; stems arising from the crown, prostrate to ascending. Vegetative parts sparsely covered with simple hairs or glabrate (younger parts and underside of leaflets more hairy). Stipules adnate at base, with 1-2 lengthy teeth in their lower part. Leaflets 8-20 mm x 5-12 mm, obovate; margin slightly serrate in its apical part; midrib ending in a minute tooth. Peduncle 6 to 15-flowered, longer than the corresponding petiole, with a short terminal cusp. Florets 5-8 mm long. Pedicel longer than the calyx tube; bract shorter than the pedicel. Calyx 2.8-3 mm long; teeth awl-shaped, shorter than the tube. Corolla yellow; standard tongue-shaped, length more than twice that of width, its lower-middle part wider than the basal and apical parts (Fig. 36,-b); wings longer than the keel. Young pod covered with appressed simple hairs or almost glabrous, protruding sideways from the calyx. Mature pod glabrate, lens or short-cylinder-shaped, greenish-gray ro nearly black, spineless. Coils 2-4, not tightly appressed, 5-6 mm in $\phi$, turning clockwise; veins 8-10, curved, anastomosing before reaching half of

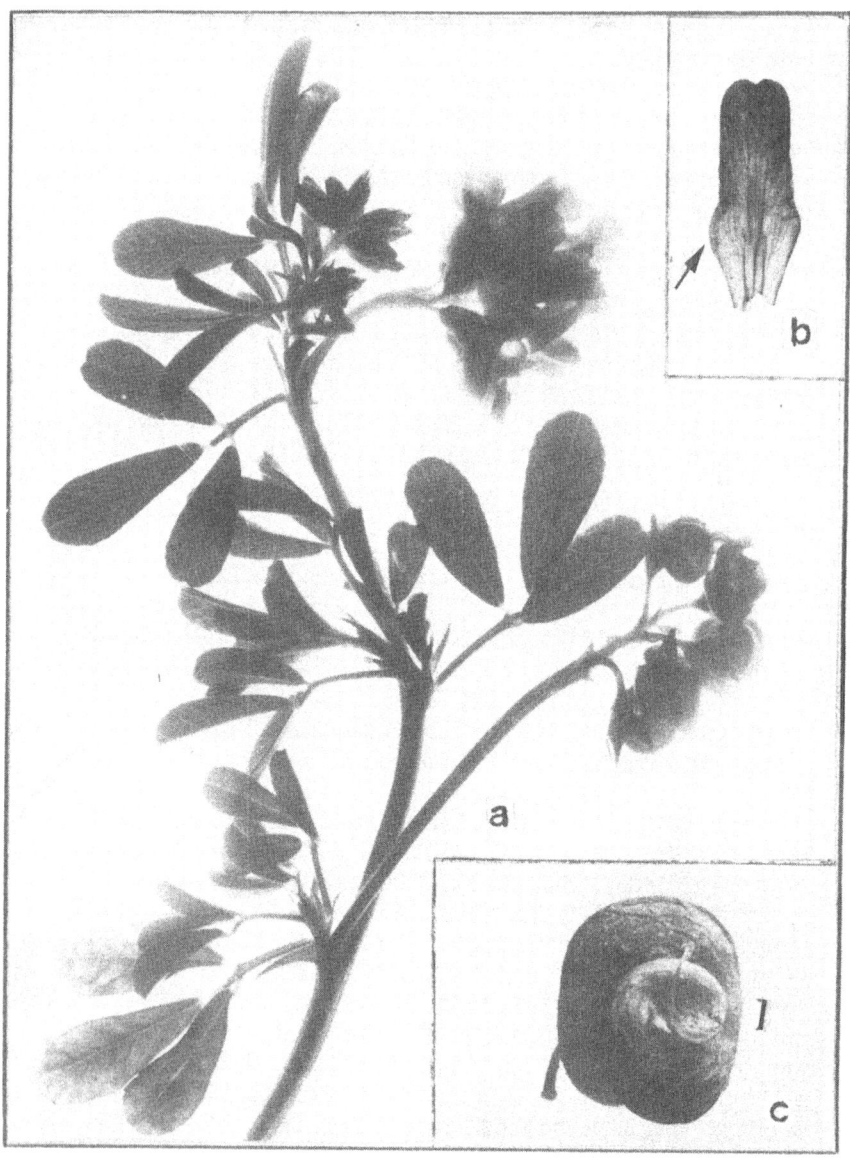

*Fig. 36*, *M. dzhawakhetica*. Branch (a), pod (c), and standard petal (b) (arrow points to widened middle part).

the radius, and entering the dorsal suture. Seeds yellowish-brown, 3 mm x 1.5 mm, 1-2 in each coil, separated by a thick wall. Seed weight 2.3-3.5 g/1000. Radicle slightly longer than half of the seed. $2n = 32$ (Lesins & Lesins, 1966); 16?

*Habitat and Distribution.* In valleys of the mid-mountain zone (1200-1500 m a.s.l.). Endemic to the mountains of Transcaucasia.

*Variation Within Species and Relationship to Other Species.* We were able to investigate material sent to us under the name of var. *timofeewii.* The seed had been collected at Akhalkalaki, Georgian SSR, from where the original *M. dzhawakhetica* had been described. Some nomenclatural confusion regarding the taxon has occurred. At the beginning Bordzilovsky (l.c.) described the form with pods glabrescent or sparsely covered with simple hairs as a separate species *M. dzhawakhetica.* Later he ranked it as *M. papillosa* var. *dzhawakhetica* (1915). Vassilczenko (1949) ascribed the character of pods almost glabrous, as well as those covered (sometimes densely) with short, as if membranous, hairs to *M. dzhawakhetica*; the main taxonomic difference thus consisting in the size of the pods: smaller (3.5-6 mm in $\phi$) characteristic of *M. dzhawakhetica*, larger (7-8 mm in $\phi$) of *M. papillosa.* We followed his classification in our earlier papers (Lesins, 1956b; 1961) and named our material *M. dzhawakhetica.* Later, when plant material with almost glabrous pods became available, it was obvious that the two taxa were different species (Lesins, 1966). Recently we intercrossed $2n = 32$ types having glabrate pods with those having membranous, articulate-hairy pods (*M. papillosa*). $F_1$ plants from maternal plants with glabrate pods had pods with some rudimentary articulate hairs, indicating their hybrid nature. On intercrossing, they appear to have greatly reduced fertility, at least under our greenhouse conditions. Further studies are required to clarify the cause of low fertility.

A unique character indicating a close relationship between *M. dzhawakhetica* and *M. papillosa* is that they both can be crossed with *M. sativa* at an unequal genome ratio: two genomes ($n = 16$) of *M. dzhawakhetica* or *M. papillosa* to one genome ($n = 8$) of *M. sativa* (Lesins, 1961; 1966). Such a character, rare in angiosperms in general, in our opinion is more weighty in assessing taxonomic relationships than many characters of more trivial nature. There is a report of a single case of *M. sativa* x *M. dzhawakhetica* where genomic ratio 1:1 has been observed (Clement, 1963). Such a cross, so far as we know, has never been obtained again, though attempts have been made by us as well as by other researchers.

*Agricultural Value.* Due to the possibility of hybridization with *M. sativa* (see also under *M. papillosa*), *M. dzhawakhetica* may supply genetic material if and when the cultivation of alfalfa has to be expanded to presently unusual environments, such as pastures and meadows in mountain terrains. At Brandon the plants survived winters without injury. Yield of forage has been about half that of commercial alfalfa. Regrowth after cutting was estimated as fair.

127

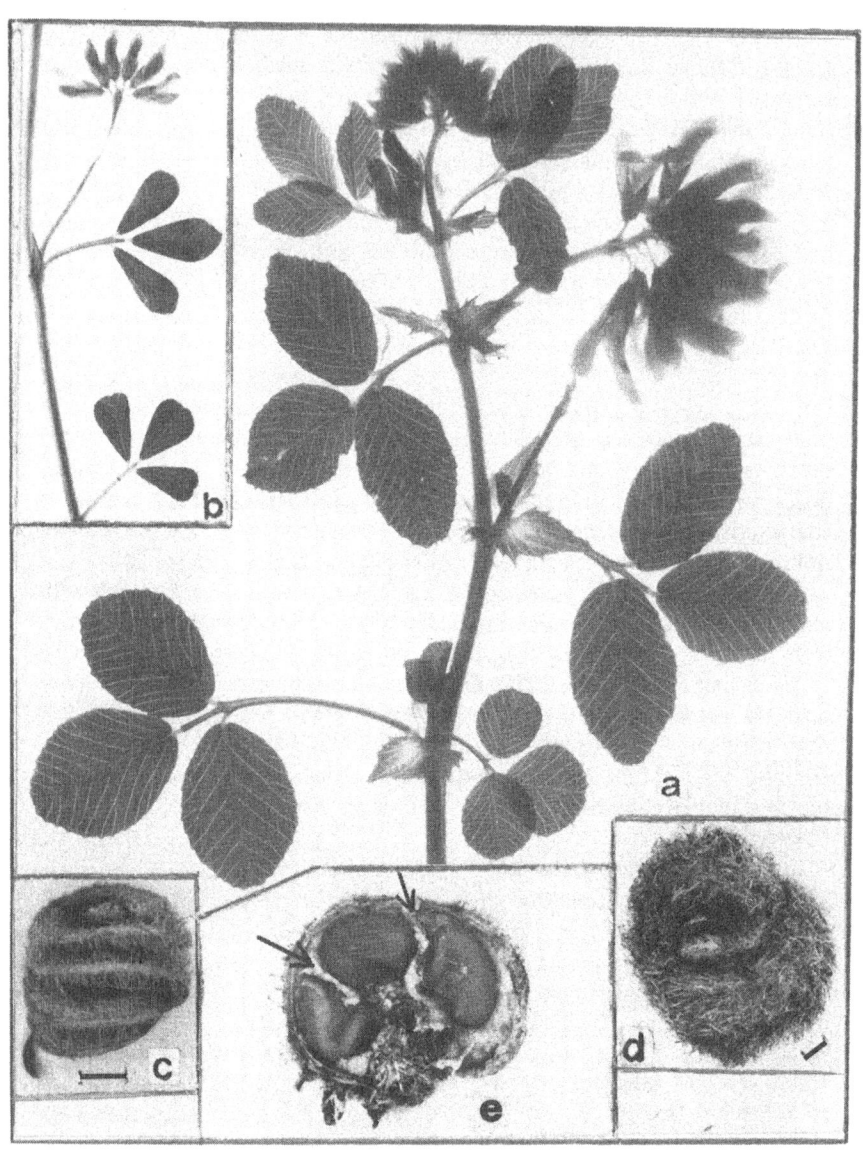

*Fig. 37, M. papillosa.* Branches: ssp. *macrocarpa* (a) (= *M. ignatzii*), magn. 1.8; ssp. *microcarpa* var. *citrina* (b), nat. size. Pods: ssp. *macrocarpa* (c) (= *M. ignatzii*); ssp. *microcarpa* (d) (= *M. papillosa*). Cross section of a pod of ssp. *microcarpa* (e); note thick partitions between seeds (arrows).

128

22. *Medicago papillosa* Boissier, Diagn. Plant. Orient., 1 (2):23 (1843). Figs. 37; 5,-31,-32.

Plants 20-60 cm long; stems prostrate to erect. Vegetative parts with simple hairs or glabrate, especially upper side of leaflets may be glabrous. Stipules triangular, adnate, narrow with a long terminal tooth, or wide uniformly tapering; toothed in basal part or around the entire margin. Leaflets 7-20 mm x 5-12 mm, broadly elliptical to narrowly obovate; margin in its apical 1/3-3/4 slightly serrate or almost entire; midrib ending in a small tooth. Peduncle with a head-shaped inflorescence, 4 to 15(18)-flowered, longer than the corresponding petiole, with a distinct terminal cusp. Florets 6-10 mm long. Pedicel distinctly longer than the calyx tube; bract shorter than the pedicel. Calyx less than half the length of the corolla, with appressed hairs to glabrate; teeth shorter than the tube. Corolla bright to pale yellow; standard usually more than twice as long as wide, tongue-shaped, somewhat wider in the lower mid-region; wings longer than the keel. Young pod protruding sideways from the calyx, densely covered with many-celled, glandular hairs. Mature pod yellowish to dark grey, covered with rough, articulate, membranous, semi-transparent hairs, their glandular tips usually broken off; spineless. Coils 1½-4(5), not tightly appressed, 4-8 mm in $\phi$, turning clockwise. On coil face 8-12 veins, somewhat curved, anastomosing shortly after leaving the ventral suture (for observation, hairs have to be cleared off). Seeds greenish-yellow to brown, 2.3-2.5 mm x 1.3-1.5 mm, 1-3 in each coil, separated by a thick partition. Seed weight 2.3-2.7 g/1000. Radicle slightly longer than half of the seed. $2n = 16, 32$.

*Habitat and Distribution.* Growing in soils of volcanic origin and on calcareous rocky mountain slopes, at and above 2000 m elevation. Endemic to the Pontus Mts. of north-eastern Anatolia and to the adjacent Caucasus Mts. of Transcaucasia. We collected the species in north-eastern Anatolia in summer pastures at Dzimil (Čimil), Istavris, Alischeri and Cigana gecidi.

*Variation Within Species.* Boissier (Flora Orient, 2:96, 1872) recorded a variety *macrocarpa* with large florets. Urban (1873) subsequently divided the species into two subspecies: *microcarpa* Urb. with 2 to 3.5 coils per pod, 5-6 mm in $\phi$, and *macrocarpa* Boiss. with 3 to 4 coils per pod, 6-7 mm in $\phi$. We note that ssp. *macrocarpa* is diploid, $2n = 16$ (Lesins & Lesins, 1963), whereas ssp. *microcarpa* with coil diameter even smaller (4 mm, Fig. 37,-d) than that given by Urban has $2n = 16$ and 32. In ssp. *microcarpa* there is a variety, $2n = 32$, which differs from the rest in having lemon-yellow corollas with broader standard petals (length:width ratio slightly less than 2:1), and having narrow, obovate to cuneate leaflets (as in Fig. 37,-b) of bluish-green color. This variety, for which we propose the name var. *citrina*, is found near Erzurum, Turkey (collected by F. Tosun). In one accession of $2n = 16$ and in one of $2n = 32$ of ssp. *microcarpa* we found some pods with a mixture of articulate and simple hairs. They may be of hybrid origin from natural intercrossing with *M. dzhawakhetica*. For the former cross there should then exist, though not reported yet, a $2n = 16$ type of *M. dzhawakhetica*. Davis (1970) mentions two plants which had characters intermediate

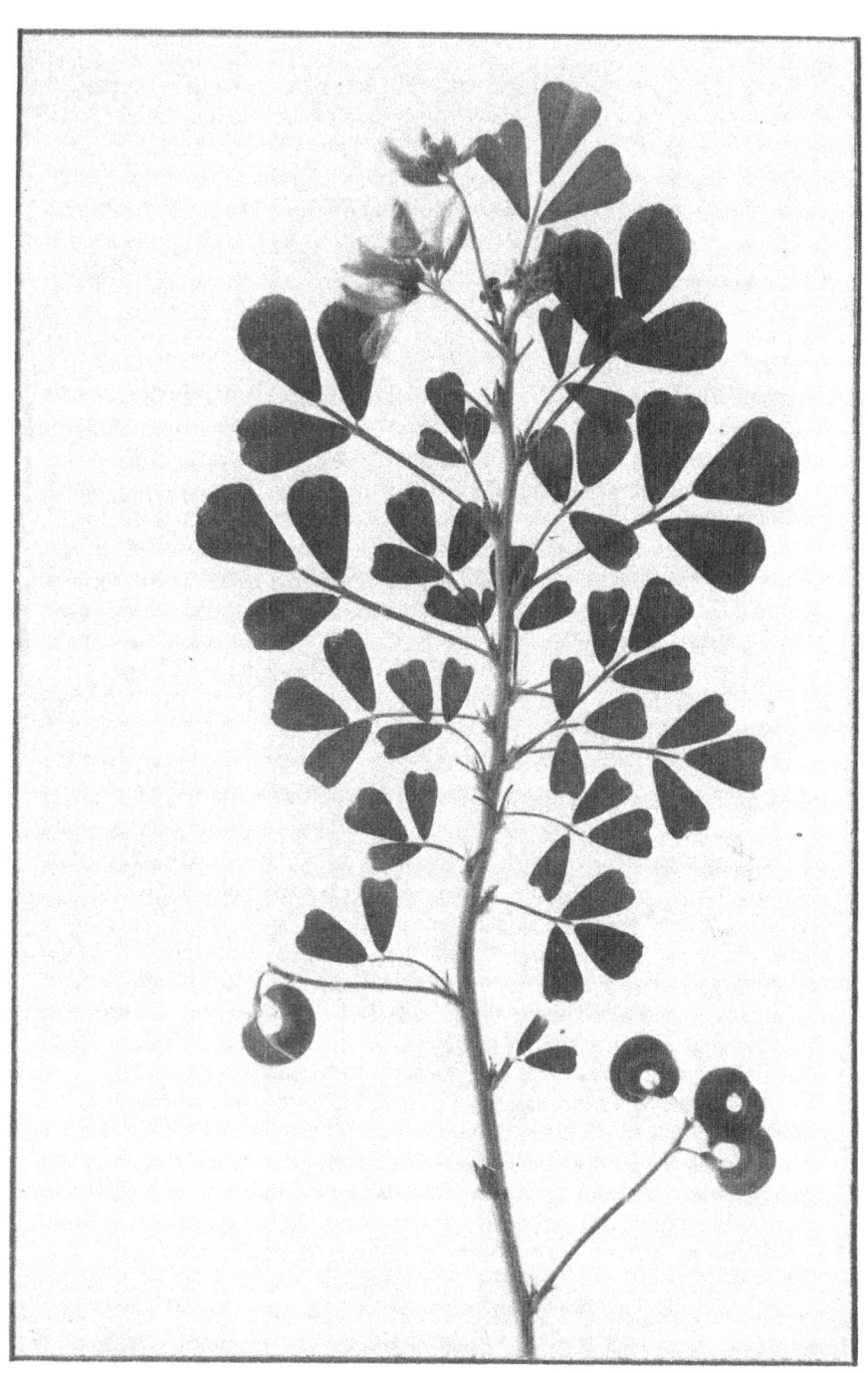

*Fig. 38, M. arborea.* Branch with racemes of flowers and pods (nat. size).

130

between *M. papillosa* on one hand and *M. varia* on the other. In this connection, it may be recalled that the species of subsect. *Papillosa* which hybridize with the *M. sativa-M. falcata* group result in triploid sterile $F_1$s (Lesins, 1961b; Lesins & Lesins, 1966). Hence, the probability of finding natural hybrids of sect. *Papillosa* with *M. sativa* s.l., beyond the $F_1$ generation, is very low.

The ssp. *macrocarpa* Boiss. certainly is morphologically well delineated from ssp. *microcarpa* Urb. in having larger florets (10 mm vs. 6-8 mm), larger pods (6-8 mm vs. 4-5 mm in $\phi$) of cylindrical vs roundish shape (Fig. 37,-c vs. 37,-d), broadly elliptical vs. obovate leaflets; and for diagnostic purposes in fruitless stage the broad, uniformly tapering, all-round toothed stipules are useful for comparison with those in ssp. *microcarpa* having stipules with a narrow terminal tooth and with a few rugged teeth at the somewhat widened base (Fig. 37,-a vs.-b). The plants we saw and from which we collected pods at Dzimil and Istavris (Pontus Mts. north of Erzurum) were almost erect in growth habit. Recently (unpubl.) it was found that $F_1$ hybrids between ssp. *macrocarpa* $2n = 16$ and ssp. *microcarpa* $2n = 16$ were sterile. Together with the above noted morphological differences then, the ssp. *macrocarpa* deserves a species rank. We propose the name M.

*23.* ignatzii (Boiss.) in honor of Ignatz Urban; the ssp. *microcarpa* then retaining the name *M. papillosa* Boiss. As mentioned before, *M. dzhawakhetica* ($2n = 32$) has been hybridized with tetraploid ($2n = 32$) *M. papillosa*. Whether the success in obtaining hybrids was in connection with the tetraploid level of the parents, or the tetraploid *M. papillosa* is indeed closer related to *M. dzhawakhetica* than to *M. ignatzii*, needs further investigations.

Agricultural Value. Though the direct, triploid hybrids with *M. sativa* were sterile, after artificially raising them to the hexaploid level the sterility barrier was overcome. A further improvement was that on backcrossing to artificial hexaploid *M. sativa*, plants were produced with 4 genomes of *M. sativa* and 2 genomes of *M. papillosa* (as *M. dzhawakhetica*, Lesins, 1961b), thus with a balanced genomic constitution. It may be concluded that the germ plasm of *M. papillosa* may be utilized in alfalfa breeding if hereditary material harbored by *M. papillosa* be required. At Edmonton during one winter, *M. papillosa* had a severe reduction in stands caused by some root rot. At Brandon, plants only occasionally have survived winters. Forage yields at Brandon have been 30% of the commercial alfalfa.

SECTION ARBOREAE (L.) *sect. nova.*

Shrubs up to 4 m high. Corolla yellow. Pods with 0.5-1.5 coils, with a large central opening. Florets with the standard shorter than or equal to the keel, wings shorter than the keel. $2n = 32,48$.

One representative, *M. arborea.*

*24.* *Medicago arborea* Linnaeus, Sp. Pl.:778 (1753). Figs. 38; 5,-3.

Shrubs 1-4 m high. Vegetative parts (except the upper side of leaves in some accessions) densely covered with appressed, silky hairs, giving the plant a greyish-green to whitish appearance. Stipules triangular, entire. Leaflets obovate to obcordate, 10-20 mm x 8-18 mm; margin entire or with scarcely noticeable wavy teeth; midrib ending in a minute tooth concealed within lengthy, whitish hairs. Peduncle 7 to 14-flowered, longer than the corresponding petiole, with a small terminal cusp. Florets 11-13 mm long. Pedicel longer than the calyx tube; bract inconspicuous, much shorter than the pedicel. Calyx half the length of the floret or less; teeth slightly shorter than the tube. Corolla yellow; standard elliptical, shorter than or as long as the keel. Pods greyish-yellow, with 0.5-1.5 coils, 10-15 mm in $\phi$, with an opening in the centre, spineless. Numerous slightly bent veins anastomose in the middle part of the pod face, thickening towards the dorsal suture. Seeds brownish, 3-4.5 mm x 2-3 mm, 2-8 in one pod, separated by spongy partitions. Seed weight 6-9.5 g/1000. Radicle equal to or slightly longer than half of the seed. $2n = 32, 48$.

*Habitat and Distribution.*  Preferring rocky hillsides, generally in dry soils. Plinius, in the first century of our era, writes that the cytisus shrub (= *M. arborea*) has been discovered on the Island Cythnus and from there has been brought to other adjacent islands of the Aegean Archipelago and to the Greek mainland. *M. arborea* at present is distributed from the Canary and Balearic Islands along southern Europe to Asia Minor; Algeria is mentioned by some authors, but Nègre (1959) does not record it for North Africa. It may have been introduced as an ornamental in some of the mentioned areas. Post (1896) records it as having been introduced in the vicinity of Jerusalem. We collected the species in Greece, southern France (Nice, Cannes), Spain, and on Sardinia.

*Variation Within Species.*  We could verify correctness of the chromosome number $2n = 48$ reported by Fernandes & Fatima Santos (1971) on seeds originally from the Balearic islet, Espartar. Mention of a diploid ($2n = 16$) chromosome number by the above authors turned out to be a misinterpretation of some literature sources (person. corresp.). Plants from Espartar Island have been described by Font Quer (1924) as var. *citrina*, differing from the usual *M. arborea* by larger pods and lemon-yellow, instead of orange-yellow, flowers. The species has ordinarily the tetraploid, $2n = 32$, chromosome number as reported by earlier researchers (Ghimpu, 1930; Mariani, 1963; Raven et al., 1965), and found in our accessions before acquisition of material from Espartar.

   *M. arborea* is the only species in *Medicago* with a woody growth habit. As noted in the General Part of this work, we consider the shrub-woody species to be older than the herbaceous ones. We also noted that *M. arborea*, at its present chromosome level, has unlikely been involved in the origin of the present herbaceous *Medicago*. Characteristic of *M. arborea* is its flower structure with the standard usually shorter than the keel. Self-pollination, which takes place in some of our *M. arborea* accessions, brings, paradoxically, *M. arborea* closer to the phylogenetically younger annual species, in which automatic tripping and selfing is the rule. We pointed to the possible explanation of this phenomenon.

*Agricultural Value.*  In the past, when areas where *M. arborea* was endemic were sparsely inhabited by man, they provided a tangible supply of fodder for his herd of goats, which were often the most valuable domestic animals. Under the name *Cytisus* it has been praised as a fodder plant by ancient Greeks and Romans. Changes that have taken place since those times have deprived the species of almost any importance in forage production, although there are experiments of growing it similarly to the basket osier and harvesting young shoots for fodder before they become woody.

At Brandon no plants have survived winter in the field.

SECTION MARINAE Grossheim, pro ser., in Kom. (ed.) Fl. USSR 11:160 (1945).

Perennial herbs; stems semi-prostrate; whole plant including pods covered with feltlike, whitish hairs. Corolla lemon yellow. Pods with 2-4 coils, not open in the centre; coils with spines, tubercles, or almost spineless.

*Fig. 39,   M. marina.* Branch (a), and pod (b).

Florets with the standard longer than the keel, wings equal to or longer than the keel. Growing exclusively in seashore sands. $2n = 16$.

The only representative of the section is *M. marina*. We have noted that its pods had some resemblance to those of *M. pironae* and *M. daghestanica* in that they all have lateral veins.

25. *Medicago marina* Linnaeus, Sp. Pl.:779 (1753). Figs. 39; 5,-25.

Plants up to 60 cm long; stems roundish in cross section, prostrate to decumbent, arising from the crown. Vegetative parts densely covered with simple, whitish hairs giving the plant a greyish appearance. Stipules entire or slightly toothed in their basal part. Leaflets thick, 8-16 mm x 7-11 mm, obovate, often truncate or emarginate; margin due to dense cover of hairs appears entire and without terminal tooth. Peduncle head-shaped, 7 to 12-flowered, longer than the corresponding petiole, with a small terminal cusp. Florets fragrant, 6-10 mm long. Pedicel shorter than the calyx tube; bract small, much shorter than the pedicel. Calyx 2-4 mm long; teeth shorter than or equal to the tube. Corolla lemon yellow; standard obovate; wings equal to or longer than the keel. Young pod rises from the calyx tube, then bends sideways from it. Mature pods yellowish-grey to dark grey usually covered with a mat of hairs, cylindrical, spiny, tubercled, or almost spineless. Coils 2-4, not tightly appressed, 4-6 mm in $\phi$, turning clockwise. From the ventral suture originate 6-7 slightly bent veins, anastomosing on the outer part of coil face, then running almost parallel to the lateral vein, finally joining it; between the lateral vein and the dorsal suture, a shallow furrow. Spines or tubercles 8-11 in each row, conical, grooved at the base, inserted slightly obliquely to the face of the coil. Seeds brownish, 1-2 in each coil, 3-3.8 mm x 1.5-2 mm, separated by a thin parenchymatous wall. Seed weight 3-4.7 g/1000. Radicle less than half of the seed length. $2n = 16$.
*Habitat.* Exclusively on seashores, usually in loose sand. The only *Medicago* of such habitat requirement, though seashore soils of more solid consistency may be inhabited by other species, e.g., *M. littoralis*.
*Distribution.* Shores of the Mediterranean and Black Seas, and the Atlantic coast of Iberia and France.
*Relationship to Other Species.* Attempted crossings with *M. sativa* and *M. pironae* and *M. daghestanica* were unsuccessful.
*Agricultural Value.* At Brandon a few plants occasionally have come through the winter; however, they showed little growth during the following summer.

SECTION SUFFRUTICOSAE Vassilczenko, pro ser., in Acta Inst. Bot. Komarovii, ser. 1/8:74 (1949).

Perennial herbs. Corolla yellow. Leaflets broad (length:width, 3:2 and wider), their upper side glabrous. Pods coiled, or almost straight, covered with glandular and/or simple hairs, or glabrous. Florets with the standard longer than the keel, wings also longer than the keel. Endemic to the Pyrenees and Corbier Mountains of the Iberian Peninsula and to the Atlas Mountains in Morocco. $2n = 16$.

Key to species in section *Suffruticosae:*

1       Pods slightly bent, flat, wide (4 mm and wider), glabrous. Endemic to Corbiers and the Eastern Pyrenees      *M. hybrida*

 —     Pods coiled; coils 1.5 to 4. Endemic to mountains of Iberian Peninsula and Morocco      *M. suffruticosa*

It was noted in the General Part that hybridization tests showed that *M. suffruticosa* and *M. hybrida* were very closely related (Lesins, 1969). These findings confirmed a previously noted similarity (Lesins & Lesins, 1965; 1966) between the chromosome complements of *M. hybrida* and *M. leiocarpa* (*M. suffruticosa* ssp. *leiocarpa*). It was further found that the taxa of section *Suffruticosae* were not closely related to the *M. sativa-M. falcata* group since no hybridization was possible between the two groups. Later, Gillies (1972c), studying pachytene chromosomes, showed that the chromosome complements of *M. hybrida*, *M. suffruticosa* and their hybrids were very similar, but distinctly different from those of *M. sativa* s.l.. Recently it was found that section *Suffruticosae* differed from all other perennial *Medicago* in having in their petals small amounts of β-zeacarotene, a carotenoid not found in any other perennial *Medicago* species (Ignasiak & Lesins, 1975). The above evidence justifies considering the *suffruticosa* group as a separate section. This section does not fit well into either *Medicago* or *Trigonella*. *M. hybrida* has often been referred to as *Trigonella hybrida*, and Gillies noted that its karyotype in some respects resembled that of a *Trigonella* species studied by Coutinho & Santos (1943). On the other hand its close relative, *M. suffruticosa*, does not fit well into genus *Trigonella* because of its coiled pods, no taxon in *Trigonella* having that feature.

Referring further to hybridization studies in *Suffruticosae*, touched upon in the General Part, it was found that inheritance of the character, coiled pods vs. straight pods was determined by six complementary factors located at different sites of the chromosome complement (Lesins, 1969). Occurrence of a similar inheritance of coiled pods vs. straight pods was found in *M. sativa* x *M. falcata* and *M. falcata* x *M. prostrata* (Lesins, 1957 and 1962, resp.) hybrids.

It may be noted that self-fertilization in *Suffruticosae*, notably in *M. hybrida*, is more common than in other *Medicago* perennials, especially in those of subsection *Falcatae*.

The occurrence of *M. suffruticosa* in both the Pyrenee Mts. of the Iberian Peninsula and in the Atlas Mts. of N. Africa adds support to the view that these continents had been connected in the past.

26.   *Medicago suffruticosa* Ramond ex DC in Lam. et DC, Fl. Franc. 4:541 (1805).

     (A) *M. suffruticosa* ssp. *suffruticosa* Urban, Verh. bot. Ver. Brand. 15:58 (1873). Figs. 40; 6,-55; 8,-4.

     Plants 30-70 cm long; stems numerous, arising from the crown with little secondary ramification, leafy, prostrate, decumbent or deflexed (if close to

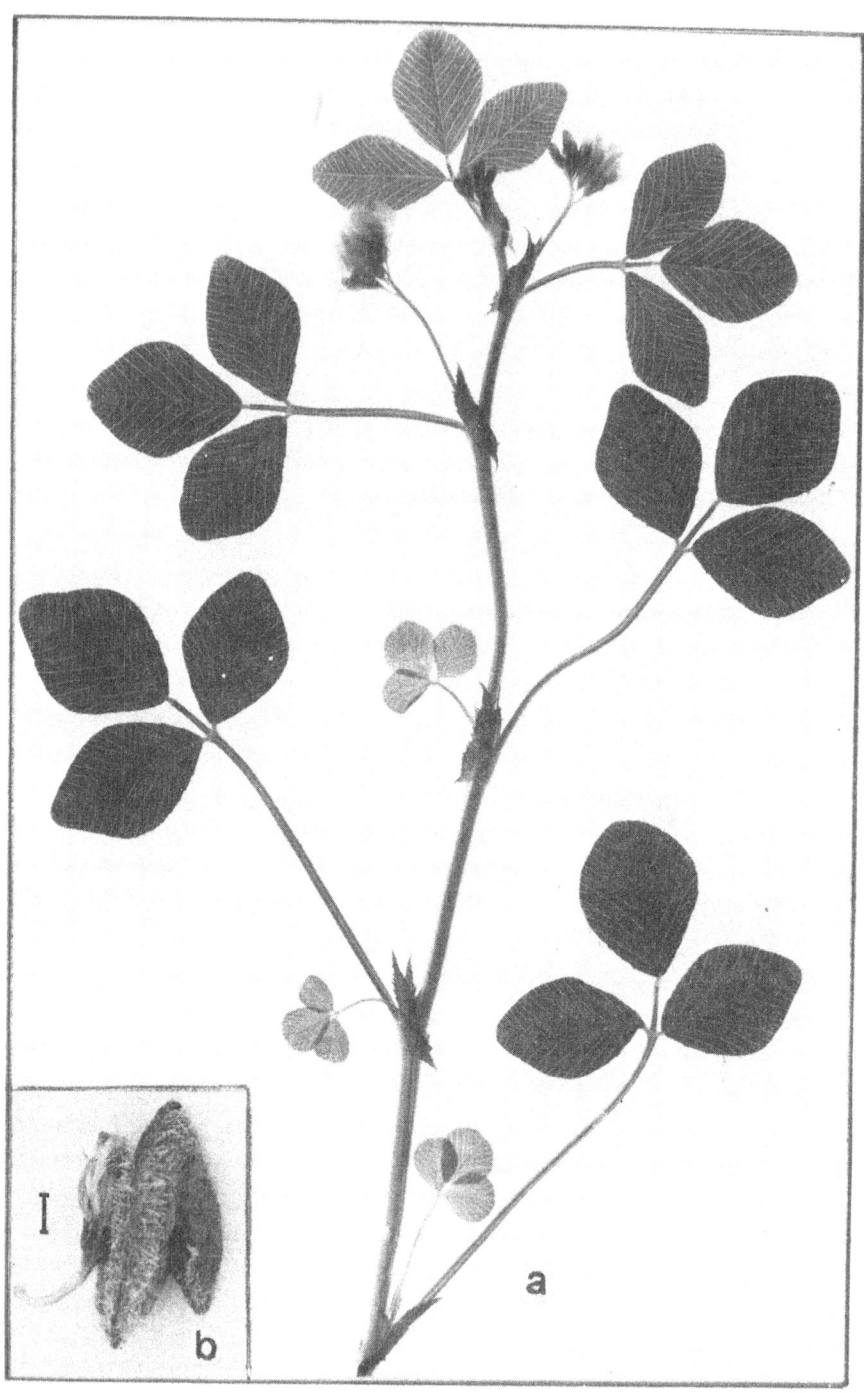

*Fig. 40, M. suffruticosa* ssp. *suffruticosa*. Branch (a), and pod (b).

ground declivities), sparsely covered with simple hairs. Stipules large, finely toothed along the margin, ending in an elongated tooth. Leaflets 12-21 mm x 10-16 mm (length not over 2 times width), rounded, broadly elliptical to obovate, glabrous on the upper side, with light green veins conspicuous against the dark green lamina; margin entire, or finely serrate in the apical part; midrib ending with or without a minute tooth. Stalk of the middle leaflet long, half the length of the corresponding lamina. Peduncle 3 to 10-flowered, usually longer than the corresponding petiole, with a terminal cusp. Florets 7.5-8.5 mm long. Pedicel usually longer than the tube; bract less than half the length of the pedicel. Calyx 3-3.5 mm long; teeth equal in length to the tube. Corolla yellow; standard broadly obovate to obcordate; wings longer than the keel. Young pod rises from the calyx, then bends sideways through the calyx teeth. Mature pod light to dark brown, disk-

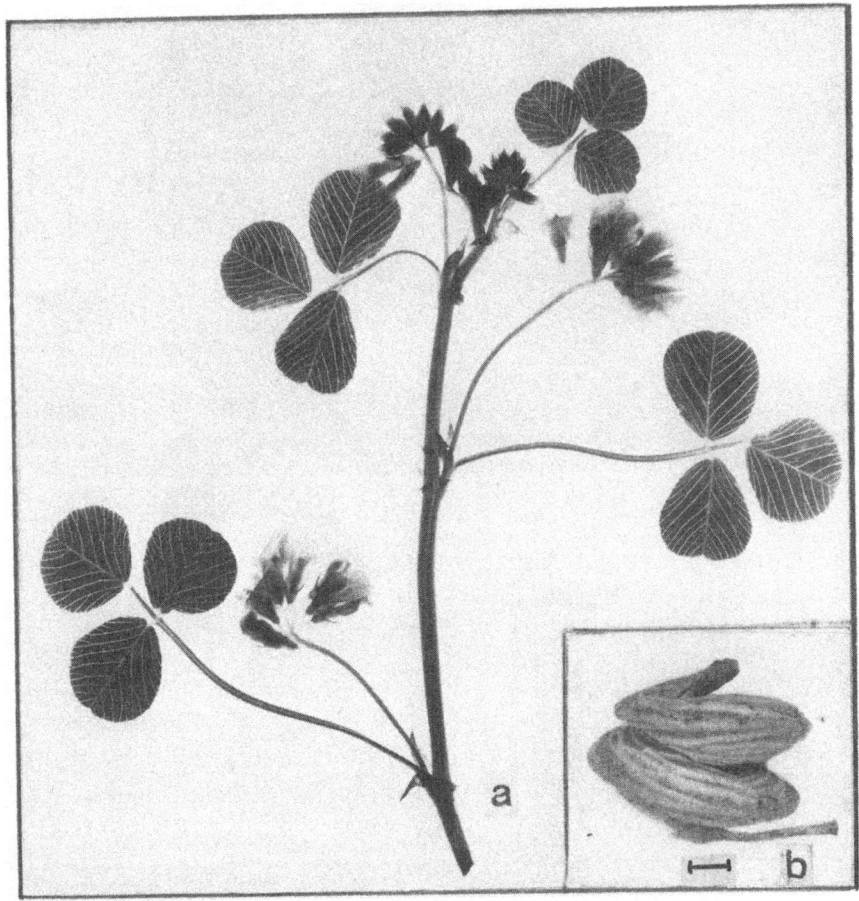

*Fig. 41,* M. suffruticosa ssp. *leiocarpa.* Branch (a), and pod (b).

shaped, with brittle walls, glabrous, or with glandular, simple, or a mixture of glandular and simple hairs; spineless; without or with a small central opening observable from the apical end of the pod. Coils 1.5-3.5, not tightly appressed, 5-7.5 mm in $\phi$, turning clockwise. On the coil face, 6-8 veins running from the ventral suture to the outer part, where they proceed almost concentrically forming a long-celled net from which individual veins pass obliquely to the dorsal suture. Seeds greenish to brownish-yellow, 2.5 mm x 1.5 mm, up to 5 in each coil, not separated. Seed weight about 2.7 g/1000. Radicle 2/3 of the seed length or more, somewhat departing from the curvature of the cotyledons (see Fig. 6,-55). $2n = 16$.

*Fig. 42, M. hybrida.* Branch (a), and pod (b).

*Habitat and Distribution.* We found *M. suffruticosa* growing on moist mountain slopes and ravines. Nègre (1956) also recorded it as growing in moist, depleted pastures at medium and high altitudes.

It is distributed in the mountains of the Iberian Peninsula and Morocco at altitudes of 800-2000 m. We collected our samples in the central Pyrenees.

*Agricultural Value.* At Brandon, spring growth and regrowth as well as forage yields were very light, and it was not fully winter-hardy. It has been observed, however, that plants tend to persist by reseeding themselves and have even become slightly bothersome in the field nursery as weeds. This type of self-propagation would be highly desirable in a pasture crop.

(B) *M. suffruticosa* ssp. *leiocarpa* Urban, Verh. bot. Ver. Brand. 15:58 (1873). Syn. *M. leiocarpa* Benth., Cat. Pl. Pyren.:100 (1826). Figs. 41; 6,-56.

Ssp. *leiocarpa* differs from ssp. *suffruticosa* in having consistently glabrous pods, usually of short-cylindrical shape; coils 2-4 vs. 1½-3½; venation on the pods more prominent, with a wider dorsal suture. Seeds larger, 3.2-3.5 mm x 1.5-1.8 mm vs. 2.5 mm x 1.5 mm, and heavier, 4.4 g/1000 vs. 2.7 g/1000. Seed radicle follows closely the curvature of cotyledons, its very tip only protruding from the depression in the cotyledons (see Fig. 6,-56). The $2n$ chromosome number is 16 (Lesins & Lesins, 1965) as in ssp. *suffruticosa*.

*Habitat and Distribution.* Ssp. *leiocarpa* prefers drier doils and grows at lower altitudes than the ssp. *suffruticosa*. We collected it in southern France on foothills of the Corbier Mts., on uncultivated strips between small fields, and in vineyards; in Morocco we found it on the foothills of the Atlas Mts. along the road Azrou-Ifrane.

*Agricultural Value.* Considering that *M. suffruticosa*, the whole species, does not hybridize with *M. sativa* (Lesins, 1969), it is not of importance for improvement of cultivated alfalfa. However, ssp. *leiocarpa*, together with other taxa of sect. *Suffruticosae*, may be of some value in pasture improvements should attention be directed to the utilization of rocky, steeply sloped land at higher altitudes not suitable for conventional crops. At Brandon plants did not come through the winter.

27. *Medicago hybrida* Trautvetter, Bull. Sci. Acad. St. Petersb. 8:271 (1841). Syn. *Trigonella hybrida* Pourr. Mém. Acad. Toul. 3:331 (1788); *M. pourretii* Noulet, Fl. Bass. Sous-Pyr.:151 (1837). Figs. 42; 5,-19; 8,-9.

Plants 20-40 cm long; stems numerous, arising from the crown, prostrate to decumbent. Vegetative parts sparsely covered with simple hairs, or glabrous. Stipules entire, or with a few small teeth along their margin. Leaflets on upper side glabrous, 14-22 mm x 9-19 mm at lower nodes rounded-obovate often obcordate, at upper ones broadly ovate to obovate; margin entire or minutely serrate, midrib ending in a scarcely perceptible tooth. Peduncle 3 to 5-flowered, longer than the corresponding petiole, with a distinct terminal cusp. Florets 8-8.5 mm long. Pedicel longer than the calyx tube; bract shorter than the pedicel. Calyx 3-4 mm long, covered with simple, appressed hairs; teeth ± the length of the tube. Corolla yellow; standard roundish, broadly obovate; wings longer than the keel. Young pod emerges from the calyx, then bends slightly sideways remaining almost straight. Mature pod slightly curved, 7.5-10 mm x 4-5 mm, light brown, spineless, glabrous. From the ventral suture 14-17 veins proceed diagonally to the dorsal suture, anastomosing in the outer 2/3 of the coil face. Seeds greenish-brown, 3 mm x 1.5 mm, 2-4 in each pod, not separated. Seed weight 4.2-4.4 g/1000. Radicle 2/3 the length of the seed, narrow in comparison to cotyledons. $2n = 16$ (Lesins & Lesins, 1965).

*Habitat and Distribution.* Endemic to the Corbier and east Pyrenean mountain ranges. Our sample was collected at about 500 m above sea level in the Corbier Mts. The air moisture at the collection site was high (in a gorge with a stream at the bottom). The taxon is the most self-fertile member of the *Suffruticosae,* and sets some seed under insect-proof greenhouse conditions. We wondered whether the high air humidity may not have been the reason for scarcity of pollinating insects, thereby compelling the species to turn to self-pollination.

*Agricultural Value.* At Brandon the plants have not survived the winter. Since the taxon is not so closely related to the *M. sativa* group as to be a donor of hereditary material to the cultivated crop, it has no agricultural value in that respect. Under certain mountain pasture conditions, however, it may have some merits, especially after hybridization with the more vigorous *M. suffruticosa.*


SUBGENUS SPIROCARPOS (Ser.) Grossheim, in Kom. (ed.) Fl. USSR 11:162 (1945).

Annuals. Corolla yellow. Pods tightly coiled leaving no opening in the centre; usually many-seeded. Seeds with their long axes more or less parallel to the ventral suture; radicle shorter than, equal to, or longer than half of the seed length. $2n = 16,32,14$.


Key to sections of subgenus *Spirocarpos*:

| | | |
|---|---|---|
| 1 | Pods with soft walls; coil wall in the central part consisting of veins with membranous connection between them. Coils thin, not appressed. Spines, if present, with two-rooted base not embedded in spongy tissue. Veins on pod face well discernible | 2 |
| — | Pods with hard or tough walls; coil wall in the central part more thick, not membranous. Coils with wide, or pergamentaceous edge, appressed or not. Spines, if present, with conical or flat base, often embedded in spongy tissue. Veins on coil face may be obscured by spongy tissue | 3 |
| 2 | Face of coils without lateral veins or veinless zone. Pods spiny. Seeds usually black | sect. *Intertextae* Urb. |
| — | Face of coils with lateral or veinless zone. Pods spiny or spineless. Seeds never black | sect. *Leptospirae* Urb. |

3       Pods with hard walls. Coils appressed, thick, turning clockwise
        or anticlockwise. Spines if present, stocky, their conical base
        often embedded in spongy tissue. Veins on coil face usually
        obscured by spongy tissue. Radicle shorter than half of the seed
        length                                           sect. *Pachyspirae* Urb.
—       Pods with hard or tough walls. Coils not appressed, turning
        clockwise; coil edge thin (pergamentaceous), or thick and con-
        spicuously transversely ridged; if spiny, spines short, inserted in
        the margin of coil edge. Veins on coil face usually clearly
        discernible. Radicle equal to or shorter than half of the seed
        length                                           sect. *Rotatae* Boiss.

SECTION ROTATAE Boissier. Fl. Orient. 2:92 (1872).

Pod walls hard or tough. Coils not tightly appressed, turning clockwise. Pods spineless, or with short spines extending from margins of the coil edge. Veins on face of coil usually well discernible. Radicle equal to or shorter than half of the seed length. $2n = 16,32$.

Key to groups of sect. *Rotatae*: .

1       Pods spiny                                              *M. rotata*
—       Pods spineless                                                    2
2       Coil edge thick, with prominent ridges obliquely or transversely
        to the dorsal suture        *M. noëana, M. rugosa, M. shepardii*
—       Coil edge thin, pergamentaceous, without ridges                  3
3       Pods biconvex. Coils imbricate, the convex ends pointing to
        both pod base and apex      *M.* x *blancheana, M. bonarotiana*
—       Pods cup-shaped. Coils imbricate, their convex ends pointing to
        pod base                                              *M. scutellata*

*Affinity Groups Within sect. Rotatae.* In the first place there is a close relationship between *M. rotata* and *M. bonarotiana.* Our investigations (Lesins et al., 1976) showed that interbreeding between the two was possible, progenies were fully viable, and inheritance of differing characters (pod shape, coil spininess, notchedness and anthocyanin color in the leaves) agreed generally with the hypothesis that one or two genes (three alleles in one case) were responsible for the determination of the characters investigated. The relationship of both taxa has been recognized early by Boissier (1872) who placed *M. rotata* together with *M. blancheana* (which, according to our study, is a product of *M. rotata* x *M. bonarotiana*) in section *Rotatae.* Urban (1873) retained this classification. Heyn (1963) also mentions that she was inclined for some time to regard *M. rotata* and *M. blancheana* as subdivisions of one species and that intermediates between them are to be found. Recently, Ponert (1973) included these two as one of subspecies in *M. rotata.* In our investigations we noted some additional similarities: block-shaped pollen grains with 4-5 germinal apertures (Fig. 7,-4, and Fig. 8,-3,

resp.), conspicuously small lateral leaflets of the first trifoliate leaf (Fig. 44,-
c), and a remarkably long after-ripening period of seeds (Lesins et al., 1976).
We decided, however, to consider *M. rotata* and *M. bonarotiana* as two

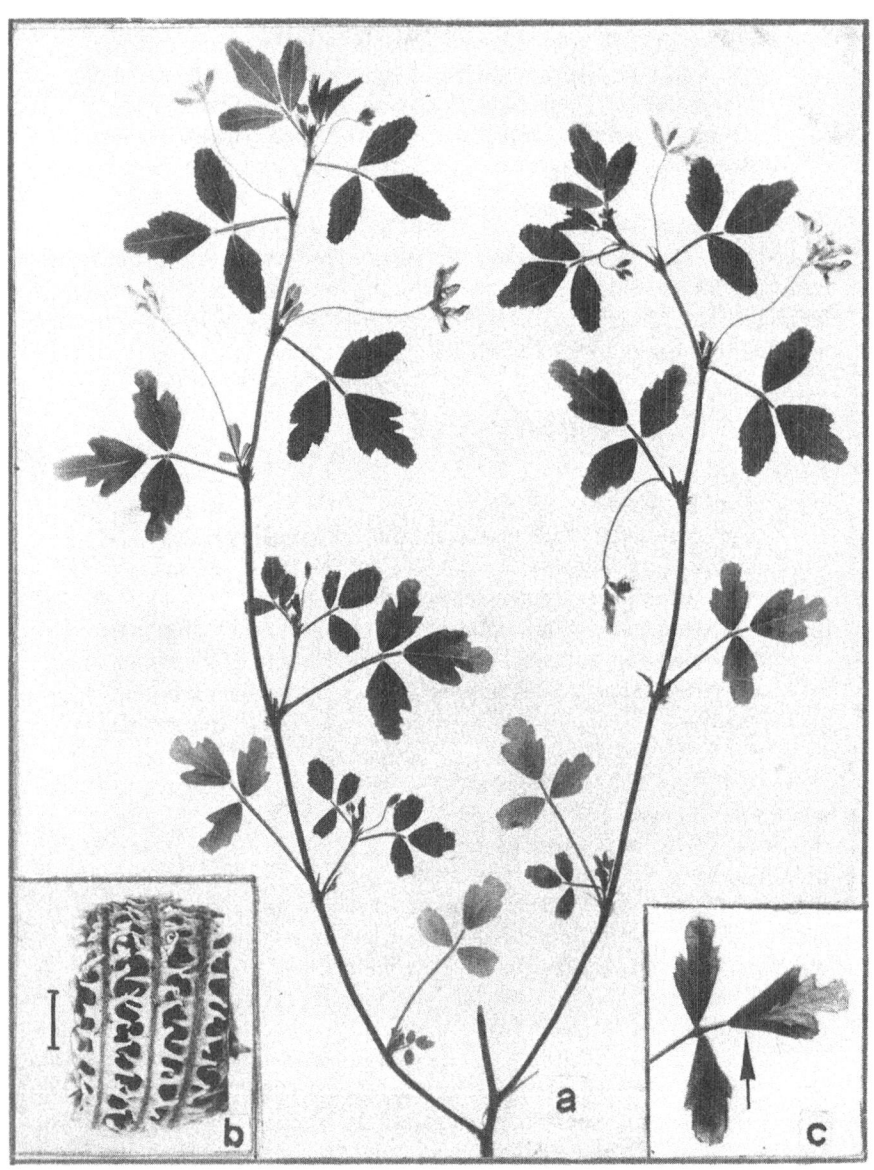

*Fig. 43, M. rotata.* Branch (a), pod (b), and leaflet (c); note anthocyanin-
colored basal part (arrow).

142

species, mainly because of their morphological differences, especially with regard to the pods.

Regarding the other *Rotatae* species, there have been groupings of *M. rugosa* (as *M. elegans*) with *M. noëana* in a separate section *Elegantes* by Boissier (1872). Post (1888) added to these two species his later described *M. shepardii*.

Urban (1873) combined *M. rugosa* with *M. scutellata* in section *Scutellatae*. It may be noted that at the time of Urban's *Medicago* monograph it was not known that these two species were the only tetraploids (2*n* = 32) among the existing close to 40 other annual species of the genus. There are other similarities between the two species: Seed radicles are shorter than half of the seed length and their tips depart considerably from the main body of the seed (Fig. 6,-43,-51); both species grow in heavy soils, and the area of distribution is common to both. However, we found differences in the pollen which in *M. rugosa* is spindle-shaped, in *M. scutellata* pyramid-shaped (Lesins & Lesins, 1963). In our crossing attempts no hybrids were obtained between the two species. In addition, the peculiar pod morphology of *M. scutellata* leads us to consider it less closely related to other members of sect. Rotatae.

The remaining three species, *M. rugosa*, *M. noëana* and *M. shepardii*, as indicated by Post (1896), have some similarity in pod appearance: the thick coil edge, with ridges running obliquely or transversely against the dorsal suture. We note that our attempts to cross *M. rugosa* with *M. noëana* were not successful.

28.  *Medicago rotata* Boissier, Diagn. Pl. Or. Nov. 1, 2:24 (1843). Figs. 43; 6,-42

Plants 30-50 cm long, ascending, branching from the main stem. Vegetative parts glabrate, or covered sparsely with simple upright or appressed hairs (on the lower side of the leaf). Stipules toothed in their basal part, ending in a long terminal tooth. Leaflets 11-24 mm x 7-15 mm, irregularly notched, or entire (at upper nodes and on secondary branches), elliptic, often obovate; margin serrate in the apical part, ending in a small tooth. Peduncle 1 to 5-flowered, usually longer than the corresponding petiole, with a terminal cusp. Florets 6-9 mm long. Pedicel usually shorter than the calyx tube; bract ± the length of the pedicel. Calyx 3-5 mm long, covered with simple hairs; teeth equal to or shorter than the tube. Corolla yellow; standard broadly obovate; wings slightly shorter than the keel. Young pod glabrate, contracted within the calyx. Mature pod brown to blackish, cylindrical, glabrous, spiny, with 3.5-5.5 coils. Coils not tightly appressed, 6-9 mm in $\phi$, turning clockwise, apical coil concave. Face of coil with radial veins, forming a distinct net. Spines 0.5-1.5 mm long, inserted at approx. 90° to the coil face, their bases united in a narrow band. Seeds yellowish to medium brown, 3.5-4.5 mm x 2-2.5 mm, 1-2 in each coil, separated (not always distinctly). Seed weight 7.5-9.9 g/1000. Seed coat yellowish to medium brown; around chalaza and extending to the hilum a darker strip (somewhat like an exclamation mark). Radicle equal to or slightly shorter

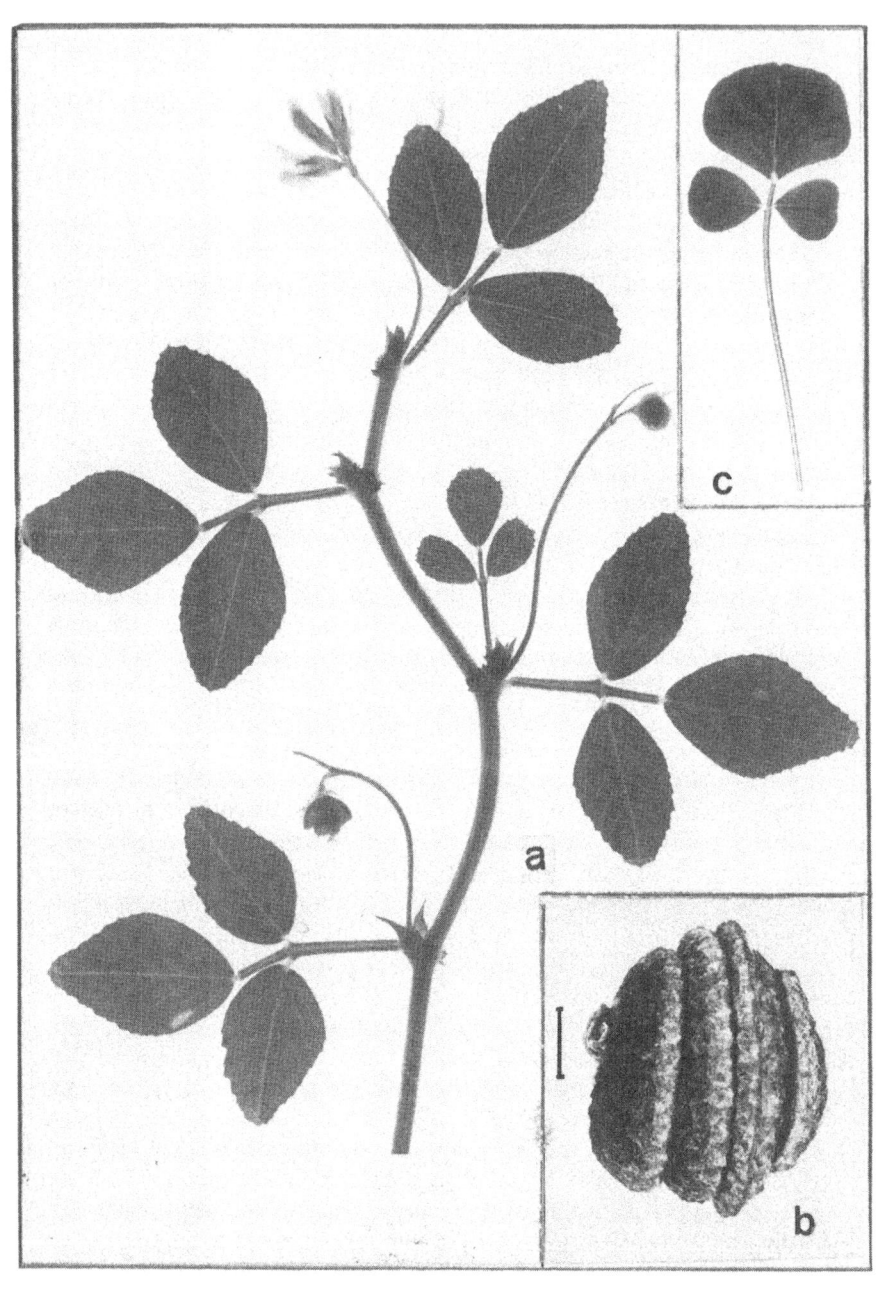

*Fig. 44, M. bonarotiana.* Branch (a), pod (b), and first trifoliate leaf (c); note the small lateral leaflets.

144

than half of the seed length. Unexpanded pollen grains irregularly block-shaped, with four or five germinal apertures. $2n = 16$.

*Habitat and Distribution.* Growing on edges of cultivated fields or as a weed in crops. We gathered pods mainly among stubble, and in fallow fields.

The species is an East-Mediterranean taxon. We collected it in Turkey and Lebanon; seeds were obtained also from Israel and Syria.

*Variation Within Species and Relationship to Other Taxa.* We devoted considerable time in investigating the various *M. rotata* in our collection and hybridizing them with *M. bonarotiana* (Lesins et al., 1976). The two taxa could be crossed readily, the pollen of $F_1$ was of good viability, $F_2$ plants could easily be established and survival was good after a suitable technique for seed treatment was worked out. Segregation ratios of different characters were explainable, assuming that one or two genes were determining the characters studied. It was found that *M. rotata* var. *rotata* as described by Heyn (1963) as well as *M. rotata* var. *eliezeri* Eig were very probably described from hybrid plants of *M. bonarotiana* x *M. rotata* origin, since we found their taxa-characteristic pod shape and coil spininess in $F_2$ plants of this cross. It should be noted that in the hybrid material pod shape and coil spininess had some variations, indicating that some modifying genetic factors were involved.

Other morphological characters usually associated with *M. rotata* are notched leaflets and anthocyanin color in their basal parts (Fig. 43,-c). These features are more clearly expressed at lower stem nodes, whereas at higher nodes and on secondary branches they may be absent. Anthocyanin coloring in higher-node leaflets in particular, depends greatly on the light conditions. It should also be mentioned that some accessions, otherwise conforming to *M. rotata* characteristics, have some glandular hairs on the calyx and young pods, probably due to introgression from *M. bonarotiana*.

29.  *Medicago bonarotiana* Arcangeli, Giorn. Botan. Ital. 8:6 (1876). Figs. 44; 5,-4; 8,-3.

Plants 15-50 cm long, procumbent to ascending, branching at some distance from the base. Vegetative parts, except upper sides of leaves, covered with glandular and simple hairs. Stipules broad, with sharply pointed teeth in their basal part, ending with a long, narrow terminal tooth. Leaflets 10-18 mm x 5-17 mm, elliptical, obovate; margin somewhat serrate in its apical half; midrib ending in a small to medium-sized tooth. Peduncle 2 to 5-flowered, longer than the corresponding petiole, bent downward, with a long cusp. Florets 7-10 mm long. Pedicel shorter than the calyx tube; bract shorter than the pedicel. Calyx 4-6 mm long, covered with glandular and simple hairs; teeth broad at the base, longer than or equal to the tube. Corolla bright yellow, usually less than twice the length of the calyx; standard broad, twice as long as the keel; wings about the length of the keel. Young pod contracted within the calyx, covered with glandular hairs. Mature pod blackish, biconvex, spineless, covered with many-celled, glandular hairs. Coils 4-6, turning clockwise, middle coil 8-13 mm in $\phi$, edge almost

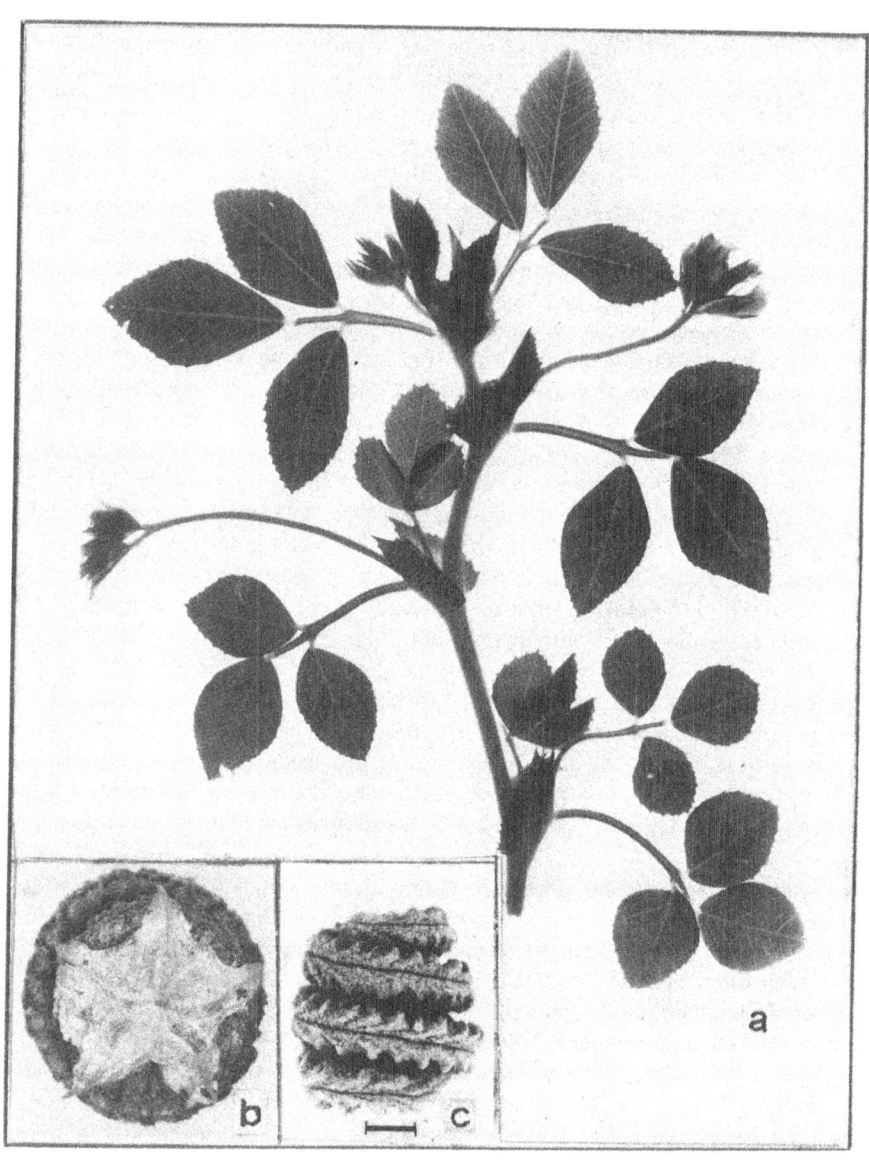

*Fig. 45, M. noëana.* Branch (a), and pods: in basal view (b) note calyx appearing as a five-point star, and in edge view (c) note dorsal suture in a narrow groove.

pergamentaceous. Coil face with radial veins anastomosing freely before entering an inconspicuous lateral vein located about 1 mm from the dorsal suture and connected to it forming almost square configurations. Seeds light brown, about 4.5 mm x 2.5 mm, separated by a thin partition. Seed weight about 10 g/1000. Radicle about half the length of the seed. Pollen grains irregularly block-shaped with four or five germinal apertures. The first trifoliate leaf with rather small lateral leaflets. $2n = 16$.

*Habitat and Distribution.* Growing in heavy, reddish clay soils, as a weed in cultivated crops, and in fallow fields.

The species is an East-Mediterranean taxon, though originally described from the vicinity of Florence, Italy. We collected it in southern Turkey and Lebanon; seeds were obtained also from Israel and Syria.

*Variation Within, and Relationship to Other Species.* Most of the variations reported in literature and observed in our collection are very probably the result of natural hybridization between *M. bonarotiana* and *M. rotata*. Thus, biconvex pods with two rows of short spines on coils are considered as characteristic for *M. blancheana* by Boissier (1856), and retained by Heyn (1963) for *M. blancheana* var. *blancheana* and by Ponert (1973) for *M. rotata* ssp. *blancheana*. We found these characters on pods in $F_2$ plants of *M. bonarotiana* x *M. rotata* (Lesins et al., 1976).

30.  *Medicago noëana* Boissier, Diagn. Pl. Orient. Ser. 2,2:10 (1856). Figs. 45; 5,-29.

Plants 10-30(60) cm long; branches angular, ascending, arising from near the base. All aboveground vegetative parts covered with simple, soft, erect hairs. Stipules large, serrate at the edges. Leaflets 14-20 mm x 11-16 mm, at lower nodes roundish-ovate, at higher nodes broadly elliptic or rhombic; margin serrate in its apical half; midrib ending in a tooth. Peduncle 2 to 7-flowered, longer than the corresponding petiole, with a terminal cusp. Florets 6-7 mm long. Pedicel shorter than the calyx tube; bract ± the length of the pedicel. Calyx longer than half of the corolla; teeth as long as the tube. Corolla yellow; standard broadly oval; wings shorter than the keel. Young pod contracted within the calyx, not turning sideways: calyx remains appressed to the base of the pod as a regular five-rayed star (Fig. 45,-b), which is in contrast to that of other *Medicago* species where calyx is appressed obliquely to the pod base. Mature pod straw-colored to brown, cylindrical, glabrous, spineless. Coils 2.5-4.5, not tightly appressed, 4-7.5 mm in $\phi$, turning clockwise, hard-walled. On coil face 14-17 veins radiating from the centre, running without noticeable branching into a weakly expressed lateral vein, from there as strong ridges obliquely to the dorsal suture which is usually located in a groove. Pod surface rough due to alveolar tissue. Seeds yellowish to brown, 1-2 in each coil, separated, 3.5-4.7 mm x 2-2.5 mm. Seed weight about 3.9 g/1000. Radicle half the length of the seed, or slightly shorter. $2n = 16$ (Lesins & Lesins, 1963).

*Habitat.* Mainly in dry, heavy, calcareous soils.

*Distribution.* Endemic to Iraq and Turkey. We collected it in Turkey in the

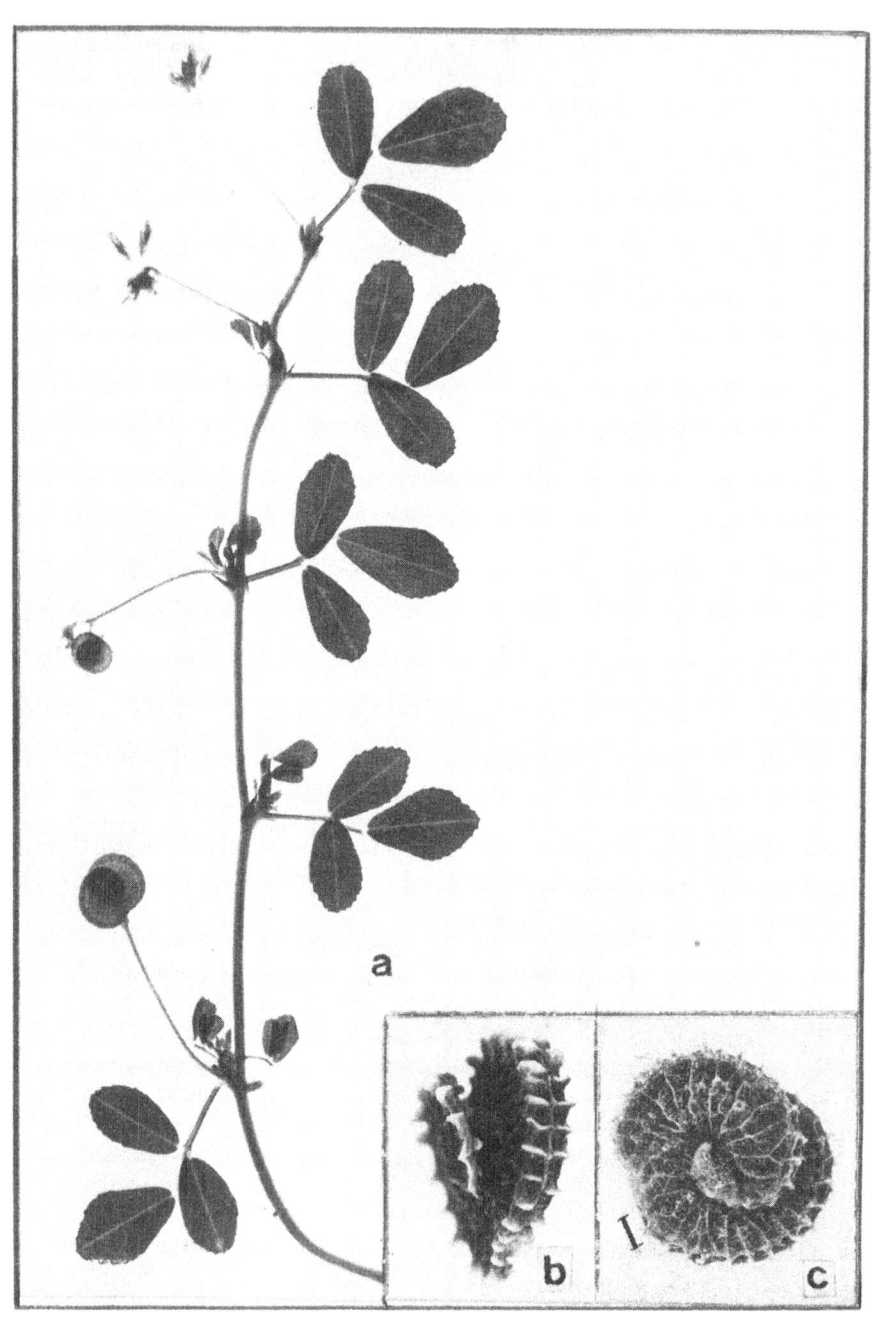

*Fig. 46,  M. shepardii*. Branch (a), and pods: in edge view (b), and in apical view (c).

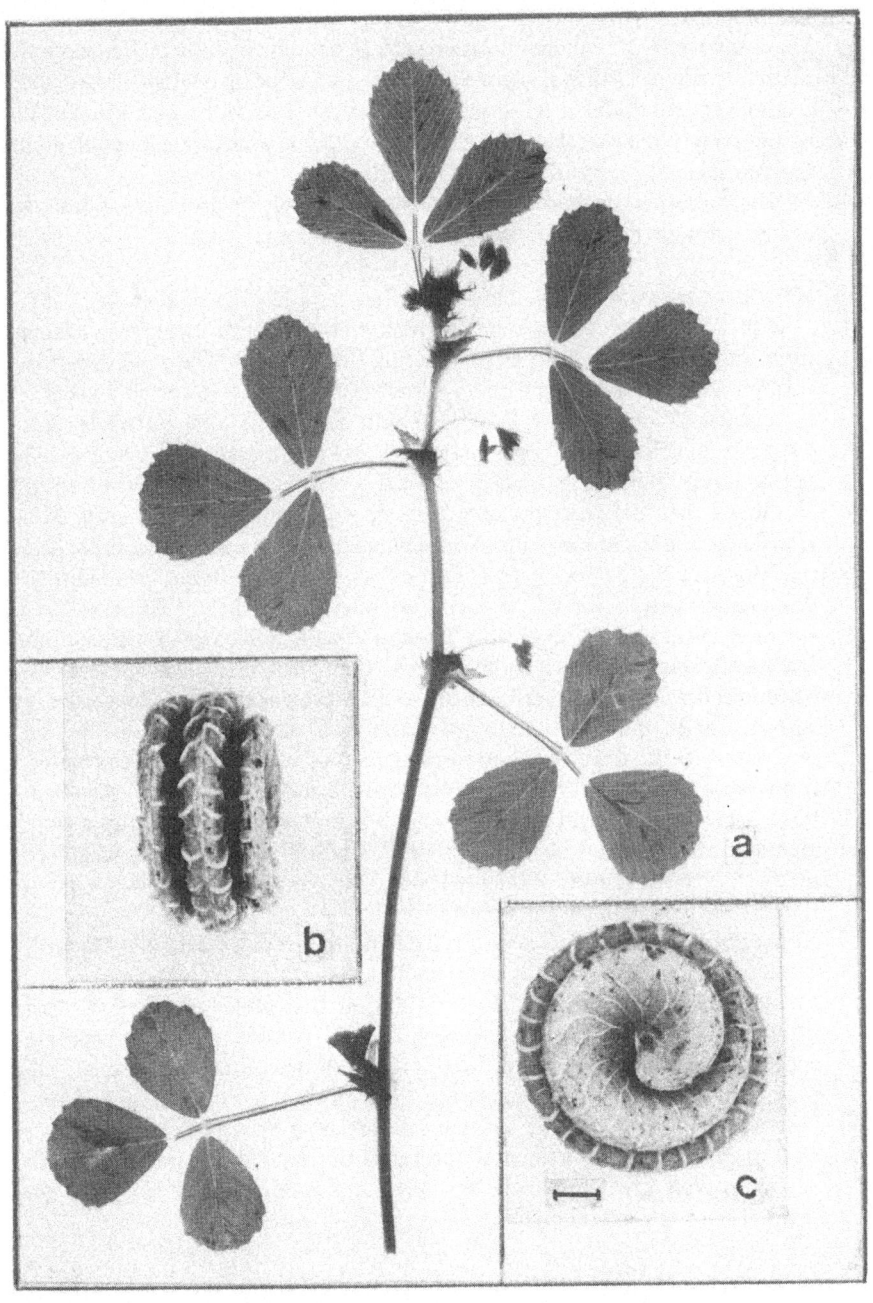

*Fig. 47, M. rugosa.* Branch (a), and pods: in edge view (b) (note elevated dorsal suture), and in apical view (c).

central (Ankara), southern (Mersin) and southeastern part (Gaziantep).

*Relationship to Other Species.* There is some similarity between the pods of *M. noëana* and *M. rugosa.* Boissier (1872) combined both in the section *Elegantes.* Heyn (1963) also considered the two as being related. To test the relationship we doubled the chromosome number of *M. noëana*, bringing it to the same ploidy level as *M. rugosa* (2n = 32), then reciprocally pollinated the two species. No hybrids were obtained.

Chromosomes in *M. noëana* differ considerably in length (2.1-3.8 $\mu$); centromeres are median to submedian (Lesins & Lesins, l.c.).

31. *Medicago shepardii* Post, J. Linn. Soc. Bot. 24:425 (1888). Figs. 46; 6,-53.
Plants 20-50 cm long; branches thin, decumbent to ascending, arising from the base. Vegetative parts, including upper sides of leaves, abundantly covered with diffuse, simple hairs. Stipules lanceolate, at lower nodes slightly toothed. Middle leaflet 5-10 (16) mm x 6-7(10) mm, lateral leaflets distinctly smaller; broadly to narrowly obovate; margin in its apical 1/3 serrate; midrib ending in a small terminal tooth. Peduncle 2 to 5(6)-flowered, longer than the corresponding petiole, with a small terminal cusp. Florets small, 3.5-4 mm long. Pedicel ± the length of the calyx tube; bract half the length of the pedicel. Calyx longer than half of the floret, covered with simple hairs; teeth shorter than or as long as the tube. Corolla yellow; standard oval; wings shorter than the keel. Young pod covered with appressed hairs, first coiling within the calyx, then turning sideways out of it. Mature pod flat, hairy to glabrescent, greyish-brown, spineless. Pods usually with 1.5 coils, thick at the edge, 3-5 mm in $\phi$, turning clockwise. On pod face distinct radial veins, anastomosing before entering the lateral veins; from there wing-like protrusions join at right angle the elevated dorsal suture. Seeds yellow, light brown, about 3.3 mm x 2 mm, 1-2(3) in a pod, separated. Seed weight about 3.4 g/1000. Radicle about half the length of the seed. 2n = 16 (Lesins & Lesins, 1963).

*Habitat.* Growing in soils of volcanic origin.

*Distribution.* Endemic to a very restricted area around Gaziantep (Antab), Turkey. We collected it at the roadside 33 km north of Gaziantep.

*Relationship to Other Taxa.* Heyn (1963) at first placed *M. shepardii* as a variety of *M. tornata*, but later (Heyn, 1970) considered it as a separate species. Our studies (Lesins & Singh, 1973) showed that there was a complete incompatibility barrier between the two taxa. Morphological differences, especially in pod shape and the interbreeding barrier, justify the retention of *M. shepardii* as a separate species. Differences in chromosome length (2.4-5 $\mu$) were greater than in any *Medicago* species examined by the authors (Lesins & Lesins, l.c.).

32. *Medicago rugosa* Desrousseaux, in Lam., Encyl. Meth. Bot. 3:632 (1792). Syn. *M. elegans* Jacq. ex Willd., Sp. Pl. 3:1408 (1802). Figs. 47; 6,-43.
Plants 30-40 cm long; prostrate to decumbent, primary branching from near the base, secondarily branching all along the main branches. Vegetative

parts, except the upper side of leaves, covered with glandular and simple hairs. Stipules ruggedly toothed along their margins. Leaflets 12-23 mm x 10-19 mm, broadly to narrowly obovate or oblanceolate; margin in its apical 1/3-1/2 serrate; midrib ending in a small tooth. Peduncle (1)2 to 5(7)-flow-

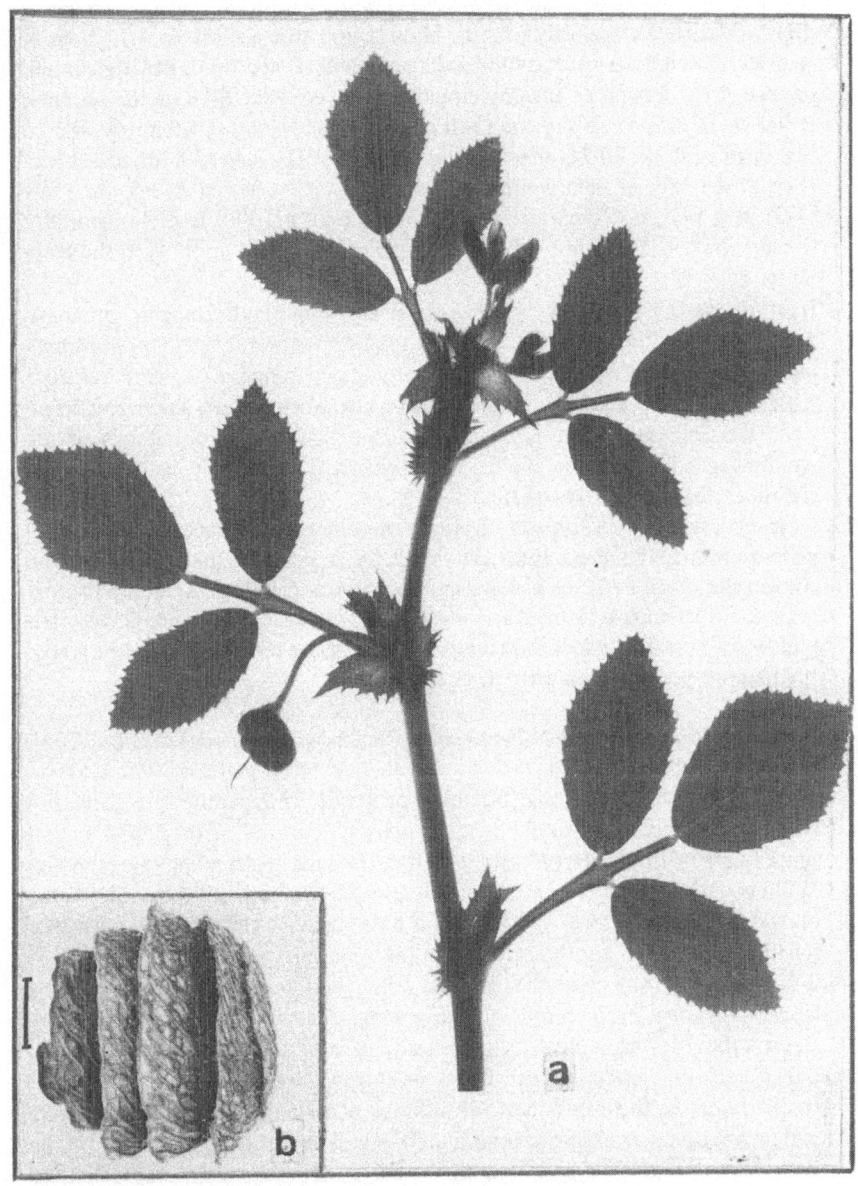

*Fig. 48,   M. scutellata*. Branch (a), and pod (b).

ered, equal in length to or shorter than the corresponding petiole, with a distinct cusp. Pedicel slightly shorter than the calyx tube; bract about equal in length to the pedicel. Florets 4-5 mm long. Calyx 2.5-3 mm long; teeth broadly triangular, shorter than or equal to the tube. Corolla yellow; standard broadly obovate; wings slightly shorter than the keel. Young pod densely covered with glandular hairs, starts coiling within the calyx, then bends sideways through the calyx teeth. Mature pod pale yellow to light brown, glabrate, discoid to short-cylindrical, spineless. Coils 2.5-5, not tightly appressed, 5-10 mm in $\phi$, turning clockwise; on coil face 8-18 slightly curved, radial veins, anastomosing soon after leaving the centre, thickening closer to the edge and as 20-24 ribs joining obliquely the elevated dorsal suture (Fig. 47,-b). Seeds light yellow to brownish, 2.5-4.5 mm x 1-3 mm, few (1-3) in a pod, not separated. Seed weight 5-14 g/1000. Radicle distinctly shorter than half of the seed, extending like a rounded hook from the main body. $2n = 32$.

*Habitat and Distribution.* M. *rugosa* is an omni-Mediterranean, predominantly east-Mediterranean, species. Its requirement for heavy clay soils may perhaps be partly responsible for its sporadic occurence. Also, spineless pods with only a few seeds in them may not be advantageous for a wider distribution. We collected it in Lebanon, Tunisia, Greece and on islands of the Archipelago; in Italy, on Sicily, Malta, Pantelleria; usually we found it at roadsides and along edges of fields.

In the search of variation in ploidy we examined 20 accessions, and all were found to be tetraploids, $2n = 32$. It is possible that the tetraploid chromosome level of this species and its low morphological variability, except for differences in pod and seed size, are causally related, since tetraploids are generally conceded to be relatively more resistant than diploids to the phenotypic variation due to gene mutations.

33. *Medicago scutellata* (L.) Miller, Gard. Dict. ed. 8, *Medicago* no. 2 (1768). Figs. 48; 6,-51.

Plants up to 60 cm long; branches prostrate to decumbent, secondarily branching along their whole length, densely covered with glandular and simple hairs. Stipules large, with pointed, irregular teeth along their margin. Leaflets glabrous on their upper side, 15-30 mm x 7-20 mm, elliptic to obovate; margin serrate almost to the base, laminal veins ending each in a separate triangular tooth; midrib ending in a larger tooth. Peduncle 1 to 3-flowered, shorter then the corresponding petiole or equal to it, with a lengthy terminal cusp. Florets 7-9 mm long. Pedicel much shorter than the calyx tube, florets almost sessile; bract usually longer than the pedicel. Calyx half the length of the floret or more, covered with glandular and simple hairs; teeth longer than the tube or occasionally equal to it. Corolla yellow to orange yellow; standard obovate; wings about the length of the keel. Young pod rises and coils in the calyx, then turns sideways; densely covered with glandular hairs. Mature pod gray to dark gray, oval or cup-shaped, spineless. Coils 5-7, imbricate like stacked bowls with the convex surfaces toward the pod base, 10-16 mm in $\phi$, turning clockwise; coil edge thin, pergamentaceous. On the pod face veins anastomose freely from the start, closer to the edge uniting in thicker strands, entering the dorsal suture at an angle opposite to the coiling direction. Seeds yellow, 5-6 mm x 3-3.5 mm, 1-2 in a coil (2-3 apical coils seedless), separated. Seed weight up to 20 g/1000. Radicle shorter than half of the seed, its tip separated from the main body of the seed. Around chalaza and extending to the hilum a brown strip shaped like an exclamation mark. $2n = 32$.

*Habitat.* Mainly in heavy soils, often in fallow fields.

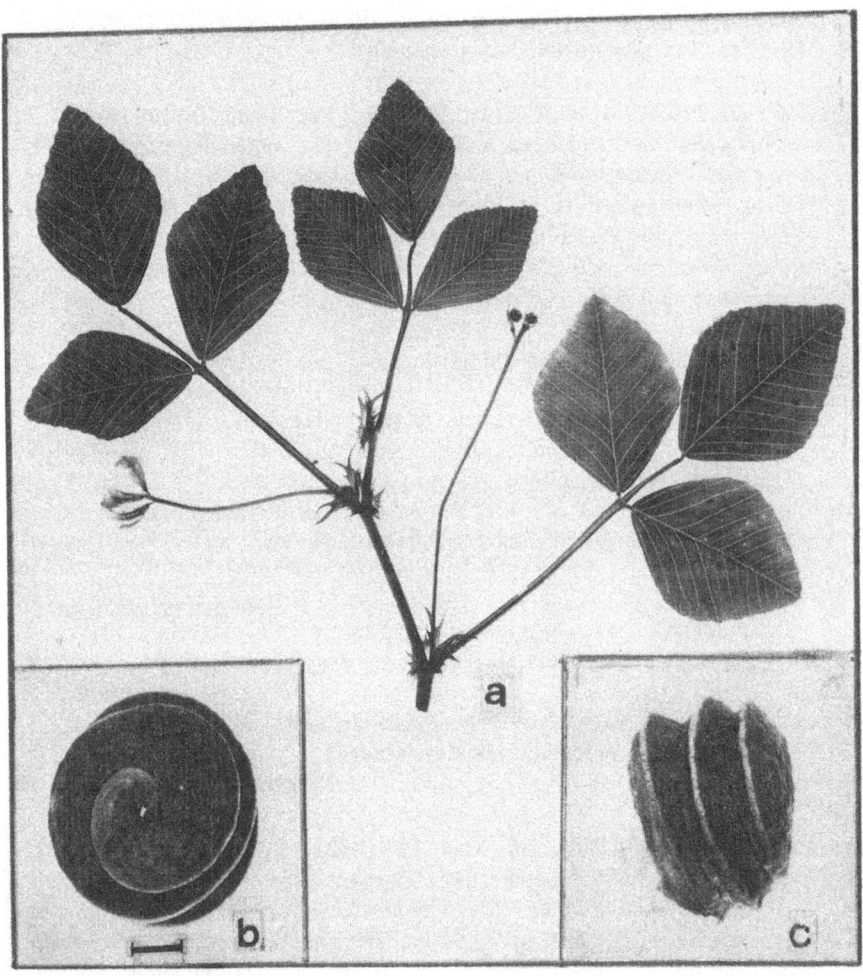

*Fig. 49, M. soleirolii.* Branch (a), note rhombical leaflets; pods: in apical (b), and side view (c).

*Distribution.* An omni-Mediterranean species, though not abundant in any growing site. We collected it in countries from Lebanon to Italy on the north, and in Algeria on the south coast of the Mediterranean.

*Variation in M. scutellata and Its Relationship to Other Taxa.* As Heyn (1963) has pointed out, the variation within the species is small considering its distribution area. The reason, in part, may be that it is a tetraploid: an occasionally mutated allele will not be readily noticed, its expression remaining masked by the three normal alleles for some generations, and eventually being lost if natural selection does not favor it. Urban (1873) included *M. scutellata* together with *M. rugosa* in one section *Scutellatae*. As pointed out in discussing sect. *Rotatae*, they have some characters in common, though pod appearance is quite different and also pollen shapes were found to be different (Lesins & Lesins, 1963). Our attempts at hybridization of the two species were not successful.

SECTION PACHYSPIRAE Urban, Verh. bot. Ver. Brand. 15:49 (1873).

Pod walls thick and hard. Coils more or less appressed, turning clockwise or anticlockwise. Spines, if present, stocky, their base thick, mostly conical, often embedded in spongy tissue. Veins on coil face often not clearly discernible because of spongy tissue, not anastomosing to any extent within half of the radius from the coil centre (except *M. soleirolii*). Radicle shorter than half of the seed length. $2n = 16,14$.

Key to groups of sect. *Pachyspirae*:

| | | |
|---|---|---|
| 1 | Lateral veins or veinless zone on pod coils absent | *M. soleirolii* |
| — | Lateral veins or veinless zone on pod coils present | 2 |
| 2 | Radial veins of coil face run into a veinless zone | |
| | | *M. murex, M. turbinata* |
| — | Radial veins of coil face run into a lateral vein | 3 |
| 3 | Lateral veins on the same level as the dorsal suture | |
| | | *M. doliata, M. littoralis, M. constricta* |
| — | Lateral veins lower than the dorsal suture | 4 |
| 4 | Groove between lateral veins and the dorsal suture absent | |
| | | *M. tornata* |
| — | Groove between lateral veins and the dorsal suture present, at least in early stages of pod development | |
| | | *M. rigidula, M. truncatula* |

34. *Medicago soleirolii* Duby, Bot. Gall.:124 (1828). Syn. *M. plagiospira* Dur., in Duchartre, Rev. Bot. 1:366 (1845). Figs. 49; 6,-54.

Plants up to 50 cm long, ascending to erect, branching mainly from the base, covered with rough, upright hairs. Stipules, pedicels, peduncle, calyx and leaflet margins covered with glandular hairs, glands often violet. Stipules deeply incised, with irregular teeth. Leaflets 19-22 mm x 15-21 mm, subrhombic in shape; apical half of margin with wide teeth, laminal veins ending in separate teeth; midrib ending in a small tooth. Peduncle 7 to 9-flowered, longer than the corresponding petiole, with a distinct terminal cusp. Florets about 10 mm long. Pedicel shorter than the calyx tube; bract slender, usually longer than the pedicel. Calyx slightly longer than half of the floret; teeth awl-shaped, twice as long as the tube. Corolla bright yellow; standard obovate; wings shorter than the keel. Young pod contracted within the calyx. Mature pod greyish to black, discoid or cylindrical, glabrate, spineless, hard-walled. Coils 2-4(5), tightly appressed, 6.5-8.5 mm in $\phi$, turning clockwise; face of coil with 5-7 indistinct veins (covered with loose cellular tissue) forming a fine network extending to the dorsal suture. Seeds about 4 mm x 2.5 mm, 1-2 in each coil, separated. Seedcoat smooth, yellowish to reddish-brown. Seed weight about 7 g/1000. Radicle slightly less than half the seed length. $2n = 16$.

*Habitat.* According to Durieu (1845), the species is growing in natural prairies surrounding Lake Houbera in eastern Algeria. We collected it in sandy soils and also on fertile hill slopes on the plains of Annaba (Bône), Algeria.

*Distribution. M. soleirolii* is a south-west Mediterranean species found mainly in the maritime region of Algeria and Tunisia (Nègre, 1959). It has also been reported growing in a few localities in Italy and in southern France, though not abundantly in any habitat.

*Distinguishing M. soleirolii from Other Taxa.* The species which has often been confused with *M. soleirolii* is the spineless 3 to 5-coiled form of *M. tornata* (Fig. 51,-i). Heyn (1963) noted that all seed samples of *M. soleirolii* which she had received from different sources have proved to be misidentified. Plants of *M. soleirolii* and *M. tornata* have similarities in branching of stems, shape and serration of leaflets, incision of stipules, ratio of length of peduncle to petiole and number of florets per raceme, and coils of *M. tornata* with indistinct lateral veins are comparable to those of *M. soleirolii* without lateral veins. There are, however, differences which on closer inspection allow separation of the two species: pods of *M. tornata* have some continuous or interrupted slits between coils, whereas coils of *M. soleirolii* have no slits between them; *M. tornata* has spiny and spineless pods, *M. soleirolii* has only spineless pods; *M. tornata* has simple hairs only, *M. soleirolii* has glandular hairs on stipules, petioles and along the edges of leaflets.

On intercrossing *M. soleirolii* with *M. tornata* a few seeds were obtained. From them some chlorophyll-deficient plantlets were grown, which survived up to 2 months. Since no chlorophyll-deficient progeny was found on selfing parental plants, it was deduced that the chlorophyll-deficients were of hybrid origin. We decided, therefore, to include *M. soleirolii* in section *Pachyspirae.*

35. *Medicago tornata* (L.) Miller, Gard. Dict. ed. 8. *Medicago* no. 3 (1768). Syn. *M. polymorpha* var. *tornata* L., Sp. Pl.:780 (1753) pro parte; *M. obscura* Retz., Obs. Bot. 1:24 (1779); *M. lenticularis* Desr., in Lam., Encycl. Method. Bot. 3:630 (1792); *M. helix* Willd., Sp. Pl. 3:1409 (1802); *M. striata*

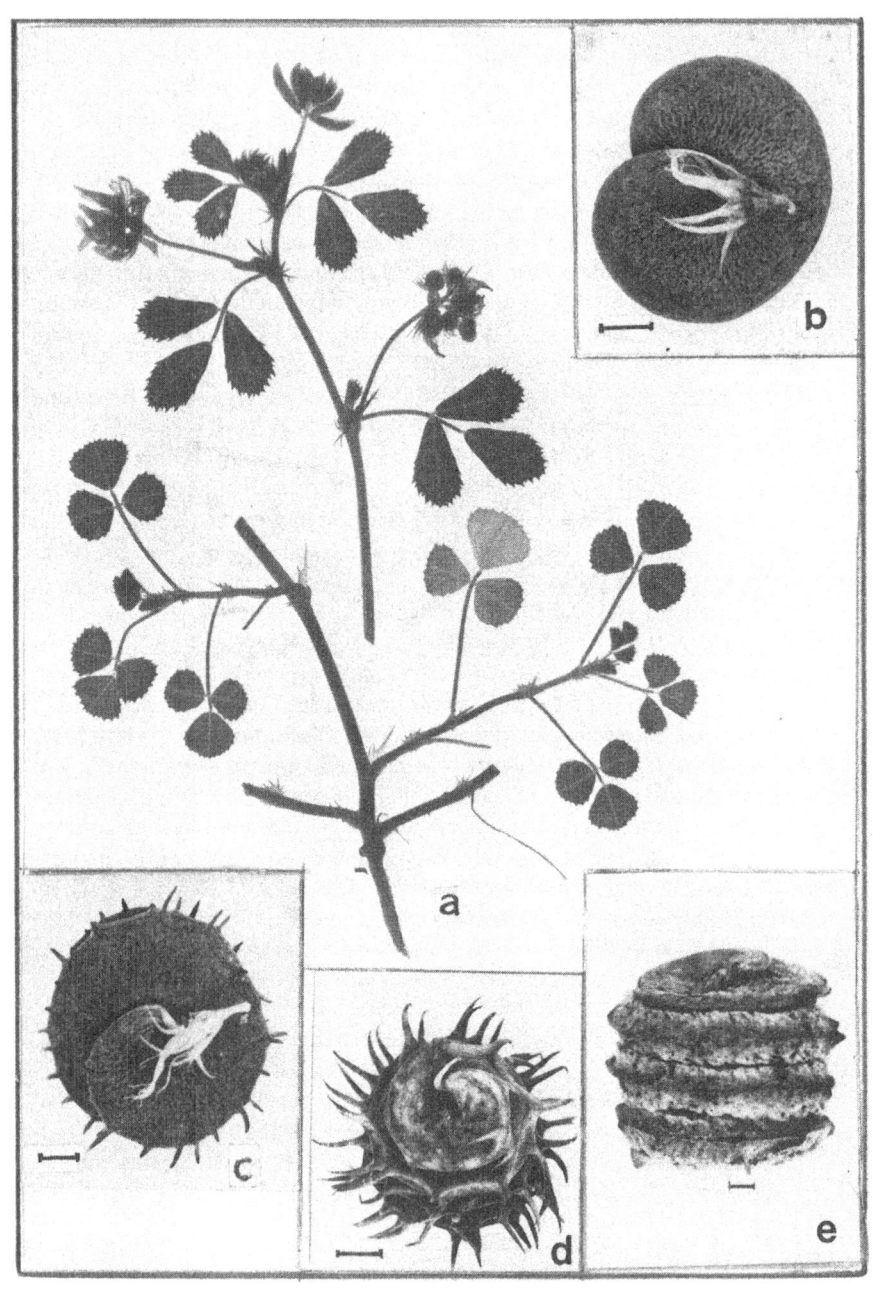

*Fig. 50, M. tornata.* Branches (a), and pods: ssp. *helix* var. *lenticularis* f.
*lenticularis* (b); ssp. *helix* var. *lenticularis* f. *aculeata* (c); ssp. *tornata* var.
*tornata* f. *muricata* (d); ssp. *tornata* var. *striata* (e).

156

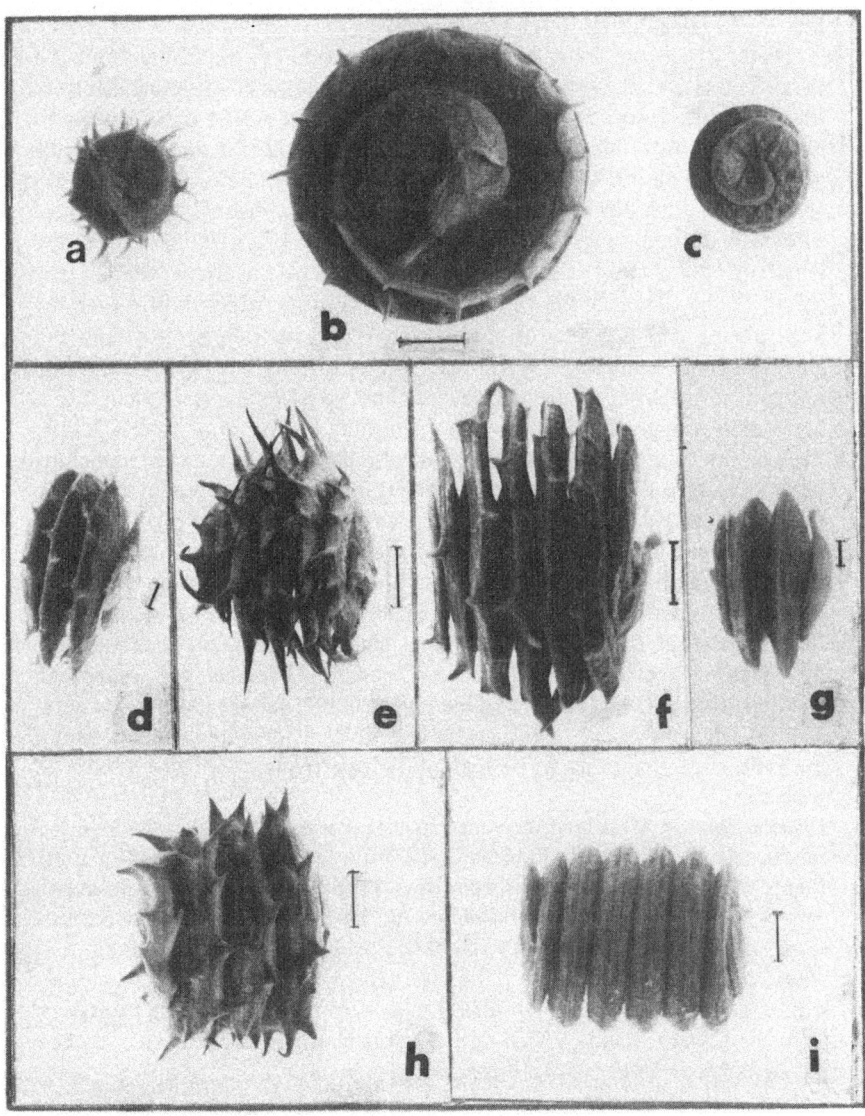

*Fig. 51,  M. tornata* pods. Upper and middle rows: ssp. *helix* var. *parvicarpa* f. *parvicarpa* (a and d), and f. *inermis* (c and g); var. *lenticularis* f. *maroccana* (b and f), and f. *spinosa* (e), note thin, centre-inclined coil edges. Lower row: ssp. *tornata* var. *tornata* f. *inermis* (i), and f. *muricata* (h), note wide edge of coils.

157

Bast., in Desv., J. Bot. 3:19 (1814); *M. italica* (Mill.) Steud. ex Fiori, Fl. Ital. 1:832 (1925). Figs. 50; 51; 6,-58,-59.

Plants 30-70 cm long, procumbent to ascending, branching from the base. Vegetative parts covered with diffuse simple hairs or almost glabrous. Stipules entire to deeply incised. Leaflets 6-18 mm x 4-14 mm, broadly to narrowly obovate, often subrhombic; apical 1/3-1/2 of leaflet margin ruggedly serrate; midrib ending in a terminal tooth. Peduncle (1)2 to 14-flowered, longer than the corresponding petiole, with a distinct terminal cusp. Florets 5-10 mm long. Pedicel shorter than the calyx tube; bract ± the length of the pedicel. Calyx ± half the length of the floret; teeth ± the length of the tube. Corolla yellow; standard obovate; wings shorter than the keel. Young pod glabrous or sparsely covered with simple, or simple and glandular hairs, contracted within calyx, later turning sideways. Mature pod ash grey to blackish, cylindrical or lenticular, spiny or spineless. Coils 1.2-8, more or less appressed, 3-10 mm in $\phi$, turning clockwise or anticlockwise. Face of the coil with 10-14 slightly curved radial veins, somewhat branching before entering the lateral vein; dorsal suture higher than the lateral veins. Coil edges either thick, or thin and slanted to the middle of the pod; sometimes puffing-up of edge margins by spongy cellular tissue may give an impression of dorsal suture being on the same level or lower. Spines, if present, 1-2 mm long, thin at the tip, usually conical at the base (sometimes flat), inserted at 135°-180° to coil face, sometimes bent and hooked. Seeds light yellow to brownish, 2-4 mm x 1-2 mm, 1-2 in each coil, separated. Seed weight 1.2-6.8 g/1000. Radicle less than half the seed length. $2n = 16$.

*Habitat.* In sandy soils, or at least a thin layer of dry, loose soil on top of the mother-rock appears to be required; most frequently in the vicinity of seashores.

*Distribution.* A West-Mediterranean species. We collected it on the Mediterranean islands of Cyprus, Pantelleria, Sardinia and on the mainland of Italy, France, Spain, Algeria and Morocco. Heyn (1963) mentions that at least two varieties are growing in Israel and one in Turkey. We did not collect any seeds of this species in Turkey. *M. shepardii* of Turkey, which Heyn (l.c.) counted as a variety of *M. tornata*, later (Heyn, 1970) as a separate species, is not interfertile with *M. tornata* (Lesins & Singh, 1973). Our accessions from the Canary Islands, which initially were classified as *M. tornata*, turned out to be *M. littoralis* (see *M. littoralis*).

*Variation in M. tornata..* *M. tornata* is the most polymorphic species in the genus. One indication of its variability is the large number of taxonomic names given to it. In addition to the more commonly used ones cited above, Heyn (1963) cites at the species rank further nine synonyms: *M. muricata* Willd., *M. canariensis* Benth., *M. plumbea* Bertol., *M. astroites* Bertol., *M. calcar* Lowe, *M. commutata* Tod., *M. laevis* Desf., *M. intermedia* Ser., and *M. corrugata* Dur. Heyn (l.c.) herself divided the species into four varieties (not counting *M. shepardii*) on the basis of spininess and number of coils per pod: 1) spineless, 1.2-2.5 coils (var. *rugulosa*), 2) spineless, 3-8 coils (var. *tornata*), 3) spiny, 1.2-2.5 coils (var. *aculeata*), and 4) spiny, 3-8 coils (var. *spinulosa*). Urban (1873) divided the species (as *M. obscura* Retz.) into three subspecies: 1) *lenticularis*, coils 1.2-1.5 with 1-2 seeds, 2) *helix*, coils 1.2-4, seeds 3-8, and 3) *tornata*, coils 4-8. Spininess was used to separate varieties and coiling direction was used to separate forms within varieties.

Nègre (1956), working in Morocco where *M. tornata* is represented most abundantly, divided the species (as *M. italica*) into three subspecies: 1) *helix*, with pods of less than 8-10 mm in $\phi$, lenticular in shape and with less than 4 coils, 2) *tornata*, with pod diameter as of the previous but with more than 4 coils per pod, and 3) *maroccana*, with pod diameter more than 8-10 mm. Later Nègre (1959) changed his classification recognizing three subspecies: 1) *corrugata*, with finely reticulate coil face, 2) *helix*, with not reticulate coil face and less than 4 coils, and 3) *tornata*, with 4-8 coils per pod. A more detailed subdivision by number of coils, pod diameter and spininess was used to delineate varieties and forms.

Examining our more than 60 *M. tornata* accessions we have been impressed by the manifold differences in the shape of the pods (lenticular vs. cylindrical), the width of the coil edge (acute vs. rather blunt), and the size of pods (from rather tiny to twice and three times larger).

Phenological observations showed that there were considerable differences in life cycle under uniform environmental conditions in the greenhouse: seeded at the same time, some accessions had already ripe pods, when others had not yet started flowering. No distinct interbreeding barriers were found within *M. tornata*, although they exist between *M. tornata* and other related species (Lesins & Erac, 1968; Lesins & Singh, 1973).

Considering recent accessions and information obtained from their examination it appears that a variety *parvicarpa* with two forms should be added to the previously known *M. tornata*. Classification and the necessary descriptions taking care of the new inclusions are as follows:

Key to taxa of *M. tornata*:

| | |
|---|---|
| Pods lens-shaped | ssp. *helix* (Willd.) Urb. |
| Pods small, 3-4.5 mm in $\phi$ | var. *parvicarpa* Les. et Les. |
| Pods spineless | f. *inermis* Les. et Les. |
| Pods spiny | f. *parvicarpa* |
| Pods larger, 4.5-10 mm in $\phi$ | var. *lenticularis* (Desr.) Nègre |
| Pods with 1.2-1.5 coils, spineless | f. *lenticularis* |
| Pods with 1.5-2.5 coils, spiny | f. *aculeata* Urb. |
| Pods with 5-6 coils, 8-10 mm in $\phi$, usually tubercled, with few (1-3) florets per raceme | f. *maroccana* Nègre |
| Pods with 3-6 coils, 4.5-8 mm in $\phi$, spiny, with more than 3 florets per raceme | f. *spinosa* Nègre |
| Pods cylindrical | ssp. *tornata* |
| Pods small, 3-4.5 mm in $\phi$ | var. *striata* Bast. |
| Pods larger, 4.5-9.5 mm in $\phi$ | var. *tornata* |
| Pods spineless | f. *inermis* Urb. |
| Pods spiny | f. *muricata* Urb. |

Ssp. *helix* (*M. obscura* Retz. ssp. *helix* (Willd.) Urb. in Verh. bot. Ver. Brandenb. 15:66, 1873) emend. (Figs. 50,-b,c; 51,-a,b,c,d,e,f,g). It is characterized by pods having the shape of a flat ($1\frac{1}{4}$-$2\frac{1}{2}$ coils) or thick (up to 6 coils) convex lens. The edges of lateral coils tend to be inclined towards the centre of the pod, and are thin as a result of the narrow angle between lateral veins and the dorsal suture. The walls of the coils may be tough, though release of seeds without crushing of the pods is possible. The spines, if present, are thin. Number of coils $1\frac{1}{4}$-6; coil diameter 3-10 mm; number of seeds per pod (1)2-7.

Var. *parvicarpa* Lesins & Lesins (Fig. 51,-a,d,c,g). Peduncle many-flowered (8-12 florets). Number of coils per pod 3-4.5; number of seeds per pod 3-5; seeds small, 2-3 mm x 1-1.8 mm; weight 1.2-2.2 g/1000. Coiling direction clockwise or anticlockwise. Both forms, f. *parvicarpa* (Fig. 51,-a,d) and f. *inermis* (Fig. 51,-c,g), were collected in Morocco at Banmansour, Lalla Mimouna and other places on the road along the canal from Kenitra to Moulay-Bousselham, north of Rabat. The growing sites were in very sandy soil. In some places, varieties with small and large pods were found together, but in sites of most barren sand only the small-podded variety was growing. Accession UAG No. 2792a is typical for f. *inermis*, and UAG No. 2681 for f. *parvicarpa*.

Var. *lenticularis* (Desr.) Nègre, Bull. Soc. Hist. Nat. Afr. Nord. :292. 1959, (Fig. 50,-b,c) comprises forms with larger pods. The spineless form, f. *lenticularis* (Fig. 50,-b), has only $1\frac{1}{4}$-$1\frac{1}{2}$ coils per pod with (1)2 seeds. The pods, thus, have the shape of a thin lens. Urban (Verh. bot. Ver. Brandenb. :66, 1873) described it as ssp. *lenticularis* var. *inermis*. He mentioned right- and left-hand coiling direction; all our accessions are with clockwise (right-hand) coiled pods. The spiny form, f. *aculeata* (Guss.) Urb. Verh. bot. Ver. Brandenb. :66, 1873, pro var. (Fig. 50,-c), usually has $1\frac{1}{2}$-$2\frac{1}{2}$ coils though one accession from Israel, (UAG No. 2032), has not more than 2 coils. With increased number of coils, the middle part of the pod assumes a more cylindrical shape. The form f. *maroccana* Nègre, Trav. Inst. Sci. Cher. Ser. Bot. 5:44, 1956, pro ssp. (Fig. 51,-b,f), in contrast to other *M. tornata*, has only 1-3 florets per peduncle. It was collected in Morocco along the road El Jadida to St. Smaill. In both forms, *maroccana* and *spinosa* Nègre, Trav. Inst. Sci. Cher. Ser. Bot. 5:44. 1956, pro var. (Fig. 51,-e), although the middle part of the pods may be cylindrical, the basal and apical coils are convex with coil edges slanted in the direction toward the middle of the pod.

Ssp. *tornata* is characterized by cylindrical or conical pod shape with both ends truncate. The edges of coils are thick, the angle between lateral veins and the dorsal suture is wide; the dorsal suture may appear on the same level or lower than the margins if the edges are puffed-up by the spongy tissue. The walls of coils are brittle or hard, sometimes must be crushed for releasing of seeds. The spines, if present, are stocky. Number of coils 3.5-7.5(8); number of seeds per pod 4-9.

Var. *striata* Bastard in Desv. J. Bot. 3:19. 1814, pro sp. (Fig. 50,-e) is

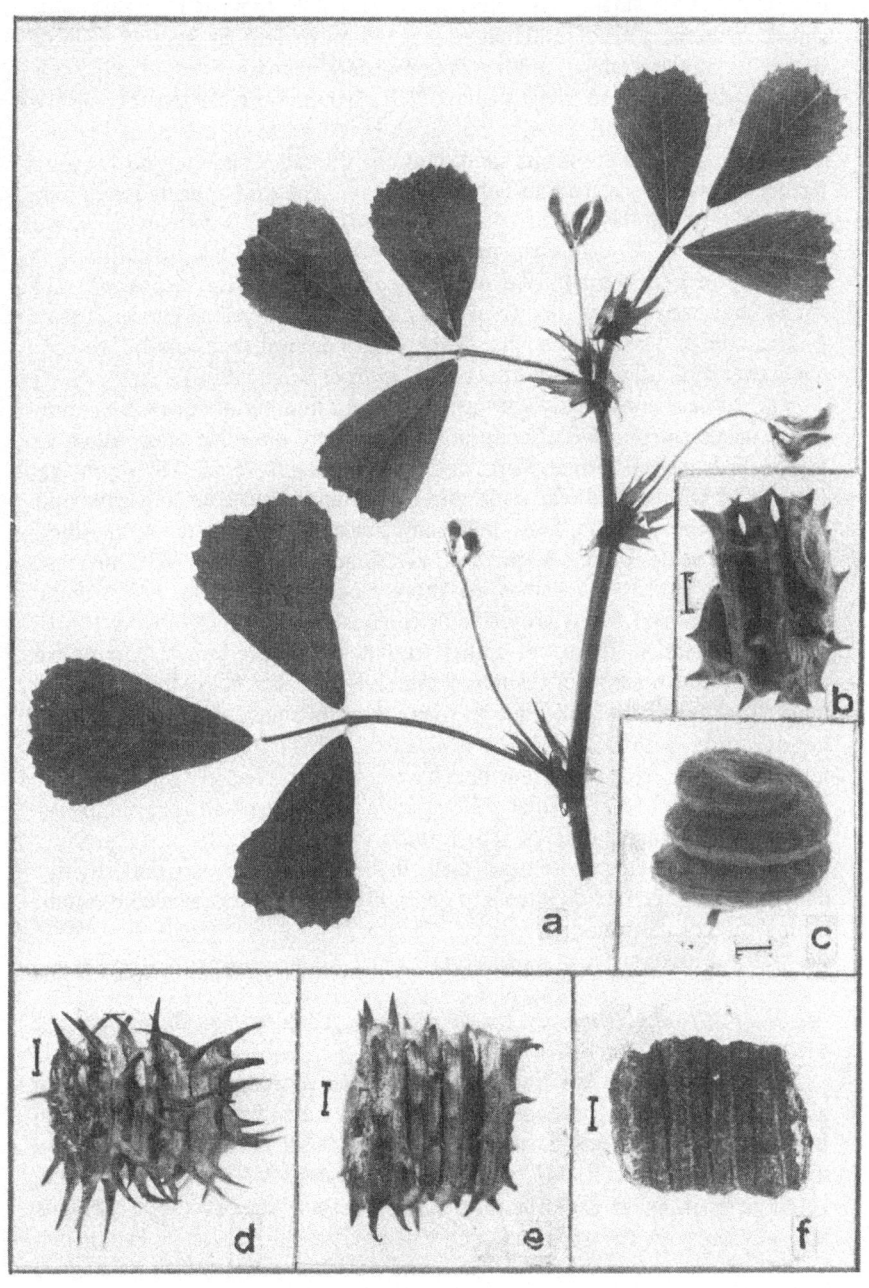

*Fig. 52, M. littoralis.* Branch (a) and pods: var. *canariensis* (b,c); other pod types (d, e, f).

characterized by small, cylindrical pods, spineless, tubercled or with tiny spines. The margins of coil edges may be puffed-up by spongy cellular tissue. In our accessions, coiling direction is clockwise. Seeds small, 2-2.5 mm x 1-1.5 mm; seed weight about 2.1 g/1000. We collected the variety between Bayonne and Biarritz on the Atlantic coast of southern France. The growing sites were moist sand, close to the sea. It is reported growing from Ollone, France, to San Sebastian, Spain. The pod appearance of var. *striata* is very similar to that of *M. littoralis*. Urban (l.c.) actually classified it as a variety of *M. littoralis*, and Heyn (l.c.) considered the possibility that it may be of hybrid origin. We intercrossed var. *striata* with *M. littoralis* and found that between the two there was a strong interbreeding barrier (Lesins & Erac, 1968). On the other hand, var. *striata* did not show any interbreeding barrier with other *M. tornata* types (Lesins & Singh, 1973).

Var. *tornata* is characterized by larger pods; number of coils 4.5-8; number of seeds per pod 5-8. The form most strongly deviating from other *M. tornata* is f. *inermis* Urban, Verh. bot. Ver. Brandenb. 15:66. 1873, pro var. (Fig. 51,-i) with cylindrical pods, often with appressed coils and wide coil edges. The pod walls are hard, and cannot be crushed by hand. In the spiny type, f. *muricata* Urban, Verh. bot. Ver. Brandenb. 15:67, 1873, pro var. (Figs. 50,-d; 51,-h), the spines are more stocky than in ssp. *helix*. Some spiny, many-coiled forms are difficult to classify whether belonging to ssp. *helix* or *tornata*. In these, the spines are inclined to the central part of the pod, suggesting a shape of the lens characteristic of ssp. *helix*, but spines are more stocky and the coil edge is wider than in typical *helix* forms. Since hybridization within *M. tornata* is possible, recombinations of characters may have arisen where different taxa have come into contact.

*Relationship to Other Species. M. rigidula* pods are often very similar to those of some accessions of ssp. *tornata* var. *tornata*. The distinguishing features are the veins on the coil face: In *M. tornata*, veins are only slightly bent, while in *M. rigidula*, strongly bent. If live material is available, establishing the chromosome number (*M. tornata* 2n = 16, *M. rigidula* 2n = 14) is the easiest way to arrive at a decision.

36. *Medicago littoralis* Rohde ex Loiseleur-Deslongchamps, Not. Fl. France:118 (1810). Figs. 52; 3,-6; 5,-23.

Plants 20-45(120) cm long, procumbent to ascending, branching from the base. Vegetative parts covered with simple hairs. Stipules incised, forming long, narrow teeth. Leaflets 8-22 mm x 4-20 mm, obovate, broadly obtriangular to obcordate (lower ones), apical part with serrate margin, often with irregular teeth; midrib ending in a triangular tooth. Leaflets sparsely hairy on the upper side, more densely on the underside. Peduncle 1 to 5-flowered, ± the length of the corresponding petiole, with a terminal cusp. Florets 5-7 mm long. Pedicel shorter than the calyx tube; bract longer than the pedicel. Calyx 2.5-4 mm long, covered with appressed or upright hairs; teeth ± the length of the tube. Corolla bright yellow; standard obovate; wings usually slightly shorter than the keel. Young pod glabrous, contracted within calyx, later protruding sideways through the calyx teeth. Mature pod straw-colored to dark gray, cylindrical, hard-walled, spiny, tubercled or spineless. Coils 3-6, appressed, 3.5-5.5 mm in $\phi$, turning clockwise or anti-clockwise; face of coil with 8-10 radial veins, somewhat curved, little branching or anastomosing only before entering the lateral vein; veins may be obscured by spongy tissue; lateral veins separated from the dorsal suture by a shallow groove which disappears at maturity; dorsal suture in ripe pods on the same level as the lateral veins (except var. *canariensis*). Spines, if present, 6-8 on each side of the coil, 1-4 mm long, conical, inserted at (90°) 130° to 180° to the face of the coil, their base often in spongy tissue. Occasionally there may be spiny and almost spineless pods on the same plant, and spines may be present or absent on different coils of the same pod. Seeds pale yellow to brownish-yellow, reniform, 2.5-3.7 mm x 1.5-2 mm, 1-2 in each coil, separated. Seed weight 2.7-3.3 g/1000. Radicle less than half of the seed length. $2n = 16$.

*Habitat and Distribution.* Growing along sandy seashores, but may penetrate deep inland in sandy soils and on rocky hillsides. An omni-Mediterranean species. We collected it in Lebanon, Turkey, Cyprus; in Greece and on its islands of Rhodes and Crete; in Italy and on Sicily, Capri and Sardinia; in France and on Corsica; in Spain, on the Canary Islands, and in N. Africa (Tunisia, Algeria, Morocco).

*Variation in M. littoralis and Its Distinction from Other Species.* Urban (1873) divided the species into three subspecies on the basis of the spininess of the pods: 1) ssp. *inermis* Urb., without spines or with only small tubercles, 2) ssp. *breviseta* DC, with spines not longer than the width of the edge of the coil, and 3) ssp. *longiseta* DC, where spines may be as long as the diameter of the coil. The first subspecies he divided according to the number of coils (this feature determining the ratio of length to breadth of the pod): (a) var. *tricycla* Urb., with fewer than 4 coils, the pod wider than long, and (b) var. *pentacycla* Urb., with 4-6 coils, the pod longer than wide. He divided ssp. *breviseta* similarly: (a) var. *depressa* Urb., the pods wider than long, and (b) var. *cylindracea* Urb., the pods longer than wide. In each variety of the above two subspecies he further distinguished forms according to their coiling direction: f. *dextrorsa* Urb., turning right (clockwise), and f. *sinistrorsa* Urb., turning left. The third, ssp. *longiseta*, he divided into forms according to coiling direction.

Heyn (1963) divided the species into two varieties: 1) var. *littoralis* with spiny pods, and var. *inermis* Moris, without spines or with tubercles. She also recorded the high variability in size of spines and number of coils. Nègre (1956) mentioned a form with lobed leaves, f. *laciniatifolia.*

We (Lesins & Erac, 1968) found that in a cross between a spineless strain and a spiny one, the spininess was dominant and was determined by a single gene. Lilienfield & Kihara (1956) found that right-hand coiling direction (anticlockwise according to our nomenclature) was dominant over the opposite coiling direction, and that the character was controlled by a single gene. It appears that these characters, though easily recognizable, are not

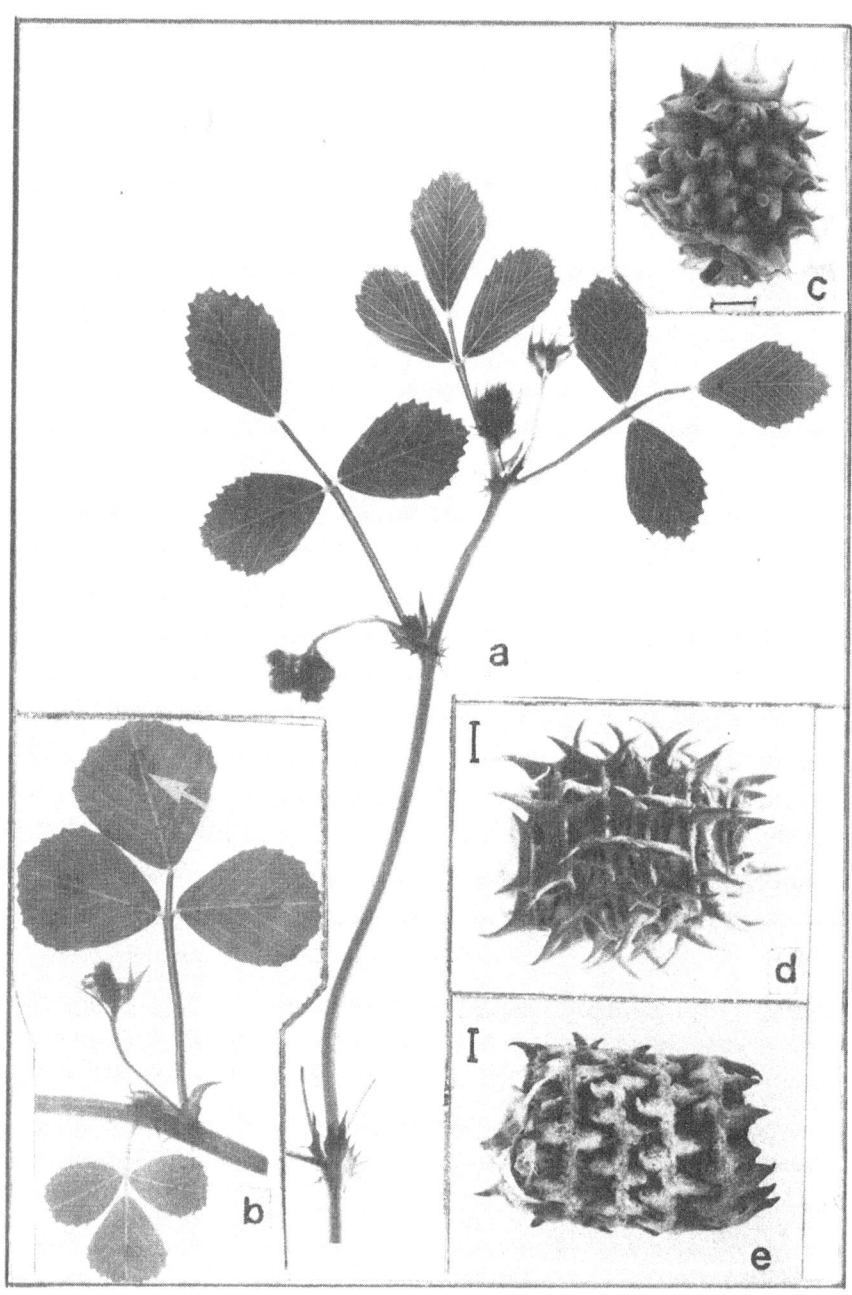

*Fig. 53, M. truncatula*. Branches: with subrhombical, obovate, sharply serrate leaflets (a); with broadly obovate, shallowly toothed leaflets (b) and anthocyanin patch in the middle (arrow). Pod types: f. *uncinata* (c), pods with spines inserted at 135° (d), and at 90° (e), note elevated dorsal suture.

deeply ingrained into the genetic setup of their carriers. We consider that differences between taxa controlled by a single gene should be classified as forms rather than at higher ranks.

Our accessions from the Canary Islands (Gran Canaria, Fuerteventura, Lanzarote and Tenerife) were puzzling in that their pods strongly resembled *M. tornata* (Fig. 52,-b,c). At first they were actually classified as *M. tornata* because lateral veins were below the dorsal suture, not on the same level as is usual in *M. littoralis* (Fig. 52,-d,e,f). Furthermore, pod coils were not as tightly appressed as is usual in *M. littoralis* and veins on coil face started to anastomose from mid-radius, not slightly before reaching the lateral vein. On the other hand peduncles had only 1-3(5) florets, generally a character-istic of *M. littoralis*. For clarification, hybridization was carried out. We knew from previous experiments (Lesins & Erac, 1968) that there was an interbreeding barrier, expressed as chlorophyll deficiency in $F_1$s if *M. litto-ralis* served as the maternal parent. The hybridization test between the Canary accessions and *M. littoralis* showed that there was no interbreeding barrier, whereas in crosses with *M. tornata*, chlorophyll deficiency in $F_1$s, and poor survival in $F_2$s was found. Hence we consider our Canary acces-sions as *M. littoralis*.

In this connection, the occurrence of *M. tornata* and *M. littoralis* in Macronesia (Canary Islands, Madeira, Porto Santo) may be discussed. For Canary Islands, Webb & Berthelot (1836) described *M. tornata* (as *M. helix* Willd. β *spinosa* Guss.) having peduncles with 2-6 florets. For Madeira and Porto Santo, Lowe (1868) described *M. tornata* (as *M. helix* W.) having peduncles with 1-5, mostly 2-3 florets, in var. *calcarata*; and 2-8, mostly 3-6 florets, in var. *inermis*. Lowe also stressed the very close similarity of *M. littoralis* Rohde (as *M. tribuloides* var. λ) with *M. tornata*. Obviously the multiflowered type which is usually associated with *M. tornata* does not occur in Macronesia. It is also probable that on testing for interbreeding, these Lowe's *M. tornata* may turn out to be part of *M. littoralis*. It is noteworthy that we did not find in the Canaries the long-spined type of *M. littoralis* which is widespread around the Mediterranean. Kunkel (1973) is probably correct in suggesting that the *M. littoralis* type of the Canaries is endemic, not introduced like most of the other *Medicago*. [He does not list *M. tornata* for the Gran Canaria Island at all.] It seems that a name in a new combination, *M. littoralis* var. *canariensis* Webb, is appropriate, taking for diagnostic purposes Webb's illustration in Webb and Berthelot Phytogr. Canar. 3, 2:table 56 (ed.2, 1850 not ed.1, 1836), where the name *M. canariensis* appears (see also Heyn, l.c. p. 87).

The species closest to *M. littoralis* is *M. truncatula*. Although the most characteristic forms of both species are easily distinguishable, there are some with intermediate characters. Data for our accessions are summarized in Table III dealing with *M. truncatula*. We found that discovering a few hairs on pods was the safest indication that an accession was *M. truncatula*, not *M. littoralis*.

37. *Medicago truncatula* Gaertner, Fruct. et Semin. 2:350 (1791). Syn. *M.tribuloides* Desr. in Lam., Meth. Bot. 3:635 (1792). Figs. 53; 3,-7; 6,-60.

Plants 15-80(120) cm long, procumbent to ascending, branching from near the base. Vegetative parts covered with simple hairs. Stipules usually deeply incised, resulting in long teeth. Leaflets 8-27 mm x 7-21 mm, obtuse, obovate (rarely obcordate), the lower ones sometimes wider than long; apical part with serrate margin (often with alternating large and small teeth), with a terminal triangular tooth. Upper side of the leaflet glabrous or sparsely hairy, underside more densely hairy. Peduncle 1 to 5-flowered, shorter (rarely equal to or longer) than the corresponding petiole, ending in a terminal cusp. Florets 5-8 mm long. Pedicel shorter than the calyx tube; bract longer than the pedicel. Calyx 2.5-4 mm long; teeth longer than the tube. Corolla yellow; standard obovate, 5-7 mm x 3-5 mm; wings shorter than the keel (rarely equal to or longer than it). Young pod contracted within calyx, more or less densely covered with simple hairs, occasionally also with glandular hairs. Mature pod light yellow to dark grey; cylindrical, hardwalled; spiny, rarely tubercled. Coils 3.5-6, appressed, 4.5-7 mm in $\phi$, turning clockwise or anticlockwise; face of coil with 6-12 radial veins, branching slightly before entering the lateral vein. Between lateral vein and the dorsal suture a groove which at pod maturity may become filled with spongy cellular tissue. The dorsal suture steeply elevated above the lateral veins (rarely on the same level). Spines 7-11 in each row, 1-4 mm long, sometimes curved, their base often broadly conical due to embedding in spongy tissue which may cover the spines to their tips; inserted at 90°-130°-180° to the face of the coil. Seeds reniform, 2.5-4.5 mm x 1.3-2.5 mm, 1-2 in each coil, separated. Seed weight 4-5 g/1000. Seedcoat smooth, pale yellow to brownish-yellow, usually with a darker patch at the tip of the radicle and around hilum. Radicle less than half of the seed length. $2n = 16$.

*Habitat and Distribution.* Urban (1873) divided the species on the basis strates. An omni-Mediterranean species. We collected it in Lebanon, Turkey, Cyprus; in Greece and on the islands of Crete and Rhodes; in Italy and on Sicily, Sardinia, and Capri; in France and on Corsica; in Spain, and in North Africa (Tunisia, Algeria, Morocco).

*Variation in M. truncatula.* Urban (1873) divided the species on the basis of the length of the spines into two varieties: 1) var. *breviaculeata* Urb., with spines as long as the width of the coil edge, inserted at a 90° angle to the face of the coil, and 2) var. *longeaculeata* Urb., with spines longer than the width of the coil edge. In this variety he included strains with hooked spines, small hairy pods, and upright, wrinkled spines. These strains have been described as separate species by some other authors. In both varieties he discerned as forms clockwise and anticlockwise coiled pod types, i.e., f. *dextrorsa* Urb. and f. *sinistrorsa* Urb.

Heyn (1963) distinguished three varieties on the basis of number of coils and the character of spines: 1) var. *truncatula* with 5-8 coils, pods longer than wide, spines more or less appressed to the pod surface, 2) var. *longeaculeata* Urb., also with 5 or more coils, length of pod equal to or less than

Table III. Pod characters in *M. truncatula* and *M. littoralis* accessions

| Species | Height of dorsal suture as to lateral veins % | | Angle of insertion of spines % | | | Hairs on pods % | | Spongy tissue on spines % | | |
|---|---|---|---|---|---|---|---|---|---|---|
| | higher | same level | 90° | 180° | inter-mediate | present | absent | none | halfway up | more than halfway up |
| *M. truncatula* | 89 | 11 | 20 | 16 | 64 | 91 | 9 | 0 | 2 | 98 |
| *M. littoralis* | 8 | 92 | 0 | 31 | 69 | 2 | 98 | 58 | 30 | 12 |

its diameter, with spines curved, not appressed to the surface of the pod, 3) var. *tricycla* Heyn, with 2.5-4 coils, length of pod less than its diameter, spines usually not appressed to the pod surface. Casellas (1962) mentioned under var. *truncatula* a microcarpous form with pods, 5-6 mm long and 3-5 mm wide, growing in mountain areas in Spain. We found a form with small pods in Sardinia, with coils 4.2-5.5 mm in $\phi$ (UAG Acc. No. 1060).

On the islands of Crete (9 km W. of Iraklion, Acc. No. 1105) and Cyprus (between Lefkoniko and Boghos, Acc. No. 1302) we collected a form which differed from the usual *M. truncatula* in having glabrous pods with the dorsal suture not higher than the lateral veins, the base of the spines inflated by spongy tissue to a bulblike structure, and with spines hooked at their tips. On superficial inspection, the pods resemble those of *M. turbinata*, but without the latter's characteristic veinless zone of the coil. The form differs from *M. littoralis* by the strongly inflated base of the spines. The name f. *uncinata* Willd. Sp. Pl. 3:1417, 1802, pro. sp. comb. nov. (Fig. 53,-c) is proposed for it.

The inheritance of spininess and coiling direction were investigated by Simon (1965). He found that spininess was dominant over the smooth (tubercled) character and was determined by one gene; similarly, clockwise (anticlockwise in our designation) coiling direction was dominant over the opposite and also was determined by one gene. It appears that these characters may be suitable for distinguishing forms, but are inadequate for delineating higher taxonomic ranks.

Lilienfeld (1962) reported on impaired fertility in crosses between two strains from Israel. Not only was seedset on $F_1$ plants less than half that of the parental stocks, but there was also a chlorophyll deficiency when one of the strains served as the female parent. Obviously, under natural conditions, gene flow in that direction would be greatly hampered. The strains Lilienfeld worked with otherwise did not differ in characters ordinarily used for distinguishing taxa.

*Distinction Between M. littoralis and M. truncatula.* As may be seen from the descriptions of these two species, the traits characterizing them have overlapping values. Urban (1873) stated that these taxa were closely related, though in general he considered them sufficiently different to be treated as separate species. The characters Urban used for distinguishing between the species were: 1) peduncle generally shorter than the petiole in *M. truncatula*, but longer than the petiole in *M. littoralis*, and 2) dorsal suture in the ripe pods at a higher level than the lateral veins in *M. truncatula*, but on the same level as the lateral veins in *M. littoralis*. In addition, Heyn (1963) stressed that in *M. truncatula* the angle of spine insertion in relation to the coil face was 90°, or close to it, vs. 180°, or slightly less in *M. littoralis*, and that *M. littoralis* has glabrous pods. According to our observations, the mature pods of *M. truncatula* have a much more pronounced development of spongy tissue on the pod surface than *M. littoralis*, embedding the base of the spines, sometimes to their very tips. In *M. littoralis*, on the other hand,

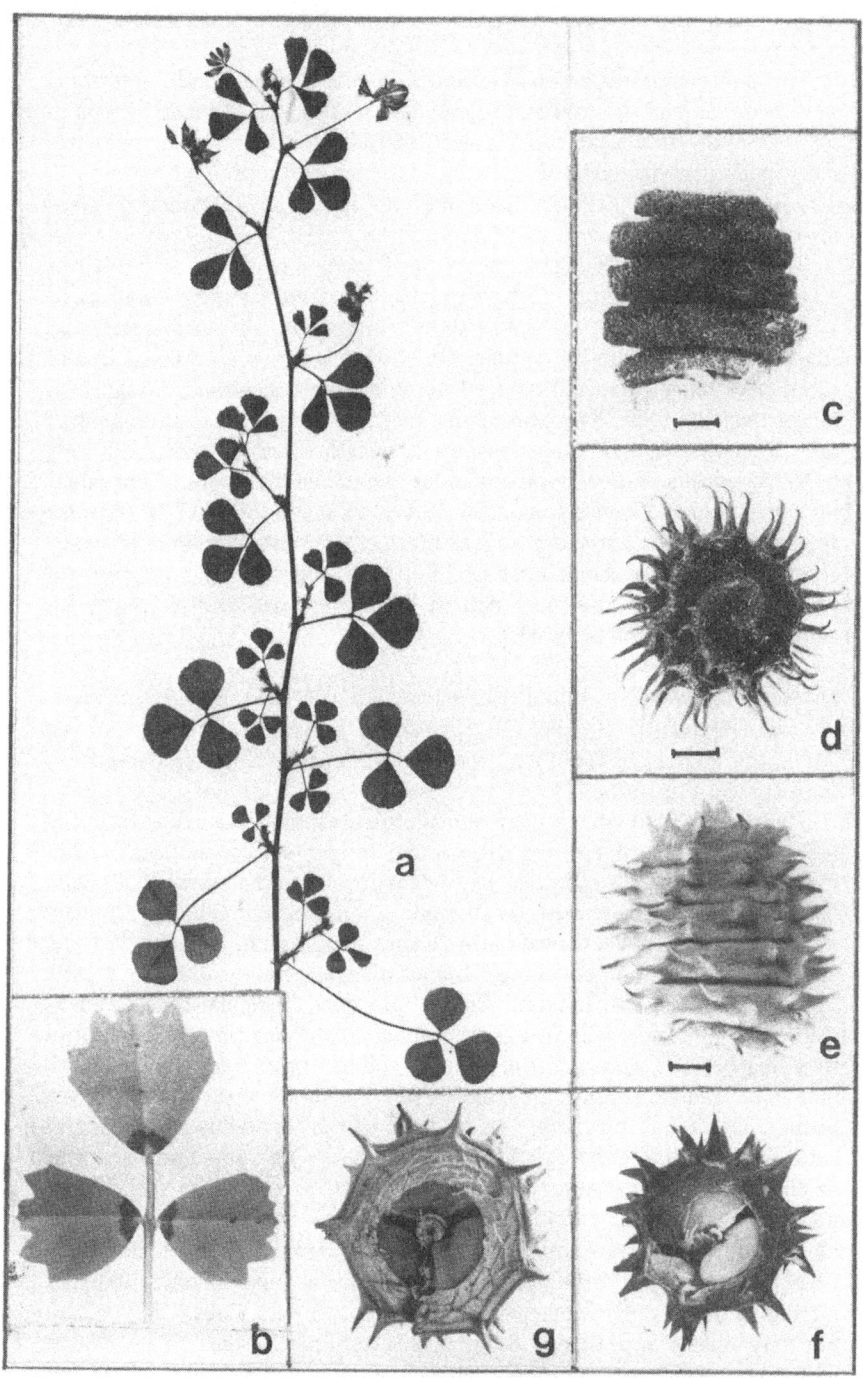

*Fig. 54, M. rigidula.* Branch (a) ½ natural size; leaf (b) with crenate leaflets and basal anthocyanin patch. Pods: spineless, glandular (c); spiny, glandular (d), and spiny, hairless (e). Transections of coils: with two seeds (f), and with three seeds (g). Note triangular seeds in three-seeded coil.

spongy tissue may be absent altogether or if present, covering the spines not more than halfway up.

The pod characters of 46 *M. truncatula* and 51 *M. littoralis* accessions were recorded and the results are presented in Table III. As may be seen in the Table, the four characters: dorsal suture higher than the lateral veins, hairy pods, narrow angle of insertion of spines, and a pronounced embedding of spine bases in spongy tissue are prevalent in *M. truncatula*, whereas the opposites are characteristic of *M. littoralis*.

The other distinguishing characters were studied on a smaller number of accessions: Of 12 *M. truncatula* accessions, 3 had the peduncle equal to or longer than the petiole; of 19 *M. littoralis* accessions, 13 had the peduncle equal to or longer than the petiole. Of 17 *M. truncatula* accessions, all had calyx teeth longer than the tube; of 15 *M. littoralis* accessions, 7 had teeth longer than the tube. The ratio of the length of the corolla to the length of the calyx was strongly dependent on the length of calyx teeth: with long teeth the corolla tended to be not more than twice the length of the calyx. Finally, 1 out of 13 accessions of *M. truncatula* and 2 out of 17 *M. littoralis* accessions had wings not shorter than the keel, contrary to what is generally considered to be a characteristic of both these species. This illustrates a general observation: the more natural populations are studied, the wider amplitude of variation is found.

38. *Medicago rigidula* (L.) Allioni, Fl. Pedem. 1:316 (1785). Syn. *M. polymorpha* var. *rigidula* L. Sp. Pl.:780 (1753); *M. gerardii* Waldst. and Kit. ex Willd., Sp. Pl. 3:1415 (1802); *M. agrestis* Ten., Cat. Pl. Hort. Neap. Append. 1:66 (1815). Figs. 54; 6,-41.

Plants (10)20-70 cm long, procumbent to ascending, branching from the base. Vegetative parts covered with simple upright hairs, sometimes in addition interspersed with glandular hairs, occasionally to the extent that plants appear greyish. Stipules with lengthy and often irregular teeth, occasionally entire (at upper nodes). Leaflets 6-20 mm x 6-19 mm, broadly obovate, occasionally retuse or obcordate. Leaflet margin serrate in the apical part, sometimes almost entire; midrib ending in a small triangular tooth. Peduncle 1 to 6(7)-flowered, longer (rarely equal to or shorter) than the corresponding petiole, with a distinct (rarely rudimentary) cusp. Florets 5.5-7 mm long. Pedicel shorter than the calyx tube; bract ± the length of the pedicel. Calyx 3-5 mm long, covered with simple, or simple and glandular hairs; teeth ± the length of the tube. Corolla yellow; standard roundish, ovate; wings shorter than or (very rarely) almost as long as the keel. Young pod contracted within calyx, glabrous to pubescent with simple or glandular hairs. Mature pod straw-colored to dark grey, glabrous, or velvety in appearance due to shorter or longer (many-celled) glandular hairs; cylindrical, discoid or barrel-shaped, hard-walled, spiny or spineless. Coils 3.5-7, usually not very tightly appressed, 4.5-9 mm in $\phi$, turning clockwise. Face of coil with 7-14 radial veins, strongly curved, occasionally S-shaped, branching little or anastomosing only on the outer part and merging with a strong lateral vein; a groove, often observable only on unripe pods, lies between the lateral vein and the dorsal suture; the dorsal suture often elevated, giving a convex shape to the edge. Spines, if present, 6-13 on each side of the coil, conical at base or with a shallow groove extending halfway up, the longest ones often hooked, inserted at (90)120°-180° to the coil face. Seeds mostly reniform, 2.5-4.5 mm x 1.3-2.5 mm, 1-3 in each coil, separated. Seed weight 2.3-6.4 g/1000. Seed coat smooth, pale yellow, darker around the hilum and at the tip of radicle. Radicle less than half the length of the seed. $2n = 14$.

*Habitat and Distribution.* Dry, rocky habitats preferred. An omni-Mediterranean species with denser representation in the eastern part. We found it to be especially abundant in Turkey.

*Variation Within Species.* Urban (1873) lists several pod forms which by Jordan (1854) have been described as separate species: 1) *M. depressa* Jord., pods wider than long, glabrous at maturity, spines very long, 2) *M. cinerascens* Jord., pods cylindrical, spines very short, 3) *M. timeroy* Jord., pods very short, covered with glandular hairs, and 4) *M. germana* Jord., pods flat, distinctly veined at maturity, with furrows between lateral veins and the dorsal suture.

Heyn (1963) dealt with considerable thoroughness with this difficult species. She divided it into several varieties: 1) var. *rigidula*, pods discoid to cylindrical, densely hairy or villose, very rarely glabrescent; coils not appressed, up to 8 mm in diameter, thin; dorsal suture protruding beyond lateral veins; spines inserted at about 180°, 2) var. *agrestis* Burn., pods discoid to cylindrical, most often glabrescent or glabrous; coils not appressed, diameter more than 8 mm, thick; dorsal suture about 2 mm thick, very slightly protruding beyond lateral veins, spines inserted at about 90°, 3) var. *cinerascens* Rouy, pods spherical to ovoid, usually densely covered with short, simple hairs; dorsal suture broad, not protruding beyond lateral veins; coils appressed, 5-6 mm in diameter, spines more or less long, slender, 4) var. *submitis* Boiss., differing from var. *cinerascens* in having coils 6-9 mm in diameter (vs. 5-6 mm), spineless or with tubercles (vs. more or less long-spined). Heyn (l.c.) admits that it is impossible to define boundaries between intraspecific entities with precision. She lists various characters which vary greatly and occur in different combinations: hairiness of vegetative parts; length of leaflets; hairiness, shape and size of pods; protrusion of dorsal suture; size, shape and number of spines per coil.

In the material examined by us, we noted that the peculiar velvety appearance of pod surface caused by glandular hairs was present in 82 accessions, in 25 accessions there were simple hairs or no hairs; no sparsely glandular hairy pods were found among our accessions. On intercrossing, however, intermediate forms were obtained.

We discovered that there was a kind of interbreeding barrier between certain *M. rigidula* strains (Lesins & Lesins, 1963): Some plants, UAG Nos. 993 and 1661 from the islands of Capri and Corsica, resp., had cylindrical pollen (Figs. 7,-5; 8,-5); some others, UAG Nos. 489 and 1743 from

171

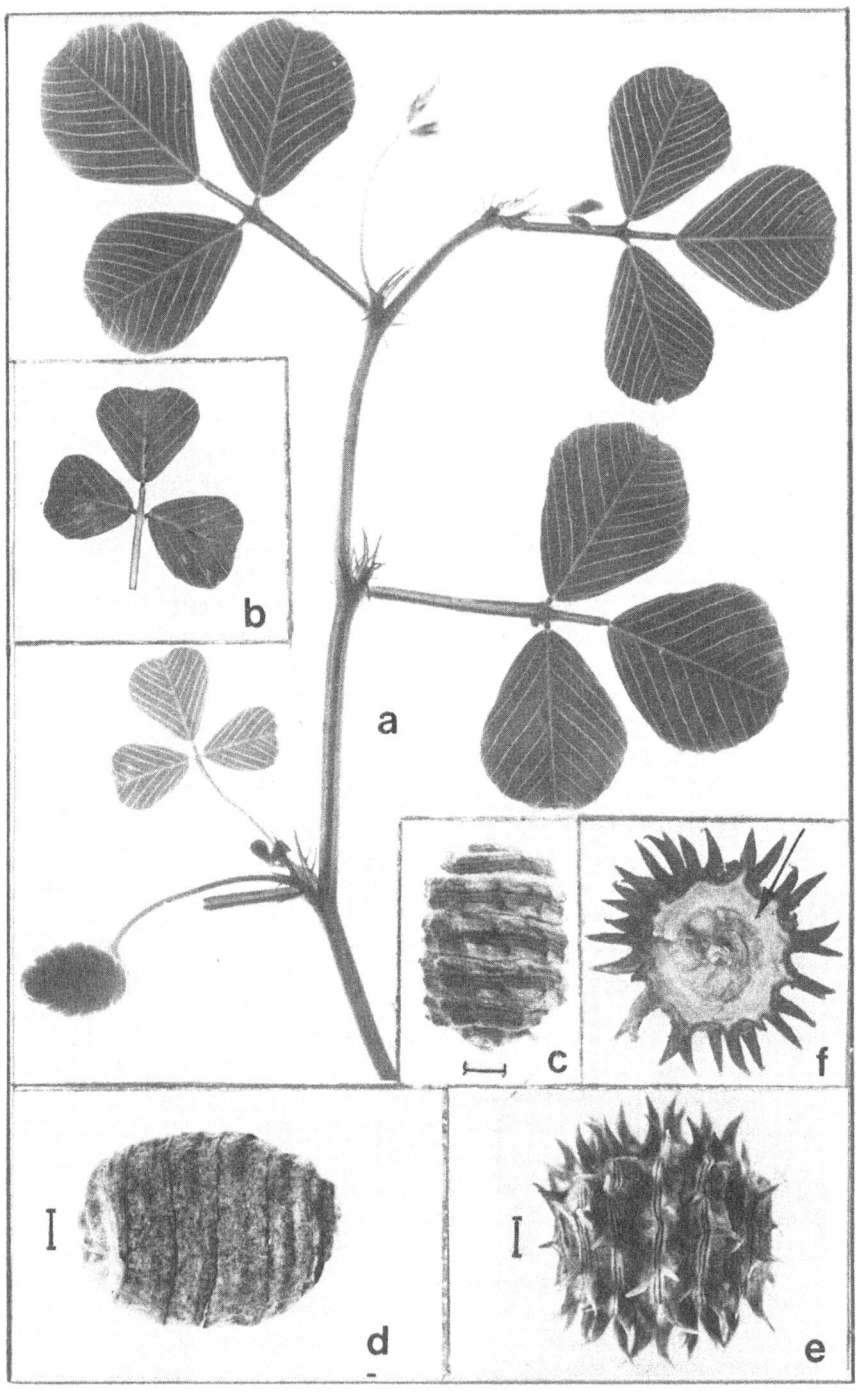

*Fig. 55, M. murex.* Branch (a) of ssp. *sphaerocarpos.* Leaf (b) of ssp. *murex,* note small white patches. Types of pods: tubercled (c); spineless and without wrinkles on coil edge, ssp. *murex* (d); spiny and with distinct wrinkles on coil edge, ssp. *sphaerocarpos* (e). Transection of pod (f) showing the veinless zone (arrow).

Iraq and southern Turkey, resp., had pyramidal-triangular pollen (Figs. 7,-6; 8,-6). In crosses of the two types, the $F_1$ had pollen viability of only 15-27%, whereas in parents it was 85-99%. Segregation for pollen types in the $F_2$ indicated two complementary genes, the pollen shape being determined by the genetic constitution of the haploid pollen (Lesins & Erac, unpubl.). It may be that under spatial isolation during long periods of time, due to genetic drift genic mutations and chromosomal rearrangements have been accumulated interfering with fertility. Or, genetic changes have been brought about under stress of the different environments.

*Marker Characters.* A good marker for intraspecific crosses is an anthocyanin-colored triangular patch at the base of the leaflets (UAG No. 489). It is a dominant character present in $F_1$s and segregating in $F_2$s in a ratio close to 3:1. Acc. No. 1743 also has the anthocyanin patch, and in addition has scalloped leaflet margins at the lower stem nodes (Fig. 54,-b). Purplish spines on pods were found in acc. No. 859.

*Distinguishing Characters Between M. rigidula and Some Other Species.* In pod appearance *M. rigidula* may resemble other hard-walled species: *M. constricta, M. murex, M. truncatula, M. doliata, M. turbinata,* and some forms of *M. tornata.* Examination of chromosome number in 55 *M. rigidula* accessions revealed invariably $2n = 14$. Since *M. truncatula, M. doliata, M. turbinata* and *M. tornata* have $2n = 16$, the separation on this basis is not difficult. Of the remaining species mentioned, *M. murex* has a marginal veinless zone on the coil face, a good character for distinguishing it from *M. rigidula,* and *M. constricta* has very tightly appressed coils, whereas in fully matyre pods of *M. rigidula* there is always a fairly distinct gap between coils. In some *M. rigidula* types, which have a dense cover of articulate hairs, the gap may be hidden and can be seen by clearing off the hairs.

Some additional distinguishing characters are given in the descriptions of the above mentioned species.

39. *Medicago murex* Willdenow, Sp. Pl. 3:1410 (1802). Syn. *M. sphaerocarpos* Bertol., Amoen. Ital.:91 (1819). Figs. 55; 5,-27.

Plants 30-90 cm long, procumbent to ascending, branches arising from near the base. Vegetative parts glabrous, or sparsely covered with simple semi-upright hairs. Stipules deeply incised, sometimes secondarily, forming long teeth; at upper nodes stipules occasionally occur as a single tooth. Leaflets 9-21 mm x 8-12 mm, obovate to obcordate; upper side glabrous, lower with a few hairs especially along the midrib; veins conspicuously lighter than the rest of the leaflet ending in fine teeth, midrib in a larger terminal tooth. Peduncle 1 to 6-flowered, with a terminal cusp, usually 2-5 times longer than the corresponding petiole, rarely only slightly longer than, or as long as the petiole. Florets 4-7 mm long. Pedicel shorter than the calyx tube; bract ± the length of the pedicel. Calyx 3-4 mm long; teeth ± the length of the tube. Corolla yellow, usually less than twice the length of the calyx, rarely equal to or more than twice its length; standard obovate; wings distinctly longer than the keel. Young pod at first contracting within the calyx, then turning sideways through the calyx teeth. Mature pod spherical to barrel-shaped, greenish-gray to black, hard-walled, spiny or spineless. Coils 6-9, appressed, 5-7.5 mm in $\phi$, turning clockwise. Face of the coil with 5-9 radial, somewhat curved veins (often indistinct under loose cellular tissue), running into a veinless outer zone, colored darker than the middle of the coil, and appearing to consist of a harder, horny substance (Fig. 55,-f). Margins of the coil edge usually on the same level as the dorsal suture. In the $2n = 14$ subspecies the dorsal suture is flanked with two deep, narrow furrows, resulting in three ridges (Fig. 55,-e). Spines, if present, 9-14 in each row, 0.5-3 mm long, with a conical or flattened base, inserted at $180°$-$240°$ (crossing over the dorsal suture) to the corresponding face of the coil; often hooked in young pods, hooks breaking off at maturity. Seeds 3.5-4.5 mm x 1.5-2.5 mm, often bow-shaped, 1-2 in each coil, separated. Seedcoat reddish-yellow to reddish-brown, darker at the tip of the radicle and chalaza. Seed weight about 7 g/1000. Radicle less than half of the seed length. $2n = 14,16$.

*Habitat and Distribution.* Usually in dry soils. The species has an omni-Mediterranean distribution. We found it to be fairly abundant on the Mediterranean isles of Cyprus, Rhodes, Crete, Sicily, Capri, Sardinia, Corsica and Pantelleria. On the mainland close to the sea we found it in Turkey, Greece, Italy, southern France and Spain. In N. Africa we collected it in Algeria. Nègre (1959) reports its occurrence also in Morocco and Tunisia. The $2n = 16$ chromosomal type seems to be very rare; we collected one sample on Sardinia, a second was sent to us from Morocco and the third is known from Israel (Heyn, 1956).

*Variation Within Species.* Urban divided *M. murex* into two varieties, var. *aculeata* Urb. and var. *inermis* Urb. These in turn he divided into forms according to the shape and size of the pods. Thus, var. *aculeata* was divided into: 1) f. *ovata* Urb., pods oval, 7-9 mm broad, 2) f. *macrocarpa* Urb., pods spherical, 7-9 mm broad, and 3) f. *sphaerocarpa* Urb., pods spherical, but only 5-7 mm broad. Var. *inermis* was divided into: 1) f. *sorrentini* Urb., pods oval, 6-9 mm broad, and 2) f. *sicula* Urb., pods spherical, 5-6 mm broad. Heyn (1963) divided the species into var. *murex*, with spiny pods, and var. *inermis* Urb., with spineless pods. The use of spininess as a basis for taxonomic classification we tested by crossing a spineless form (UAG No. 1982) with a spiny one (UAG No. 1075). The crossing was achieved quite easily; the $F_1$ plants had the spiny pods of the pollen parent indicating dominance of the spiny character. The $F_2$ segregated in 43 spiny : 11 spineless, indicating a one-gene determination of spininess (Lesins et al., 1970). It appears then that in this instance spininess is not a sufficient basis for setting up ranks higher than forms. On the island of Pantelleria we collected a form with distinctly shorter spines (UAG No. 2064. Fig. 55,-c) than in other spined accessions.

It was surprising to find that *M. murex* has two basic chromosome complements, viz. $2n = 14$ and 16 (Lesins et al., l.c.). From the 59 accessions investigated cytologically, only 4 had $2n = 16$, the rest being $2n = 14$. This was contrary to the generally held view that the species has $2n = 16$ (Heyn, l.c.; Clement, 1962; Simon & Simon, 1965). An attempt to intercross the two chromosomal types did not succeed. The morphological differences between the two types are very subtle: 1) the $2n = 16$ plants appear to have longer but thinner stems than the 14 chromosome type, 2) the leaflets are somewhat smaller and their shape more obcordate in the $2n = 16$ type than those in the 14-chromosome type, 3) the edge of the pod coil has either no ridge, one ridge, or three not distinct ridges in the 16-chromosome type, whereas three clearly expressed ridges are present in the 14-chromosome type, and 4) leaflets of the $2n = 16$ type have small whitish patches (Fig. 55,-b), which are not to be found in the $2n = 14$ type.

The $2n = 14$ type is thought to have originated from $2n = 16$ by transfer of almost all the chromosomal material from the 8th (shortest) chromosome to the 3rd chromosome, making it the longest in the $n = 7$ chromosome complement (Lesins et al., 1970; Gillies, 1971; Lesins & Gillies, 1972). Such a chromosome rearrangement coupled with a complete interbreeding barrier makes the $2n = 14$ type appear to have originated by a rare step of evolutionary saltation. Further investigations involving all the available *M. murex* material are needed to obtain insight into the origin of this and other $2n = 14$ *Medicago* taxa.

A decision on the nomenclatural rank of the two types is not easy. On the one hand their gross morphological appearance is rather similar, but on the other they do not interbreed. Willdenow's original plants probably are of the $2n = 16$ type, as ridges on coils are very indistinct (Dr. H. Scholz, Bot. Museum, Berlin-Dahlem, person. corresp.). It seems that the rank of subspecies would be appropriate for the different types and that the $2n = 16$ type should be *M. murex* ssp. *murex*, and the $2n = 14$ type *M. murex* ssp. *sphaerocarpos* (Bertol. pro. sp., l.c.).

*Distinguishing M. murex from Other Species.* At first sight it is difficult to distinguish *M. murex* from *M. constricta*, especially because of the great morphological similarity of their pods. On closer inspection, however, these taxa are easily differentiated by several characters: 1) leaflets of *M. murex* are glabrous on the upper side, of *M. constricta* both sides of leaflets are hairy, 2) the veinless outer zone on face of coils of *M. murex* pods is well delineated and darker colored, whereas in *M. constricta* it is absent, but, if appearing to be present, clearing off the loose cellular tissue will disclose some facial veins reaching the lateral vein; moreover, coil face is uniformly colored, 3) veins on face of coils in pods of *M. murex* run radially to the veinless zone, in *M. constricta* they are more strongly curved, sometimes in the distal part almost concentric to the edge, 4) wing petals in *M. murex* are longer than the keel, in *M. constricta* they are shorter, 5) in *M. murex* the young pods from the outset protrude through the calyx teeth, in *M. constricta* they are first concealed between them, 6) calyx teeth in *M. murex* usually are glabrous, in *M. constricta* they are hairy, 7) the spines of mature pods in *M. murex* usually are not hooked, in *M. constricta* they often are, 8) characteristic for *M. murex* ssp. *sphaerocarpos* is the coil edge with two furrows and three ridges, not present in *M. constricta*.

40. *Medicago constricta* Durieu, Act. Linn. Soc. Bord. 29:15 (1873). Syn. *M. globosa* Urb., Verh. bot. Ver. Brand. 15:71 (1873); *M. globosa* Presl, Del. Prag.:45 (1822). Figs. 56; 5,-8.

Plants 20-50 cm long, procumbent to ascending, branching from the base. Vegetative parts, including upper sides of leaves, covered with simple, upright hairs. Stipules deeply incised forming long teeth (at upper nodes

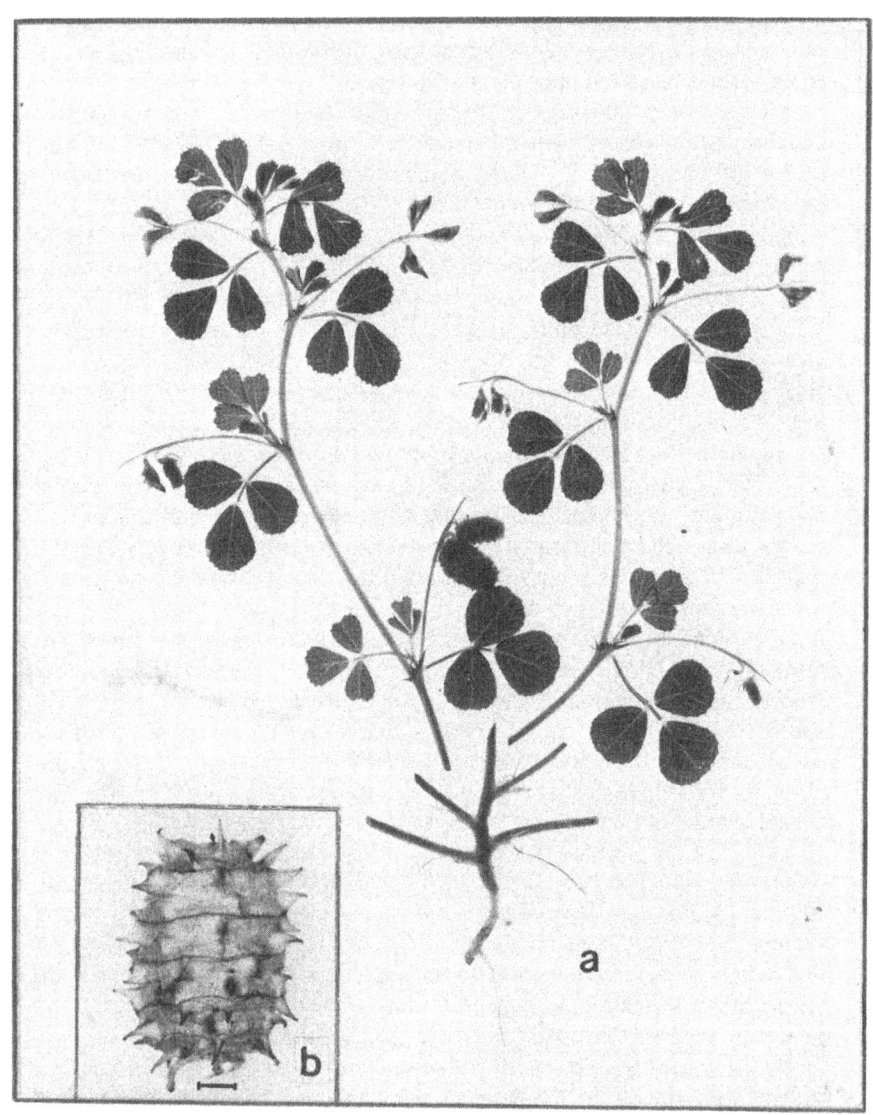

*Fig. 56, M. constricta.* Parts of herb (a), and pod (b).

stipules may be almost entire). Leaflets 7-18 mm x 7-12 mm, broadly obo-vate, rarely obcordate, margin of their apical 1/3 slightly serrate; midrib ending in a small tooth. Peduncle 1 to 5-flowered, longer than the corresponding petiole, ending with or without a terminal cusp (on the same plant peduncles with a smaller number of florets may have a cusp, those with a larger number may not). Florets 4.5-6 mm long. Pedicel shorter than or as long as the calyx tube; bract usually shorter than the pedicel. Calyx 2.5-4 mm long; teeth longer than or equal in length to the tube. Corolla yellow (buds may have a purplish tinge); standard roundish, obovate; wings shorter than the keel. Young pod contracted within the calyx, then turning sideways out of it. Mature pod straw-colored to dark brownish-grey, usually cylindrical in the middle part, hard-walled, spiny. Coils 4.5-8.5, turning clockwise, 4.5-8 mm in $\phi$, tightly appressed, the margins of coil edge may be elevated, so that the edge may appear concave. Face of coil with 7-10 strongly curved veins, almost concentric toward the coil edge, where they may be obscured by loose cellular tissue (after clearing this off, some veins are seen to reach the lateral vein). Spines 8-10 in each row, 0.5-3 mm long, inserted at 180°-210° to the corresponding coil face, often hooked. Seeds yellow, reniform, 3.5-4.5 mm x 1.3-2.5 mm, 1-2 in each coil, separated. Seed weight about 6 g/1000. Radicle less than half of the seed length. $2n$ = 14 (Lesins & Lesins as $M.\ globosa$, 1963).

Heyn (1963) pointed out that the name $M.\ globosa$ given by Presl (1822) was probably based on material not now considered to be $M.\ constricta$.

*Habitat and Distribution.* Hayn (1963) reported from Israel that the species habitat was mainly sandy clay. We also found it growing in soils of the dry type. It is an East-Mediterranean taxon. We collected it in Lebanon, Turkey, Greece and on the islands of Cyprus, Rhodes, Crete, and Karpathos.

*Variation Within the Species.* In some accessions (UAG Nos. 1212, 1222, 1226, 1471) we observed 3-5 florets per peduncle, whereas Urban (1873) and Heyn (l.c.) have reported only 1-2. On analyzing our material for ten different plant characters, J.L. Fyfe (Scott. Pl. Breed. Sta. Pentlandfield, unpubl. results) found that the accessions from Cyprus constituted a separate group excelling, especially in earliness, all accessions from other areas. One form from Cyprus (UAG No. 1274) has long, thin pods (10 mm long, 5 mm in $\phi$), with some of the coils distinctly spined, others tubercled, and still others spineless on the same pod. As a marker character, the cherry red color of spines on young pods (UAG No. 1212) was useful in crosses; the $F_2$ segregation ratio from crosses with another accession with non-red spines was 3:1, red spines being dominant (unpubl. results).

*Distinguishing M. constricta from Other Species.* Pods of $M.\ constricta$ are very similar to those of spiny $M.\ murex$ and $M.\ doliata$, and some forms of $M.\ rigidula$. Distinguishing characters from $M.\ murex$ are listed in discussion of that species. To differentiate $M.\ constricta$ from $M.\ doliata$ the chromosome number is the most reliable character: $M.\ doliata$ has $2n$ = 16, $M.\ constricta$, $2n$ = 14 (Lesins & Lesins, l.c.). $M.\ rigidula$ pods have some gaps between coils, while $M.\ constricta$ have none.

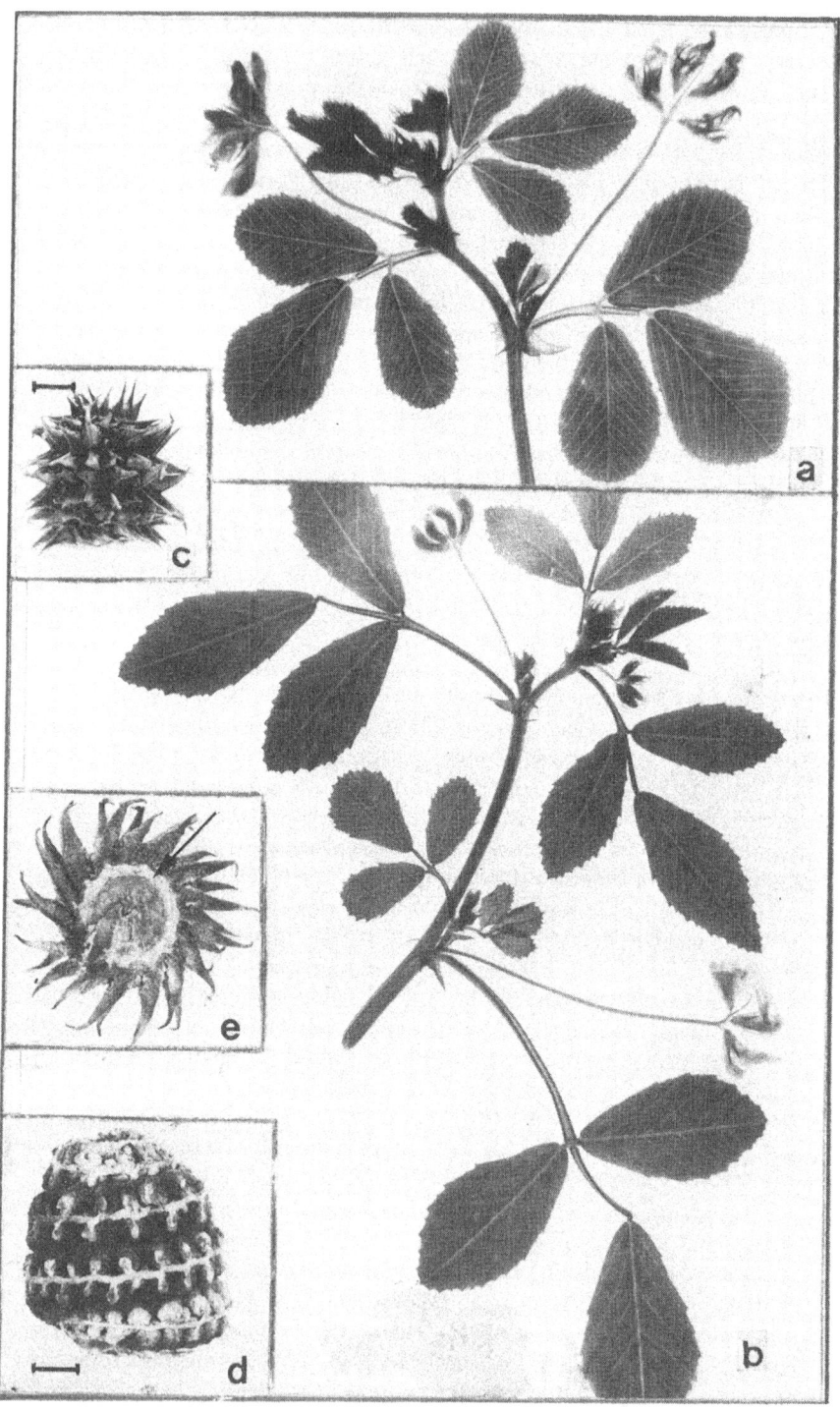

*Fig. 57, M. turbinata.* Branches, (a) and (b), with different petiole lengths, and different form and serration of leaves. Pods: spiny (c), tubercled (d). Transection through pod (e) showing veinless zone (arrow).

Hybridization attempts with the $2n = 14$ type of *M. murex* were not successful. It may be noted that the pachytene chromosome complement of *M. constricta* lacks the long chromosome characteristic of other $2n = 14$ *Medicago* species (Gillies, 1971; Lesins & Gillies, 1972). It is possible that *M. constricta* may have evolved either by further rearrangement of the *M. murex* chromosome complement, or from an extinct, or not yet discovered $2n = 16$ ancestor.

*41.* **Medicago turbinata** (L.) Allioni, Fl. Pedem. 1:315 (1785). Syn. *M. polymorpha* var. *turbinata* L., Sp. Pl.:780 (1753); *M. tuberculata* Willd., Sp. Pl. 3:1410 (1802). Figs. 57; 6,-61.

Plants 30-50 cm long, procumbent to ascending, branching from the base. Vegetative parts covered to varying degrees with simple upright hairs. Stipules toothed. Leaflets 11-22(35) mm x 7-16(30) mm, broadly obovate (at lower nodes) to narrowly oval, elliptic (at upper nodes); apical 3/4 of leaflet margin serrate; midrib ending in a small triangular tooth. Peduncle 1 to 8-flowered, longer than the corresponding petiole, with a terminal cusp. Florets 6-8 mm long. Pedicel shorter than the calyx tube; bract shorter or longer than the pedicel. Calyx 3.5-4 mm long, glabrate or covered with simple, or simple and glandular hairs; teeth longer than the tube. Corolla bright yellow; standard obovate, 6-6.5 mm x 4-4.5 mm; wings shorter than the keel, rarely as long as or longer. Young pod glabrous, at first contracted within the calyx, then protruding sideways through the calyx teeth. Mature pod ash grey to blackish, cylindrical-truncate to barrel-shaped, glabrous, spiny, tubercled, or spineless. Coils 4-6(7), usually appressed, 5-5.8 mm in $\phi$, turning clockwise or anticlockwise; coil face with 7-10 slightly curved, often indistinct veins radiating from the ventral suture and ending in a veinless zone, which occupies a space 1/5 - 1/3 of the radius of coil face. A deep, narrow groove is present between the coil margin and the dorsal suture. Dorsal suture, especially in young pods, steeply elevated in the middle of the coil edge. Spines or tubercles, if present, 12-16 in each row, 1-4 mm long, with broadly conical base embedded in alveolar tissue. Spines inserted at 180° to 225° (= arching over the dorsal suture) to the corresponding coil face. In some forms the spines are inclined in the opposite direction to that of pod coiling. This character may have provided the specific epithet 'turbinata'. Seeds light to brownish-yellow, 4.5-4.7 mm x 2.2-2.5 mm., 1-2 in each coil, separated. Seed weight 8.5-10.5 g/1000. Radicle less than half the length of the seed, darker colored at the tip and hilum. $2n = 16$.

*Habitat and Distribution.* Growing in dry clay soils (terra rossa, Heyn, 1963). Distributed in eastern and northern Mediterranean countries. We collected it in Lebanon (often), Turkey, Greece, Italy and on the islands of Cyprus, Rhodes and Crete.

*Variation Within Species.* Urban (1873, as *M. tuberculata*) distinguished four varieties on the basis of spininess: 1) var. *vulgaris* Moris, with tubercles not higher than the dorsal suture; this variety he divided into three forms: a) f. *sinistrorsa* Urb., pods coiling anticlockwise, b) f. *dextrorsa* Urb., pods

179

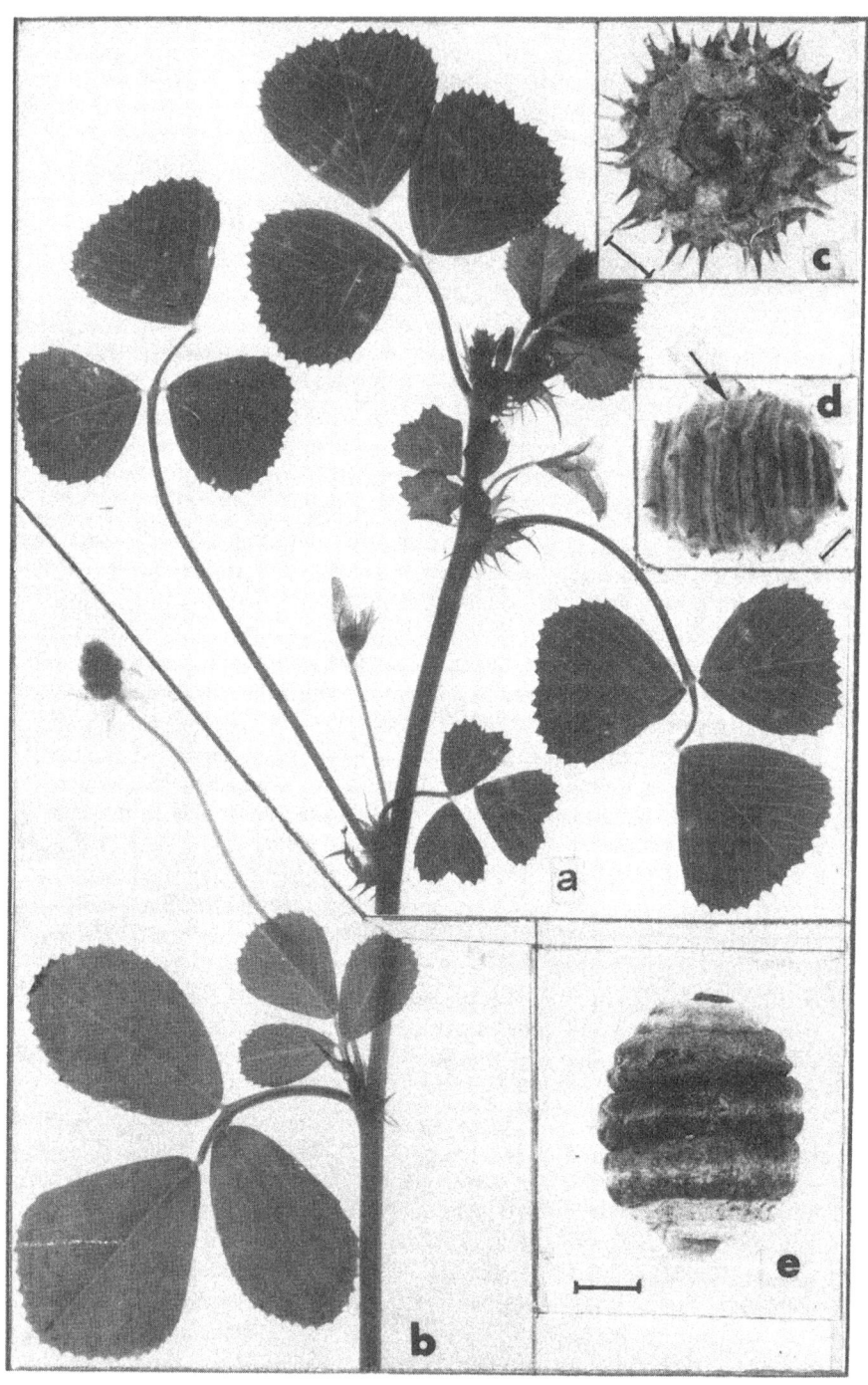

*Fig. 58*,  *M. doliata*. Branches: with peduncle shorter than the petiole (a), and longer (b). Pods: spiny (c), f. *terniana* (d) with elevated coil margins (arrow), and spineless (e).

coiling clockwise, c) f. *pubescens* Urb., with long, upright hairs giving the plant a felty, whitish appearance; 2) var. *apiculata* Urb., with short spines; 3) var. *aculeata* Moris, with spines as long as the coil radius, and 4) var. *chiotica* Urb., with spines inserted obliquely.

Heyn (l.c.) divided the species into three varieties on the basis of spininess: 1) var. *turbinata*, without or with short, appressed spines which would not aid in dispersal of the pods by adhering to animal furs, 2) var. *apiculata* (Urb.) Heyn, with slender spines, their bases close to the margin of the coil edge, with (3)5-8 florets per raceme, and 3) var. *aculeata* (Moris) Heyn, with thicker spines than in the previous variety, their bases off the margin of the coil edge; with 1-2(3) florets per raceme (vs. 3-8 florets in the other two varieties).

*Features Distinguishing M. turbinata from Other Species.* The species most similar to *M. turbinata* is *M. doliata*. Young pods in *M. turbinata*, though contracted within the calyx, soon turn sideways through its teeth, whereas in *M. doliata* young pods are contracted and concealed within the calyx (Fig. 3,-1). In *M. turbinata* the veins run into a veinless zone in the outer part of the coil face, moreover the veinless zone is darker colored than the inner part of the coil face, whereas in *M. doliata* this zone is absent, or if appearing to be present, the whole face of the coil is uniform in color. In *M. turbinata* the peduncle ends in a distinct cusp, whereas in *M. doliata* the cusp is minute or absent. In *M. turbinata* the dorsal suture is sharply elevated, with a deep, narrow groove between it and the coil margin, whereas in *M. doliata*, even if the dorsal suture occasionally may be elevated, there is no such groove. There are forms in *M. turbinata* not present in *M. doliata*, e.g., pods with spines inserted obliquely, opposite to the coiling direction, or with regularly 3-8 florets on a peduncle. Attempts to hybridize *M. turbinata* with *M. doliata* failed. *M. rigidula*, *M. constricta* and some forms of *M. murex* are somewhat similar to *M. turbinata* in pod shape. They have the chromosome number $2n = 14$, in contrast to $2n = 16$ in *M. turbinata*. In the $2n = 16$ form of *M. murex* (as in all *M. murex*), the upper side of the leaflets is completely glabrous, whereas in *M. turbinata* at least some sparse hairs are found on the upper side of leaflets. Pods of certain forms of *M. truncatula* may be quite similar to those of some *M. turbinata* forms (comp. Figs. 53,-e and 57,-d). They can be distinguished by the veins on the face of the coils: there is a veinless zone in *M. turbinata*, whereas on the coil face of *M. truncatula* no such zone is present. Sterile hybrids between *M. turbinata* and *M. truncatula* were obtained (unpubl.).

42. *Medicago doliata* Carmignani, Giorn. Pisano 12(N.32):48 (1810). Syn. *M. aculeata* Gaertn., Fruct et Sem. 2:349 (1791). *M. turbinata* Willd., Sp. Pl. 3:1409 (1802). This species is commonly known under its synonymic names. Figs. 58; 5,-1.

Plants 30-50 cm long, procumbent to ascending, branching from the base. Vegetative parts covered with diffuse simple hairs, occasionally with simple and glandular hairs. Stipules toothed, usually with a number of long, thin teeth, or occasionally, fewer broad teeth. Leaflets (8)11-16(25) mm x (6)10-14(20) mm, broadly obovate to oblanceolate-obovate, occasionally obcordate; the apical part of margin serrate, midrib ending in a small triangular tooth. Peduncle 1 to 2(4)-flowered, longer than, equal to, or shorter than the corresponding petiole, without or with a minute terminal cusp. Florets 4.5-7 mm long. Pedicel shorter than the calyx tube; bract ± the length of the pedicel. Calyx 2.5-4 mm long, covered with simple or simple and glandular hairs; teeth narrowly triangular, about the length of the tube, sometimes a little longer or shorter. Corolla yellow; standard obovate, 4-6 mm long, 2-4 mm wide; wings shorter than the keel. Young pod hairy, contracted within the calyx (Fig. 3,-1), contraction and concealment within calyx is more conspicuous than in any other species of section *Pachyspirae*. Mature pod light gray to black, spherical to oblong-oval, spineless, tubercled or spiny; glabrate, or covered with simple or glandular hairs, or with both types. Coils 5-9, turning clockwise or anticlockwise, 6-8.5 mm in $\phi$, tightly appressed, hard-walled. Coil face with 8-10 slightly curved, often indistinct veins; lateral veins on the same level as the dorsal suture, lower or higher than it. Spines, if present, 1-4 mm long, 10-14 in each row, inserted at 150°-180° to the face of the coil; their broadly conical base usually embedded in alveolar tissue, their tips thin and often hooked. Seeds pale yellow to brownish-yellow, strongly curved, (3)4.3-5 mm x (2)2.5-3 mm, 1-3 in each coil, separated, varying often considerably in size within a single pod. Seed weight 8.4-16 g/1000. Radicle less than half the seed length. $2n = 16$.

*Habitat and Distribution.* We found the species in clay soils. Heyn (1963) indicates that in Israel the species is confined to moist, heavy soils. Nègre (1956) also mentions moist growing sites. Carbonell (1962) considers dry soils as its usual habitat. While growing in our greenhouses, *M. doliata*, unlike other *Medicago* species, did not tolerate heavy watering; it turned yellowish and some plants died if watering was not reduced.

It is an omni-Mediterranean species. We collected it in Lebanon, the mainland of Italy, on the islands of Sicily and Sardinia, in France, Spain, Morocco, Algeria (where it was very common), and in Tunisia (less common).

*Variation Within Species.* Urban (1873) divided the species (as *M. turbinata* Willd.) into two varieties on the basis of presence or absence of pod spines: 1) var. *aculeata* Moris, with spines, and 2) var. *inermis* Ascher., without spines. Both varieties he in turn divided into two forms on the basis of pod coiling direction: (a) f. *dextrorsa* Ascher., turning clockwise, and (b) f. *sinistrorsa* Ascher., turning anticlockwise.

Heyn (1963) likewise based her division (as *M. aculeata* Willd.) primarily on spininess, and also used pubescence of the pods as a distinguishing character: 1) var. *aculeata* with spiny, pubescent pods, and 2) var. *inermis* Heyn, with spineless, pubescent ± glabrescent pods.

We found a form (UAG No. 2808) at Terni (vicinity of the town Tlemcen), Algeria, which differs from the usual type in a number of characters: pods, which are profusely covered with simple and multicellular glandular hairs, are small, 6-7 mm in diameter (not reaching 8 mm) and the coil edge margins in mature pods are higher than the dorsal suture, not lower than it. Peduncle is shorter than the corresponding petiole, not longer. Seed weight not exceeding 8 g/1000. In a cross between the type from Terni and one from Sardinia (UAG No. 1101) the $F_1$ had 80% viable pollen as compared to 98-100% in parents; also, seeds per pod were 3.7 as compared to 4.8-5.5 in parents. It is proposed to name the type from Terni: f. *terniana* (Fig. 58,-d).

*Features Distinguishing M. doliata from Other Species.* The species most similar in appearance to *M. doliata* is *M. turbinata.* The main distinguishing character is the veinless zone on pod face which is present only in *M. turbinata.* Further distinguishing features are dealt with in the discussion of *M. turbinata.* The *M. doliata* pods may be similar to those of *M. rigidula, M. constricta* and *M. murex,* but the chromosome number of those species is $2n = 14$ (except one ssp. of *M. murex*). Thus, if living material is available, they may be distinguished safely from *M. doliata* which has $2n = 16$. Though the pods of *M. doliata* and *M. murex* ($2n = 16$ ssp.) show considerable similarity, their leaflets differ: upper sides of *M. murex* are glabrous, those of *M. doliata* at least sparsely haired. Finally, the pods of some forms of *M. truncatula* are almost indistinguishable from *M. doliata.* However, at least in early stages of pod development, there are grooves between the lateral veins and the dorsal suture in *M. truncatula,* but not in *M. doliata.*

SECTION LEPTOSPIRAE Urban, Verh. bot. Ver. Brand. 15:50 (1873).

Pods with thin, soft walls. Coils not tightly appressed, turning clockwise. Spines, if present, two-rooted: one root from the lateral vein, the other from the pod edge, their bases not embedded in spongy tissue. Veins on face of coil well discernible. Radicle longer than or equal to half of the seed length. $2n = 16,14$.

Key to species or groups of section *Leptospirae*:

| | | |
|---|---|---|
| 1 | Pods smooth; surface densely covered with long, cottony hairs | |
| | *M. lanigera* | |
| — | Pods spiny, tubercled or wrinkled; surface not covered with cottony hairs | 2 |
| 2 | Radial veins on coil face run into a veinless zone | |
| | *M. disciformis, M. tenoreana* | |
| — | Radial veins on coil face run into a lateral vein | 3 |
| 3 | Coil edge wide, completely or almost completely covering the lateral vein and the groove between it and the dorsal plate | |
| | *M. coronata, M. praecox* | |
| — | Coil edge narrower, lateral vein and the groove between it and the dorsal suture observable from both the edge and face | 4 |
| 4 | Dorsal suture in a groove. Three grooves and 4 ridges observable on coil edge | *M. arabica* |

*Fig. 59, M. sauvagei.* Branch (a), and pods: in edge view (b), note spineless apical coil (arrow); in apical view (c), note little-protruding lateral vein on the coil face (arrow).

| — | Dorsal suture elevated above lateral veins | 5 |
| 5 | Stipules entire or slightly toothed | *M. minima* |
| — | Stipules deeply incised | 6 |
| 6 | Flowers with wing petals longer than the keel | *M. polymorpha* |
| — | Flowers with wing petals shorter than the keel | 7 |
| 7 | Apical coil spineless, lateral veins at 1/3 of coil radius from the dorsal suture, protruding only slightly from the plane of coil face (Fig. 59,-c) | *M. sauvagei* |
| — | Apical coil spiny; lateral veins closer to the dorsal suture (1/6-1/5 of the radius), protruding conspicuously from the plane of coil face as shoulders at 90° (Fig. 60,-d,f). | *M. laciniata* |

*43.* *Medicago sauvagei* Nègre, Comptes Rendus Soc. Sci. Nat. Maroc 7:175 (1954). Figs. 59; 6,-48.

Plants up to 40 cm long, branches decumbent to ascending, glabrous or nearly so. Stipules deeply incised (to 2/3 of their length). Leaflets 9-20 mm x 5-10 mm, obovate to cuneate; margin in its apical 2/3 coarsely serrate; at the truncate apex, the midrib tooth and the adjacent marginal teeth give an impression of a three-toothed apex. Peduncle 1 to 3-flowered, ± the length of the corresponding petiole, with a distinct terminal cusp. Florets 5.5-7.5 mm long. Pedicel slightly shorter than or as long as the calyx tube; bract short, half the length of the pedicel. Calyx slightly shorter than half of the corolla, sparsely covered with appressed hairs; teeth shorter than the tube. Corolla yellow; standard oval; wings considerably shorter than the keel. Young pod glabrous, emerges from the calyx, then turns sideways from it. Mature pod straw-colored to grey, discoid, spiny, glabrous. Coils 4-6, loose, up to 10 mm in $\phi$, turning clockwise, the apical coil spineless. Coil face with 12-15 strongly curved, often S-shaped veins, branching but little before entering the lateral vein which lies on the face of the coil distinctly below the dorsal suture. Spines up to 23 in a row, 1-2 mm long, straight, with terminal hooks, inserted at 135°-180° to the face of the coil, somewhat slanted towards the basal end of the pod. Seeds dark yellow to yellow brown, 2.5 mm x 1.5-2.5 mm, 1-3 in a coil, not separated, or separated by a slight partition only. Seed weight about 5.5 g/1000. Radicle up to 2/3 of the seed length. $2n = 16$ (Lesins & Lesins, 1961).

*Habitat.* Nègre (1956) reported soil type as reddish brown clay, and growth sites at altitudes above 400 m. We also found it growing in reddish clay on a hillside overlooking a deep valley.

*Distribution.* The species is endemic to Morocco. Nègre (1959) indicated that no new growing sites have been found except for the two he earlier reported (1956). We found it in a third site enroute from Rabat to Rommani, growing quite abundantly along the roadside, 3-4 km before reaching Nkhella, at about 200 m altitude. Here at one spot *M. sauvagei* was growing together with *M. laciniata*, with which it can produce hybrids (see below). *Relationship of M. sauvagei to Other Taxa.* *M. sauvagei* could be crossed with *M. laciniata* (Singh and Lesins, 1972). Gene flow from *M. laciniata* to

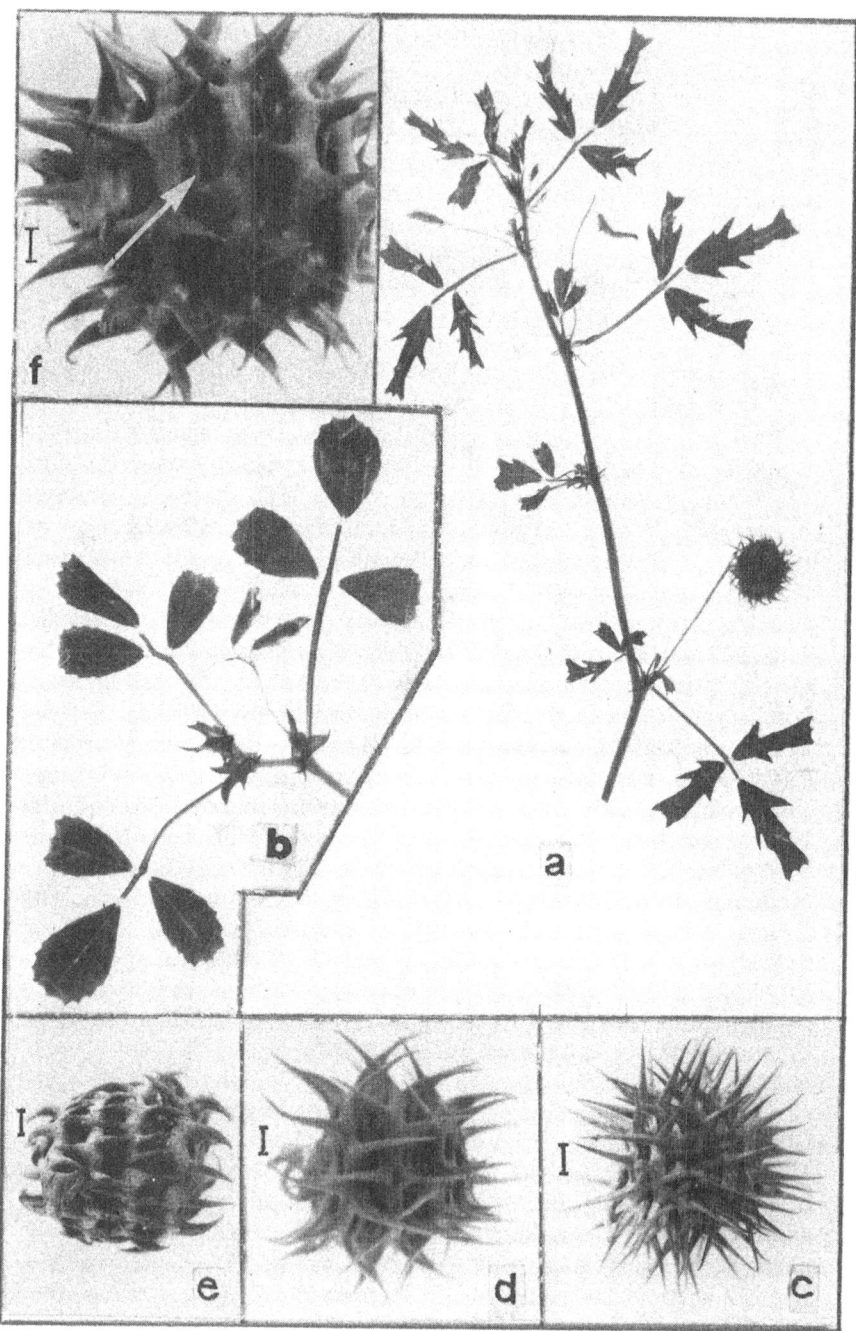

Fig. 60, *M. laciniata*. Branches: var. *laciniata* (a) nat. size, and var. *brachya-cantha* (b) magn. 1.3. Types of pods: var. *laciniata* (c); var. *brachyacantha* (d) and (f), note at (f) the characteristic position of lateral veins (arrow) between the elevated dorsal sutures. A form (UAG No. 40) with appressed spines (e).

*M. sauvagei* seems, under natural conditions, to be effectively prevented by a chlorophyll deficiency of the $F_1$ hybrids. In the opposite direction, although the transfer of hereditary material is possible under artificial conditions, it may be rare under natural conditions because both species are self-fertilizers, $F_1$ hybrids have poor pollen, seed set is low, and there is a high mortality in $F_2$ seedlings. The almost normal meiotic stages in $F_1$ plants indicate that the two species have a close evolutionary relationship. The pods of *M. sauvagei* are superficially very similar to those of *M. disciformis*, in that the apical coil in both taxa is spineless. However, pods of *M. sauvagei* have a lateral vein, whereas pods of *M. disciformis* have a veinless zone instead. Hybridization attempts between the two species were not successful (K. Lesins, unpubl.).

Another species which in appearance, especially of pods, may be confused with *M. sauvagei* is *M. polymorpha*. The latter's spiny, middle to large-sized pods may be mistaken for those of *M. sauvagei*, especially when field-aged pods of *M. polymorpha* are with considerably abrased pod spines. Besides the spineless apical coil of *M. sauvagei*, on close inspection it will be seen that the lateral veins of *M. polymorpha* are only half as far from the dorsal suture as those of *M. sauvagei*; furthermore, the lateral veins on the coil face in *M. polymorpha* are protruding much more from the plane of the coil face than in *M. sauvagei*, hence the grooves between the lateral veins and the dorsal suture are deeper in *M. polymorpha* than in *M. sauvagei*. In addition, the radicle in *M. sauvagei* is longer than that in *M. polymorpha* (Figs. 6,-48 and 5,-35, resp.). If living material is available, the distinction is easy: *M. polymorpha* has $2n = 14$, *M. sauvagei* $2n = 16$ (Lesins & Lesins, l.c.).

44. *Medicago laciniata* (L.) Miller, Gard, Dict. ed. 8, *Medicago* no. 5 (1768). Syn. *M. polymorpha* var. *laciniata* L., Sp. Pl.: 781(1753); *M. aschersoniana* Urb., Verh. bot. Ver. Brand. 15:77 (1873). Figs. 60; 3,-4; 5,-21; 9,-7.

Plants 15-35 cm long, few branches arising from the base, procumbent to ascending. Vegetative parts glabrous, or sparsely covered with simple hairs. Stipules feather-edged, or with one to two long teeth at the base. Leaves glabrous on the upper side, sparsely covered with simple hairs on the underside. Leaflets 4-15 mm x 2-7, obcordate, cuneate, truncate, with a broad terminal tooth; margins laciniate, or entire toward the base and serrate in the upper third. Peduncle 1 to 2-flowered, longer than, equal to, or shorter than the corresponding petiole, with a terminal cusp. Florets 4-6.5 mm long. Pedicel shorter than the calyx tube; bract about as long as the pedicel. Calyx 2-3 mm long, sparsely covered with appressed hairs; teeth shorter than the tube (1:2). Corolla yellow, more than twice the length of the calyx; standard ovate; wings shorter than the keel. Young pods glabrous (rarely with sparse hairs), first emerging straight from the calyx (Fig. 3,-4), later bending sideways. Mature pods yellow brown to greyish, spherical, ovoid, cylindrical or cone-shaped, spiny. Coils 3-7, turning clockwise, 3-5 mm in $\phi$; on coil face 8-16 S-shaped veins running into a protruding lateral vein; lateral veins adjoining the elevated dorsal suture as shoulders at right angles. Spines 8-16 in a row, 1-4 mm long, grooved in their basal part, inserted at 90°-180° to the coil face, often ending with a hook. Seeds pale yellow to yellow brown, 2.3-3.0 mm x 1.2-1.5 mm, 1-2 in each coil, not separated. Seed weight 1.4-2 g/1000. Radicle up to 2/3 of the seed length. At the tip of the radicle and around the chalaza, a darker color than on the rest of the seedcoat. Cotyledons after germination long and narrow (10-16 mm x 1.5-2.5 mm), their petiolar part uniformly narrow for about 1/3 of the total cotyledon length (Fig. 9,-7). $2n = 16$.

*Habitat and Distribution.* The natural habitat of *M. laciniata* is in dry, sandy or stony desertlike environments, where it is often the only *Medicago* species that survives.

The species appears to be native to the countries of the southern coast of the Mediterranean sea. It has been reported growing from the Canary Islands, over N. Africa to Asia, extending to the dry regions of Pakistan and India. We did not encounter it in the northern Mediterranean countries from Lebanon to Spain, but collected it in abundance on the Canary Islands and in N. Africa. Adventitiously it has been found as far north as Holland, probably brought there with wool and deposited with waste in the vicinity of wool processing factories (Oostroom & Reichgelt, 1958).

*Variation Within Species.* The characters used by different authors for establishing separate taxonomic units have been: laciniation of leaflets and stipules, length of peduncles, size of pods, number of veins on coil face, and number, length and angle of insertion of spines. The lack of laciniation in leaflets has been the main character for establishing a separate species, *M. aschersoniana*, by Urban (l.c.). Later he may have changed his views regarding the taxonomic rank of *M. aschersoniana* by not retaining it as a separate species (Heyn, p. 60, 1963). Heyn (l.c.) lists as synonyms more than ten taxa of *M. laciniata* at the varietal rank. She herself recognized two varieties: 1) var. *laciniata* with stipules and leaflets laciniate, peduncle longer than the petiole, pods with 5-7 coils and long spines, 2) var. *brachyacantha* Boiss. with stipules and leaves serrate, peduncle equal or shorter than the petiole, pods with 2.5-4.5 coils and variable length of spines. A form from Iraq (Rechinger's No. 14314; UAG No. 40, Fig. 60,-e) has insertion of spines at 90° to the face of the coil.

In hybridization experiments (Lesins & Erac, 1968) involving two types of var. *laciniata* and one of *aschersoniana*, the following segregation ratios were found: 1) laciniate vs. entire leaflets, 3:1; 2) peduncle:petiole length, 15:1; and 3) spherical vs. conical pod shape, 3:1. In addition, $F_1$ pollen fertility, and survival of $F_2$ plants were good. Therefore, the types tested should be considered varieties or even forms of *M. laciniata*. Lilienfeld (1959) also found a monohybrid segregation ratio of laciniate vs. entire leaf, the first being a dominant character. Probably there are two different alleles responsible for the laciniate-type leaves (Lesins & Erac, l.c.).

*Distinguishing M. laciniata from Other Taxa.* Some species from section *Leptospirae*, especially *M. polymorpha* and *M. minima*, may be misidenti-

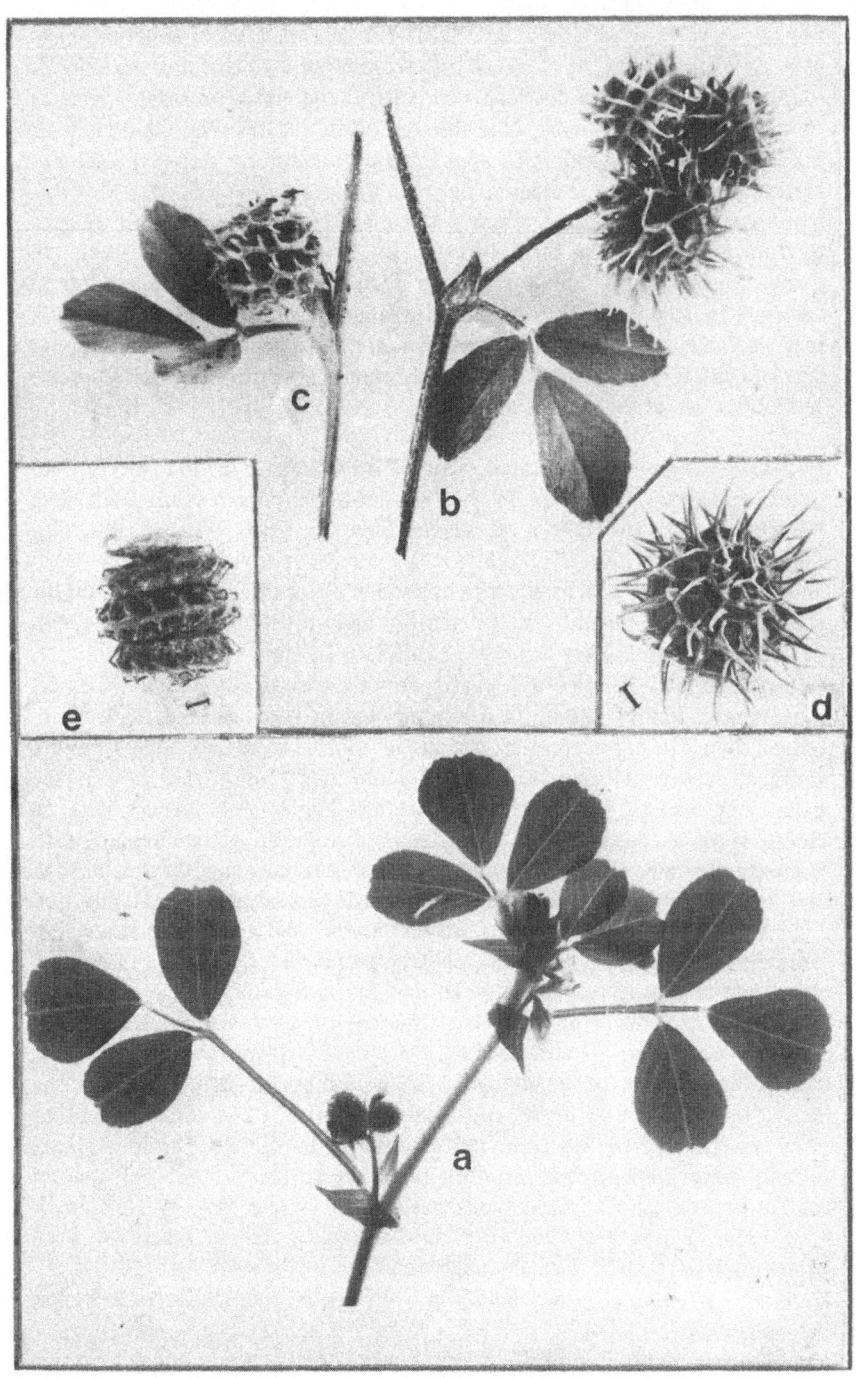

*Fig. 61, M. minima.* Branch (a), note entire stipules. Peduncles: longer than the petiole (b), shorter (c). Pod types: long-spined (d), spineless (e), short-spined (at c).

fied as *M. laciniata*. Examination of live plants for their chromosome number can solve the question of identity between *M. laciniata* and *M. polymorpha*: *M. laciniata* has $2n = 16$, *M. polymorpha* $2n = 14$ chromosomes. In *M. laciniata*, looking down between two coils, the lateral veins often appear as shoulders siding at $90°$ angles the elevated dorsal sutures (Fig. 60,-d,f); in *M. polymorpha*, the lateral veins appear to lie on a more sloped plane. For further distinguishing characters between *M. laciniata* and *M. polymorpha*, see discussion of the latter. Some of the characters distinguishing *M. laciniata* from *M. minima* are the following: In *M. laciniata* the lateral veins are about 1/5-1/6 of the radius from the dorsal suture; in *M. minima* lateral veins are located 1/3-2/5 of the radius from the dorsal suture; furthermore, in *M. laciniata* stipules usually are laciniate or deeply toothed, and the upper side of leaflets is glabrous; in *M. minima* stipules are entire or slightly serrate and both sides of the leaflets are hairy.

*45.* *Medicago minima* (L.) Bartalini, Catal. Piante Siena : 61 (1776). Syn. *M. polymorpha* var. *minima* L., Sp. Pl.:780 (1853); *M. meyeri* Grun., Bull. Soc. Nat. Mosc. 40:416 (1867); *M. sessilis* Peyr. ex Post., Fl. Syr. Pal. Sin. Suppl.:10 (1896). Figs. 61; 5,-26.

Plants 20-60(90) cm long, more or less densely covered with simple and/or glandular hairs; branches arising from the base, prostrate to ascending. Stipules entire or minutely toothed. Leaflets 6-14 mm x 4-8 mm, obovate, rarely emarginate to obcordate (at lower nodes); margin in its 1/4-1/3 apical part serrate; midrib ending in a terminal tooth. Peduncle 1 to 7-flowered, longer than the corresponding petiole, or shorter to sessile, with a minute terminal cusp or without it. Florets 3-6 mm long. Pedicel shorter than the calyx tube; bract ± the length of the pedicel. Calyx ± half the length of the floret; teeth as long as the tube or longer. Corolla bright to lemon yellow; standard obovate, sometimes emarginate; wings usually slightly shorter than the keel. Young pod turns sideways through the calyx teeth. Mature pod light to dark brown, glabrescent, or with simple and glandular hairs, cylindrical or oval, spiny, with short prickles, or spineless. Coils 3-5, thin-walled, 2.5-4 mm in $\phi$, turning clockwise. On pod face 5-8 S-shaped veins entering a lateral vein at 3/5-2/3 of radius from the centre; between it and the dorsal suture a wide groove transversed by the bases of spines, these dividing the groove in portions of quadrangular shape. Spines, from up to twice the length of the diameter of the coil to about 1/10 of it, or absent; if present they are grooved to 2/3 from the base or up to the tip; the long spines usually have hooked tips; insertion of spines at $180°$ to the coil face or obliquely to it (short spined types). Seeds pale yellow, small, 1.7-2.5 mm x 1-1.3 mm, 1-2 in each coil (first and last coil seedless), separated. Seed weight 0.9-1.4 g/1000. Radicle longer than half of the seed. $2n = 16$.

*Habitat.* *M. minima* grows mainly in dry soils, on rocky hillsides as well as in sand.

*Distribution.* This is one of the most widely distributed annual *Medicago* species. It is onmi-Mediterranean, extending to the north up to the Isle of

Öland (Sweden), to the west to the Canary Islands, and adventitious in many parts of the world. It easily spreads because of the hooked spines which attach readily to the fleece of sheep and other furred animals. It has often been found growing temporally in the vicinity of wool factories.

*Variation Within Species.* The species is rather variable. Urban (1873) listed four varieties on the basis of the length of the spines: 1) *longiseta* DC, spines as long as or longer than the diameter of the middle coil, 2) *vulgaris* Urb., spines as long as or longer than the radius of the coil, 3) *brachyodon* Reichb., spines shorter than the radius of the coil, 4) *pulchella* Lowe, spines shorter than or scarcely as long as the thickness of the coil edge, or absent. In addition, he lists three forms on the basis of plant hairiness: 1) *pubescens* Webb, the plant more or less hairy, but not to the extent of being felty or having glandular hairs, 2) *mollissima* Urb., with felty hairiness, especially on young leaves and stems, 3) *viscida* Koch, with glandular hairs on stems, leaves, and pods to the extent of being sticky.

Heyn (1963) distinguished two varieties on the basis of the length of spines and the shape of pods: 1) *minima* (syn. *M. hirsuta* Bart., *M. recta* Willd., *M. mollisima* Roth, *M. graeca* Hornem., among others) with spines longer than the radius of coils and with discoid pods, 2) *brevispina* Benth., with spines shorter than half of the radius of coils, and with ovoid pods. Synonyms of this latter variety at species level listed by Heyn (l.c.) are: *M. pulchella* Lowe, also *M. pulchella* Tod., *M. meyeri* Grun., *M. brachyacantha* Kern., *M. sessilis* Peyr., and *M. inconspicua* Nevsky.

We observed variations of characters in different combinations on intercrossing two different morphological forms: 1) long-spined (length of spines more than diameter of the coil), long-peduncled (peduncle longer than the petiole), with 2) short-spined (length about 1/5 of coil radius), sessile (peduncle very short, pods sitting in the axle of the corresponding petiole). It was found in the $F_2$ that for long peduncle : short peduncle the segregation ratio was 3:1, thus a simple monogenic inheritance; whereas in 52 plants none was with short spines, indicating that the character was inherited as a recessive and that two or more genes were involved (Lesins et al., unpubl.). Under natural conditions (at Sofari, Lebanon) we observed both types growing intermixed side by side. We noted a conspicuous physiological variation between accessions: under the same growing conditions and seeded at the same time a spineless accession (as *M. meyeri*) reached only bud stage, while a spiny var. *minima* already had ripe pods.

*Distinction Between Species.* Heyn (l.c.) indicates that *M. minima* has often been confused with *M. laciniata*. Both species have forms with long spines as well as with short ones; the main distinguishing characters between the two we noted in discussing *M. laciniata*. The characters distinguishing *M. minima* and *M. polymorpha* are discussed when dealing with the latter species.

*46.*    **Medicago praecox** de Candolle., Cat. Plant. Hort. Monsp.:123 (1813). Figs. 62; 6,-36.

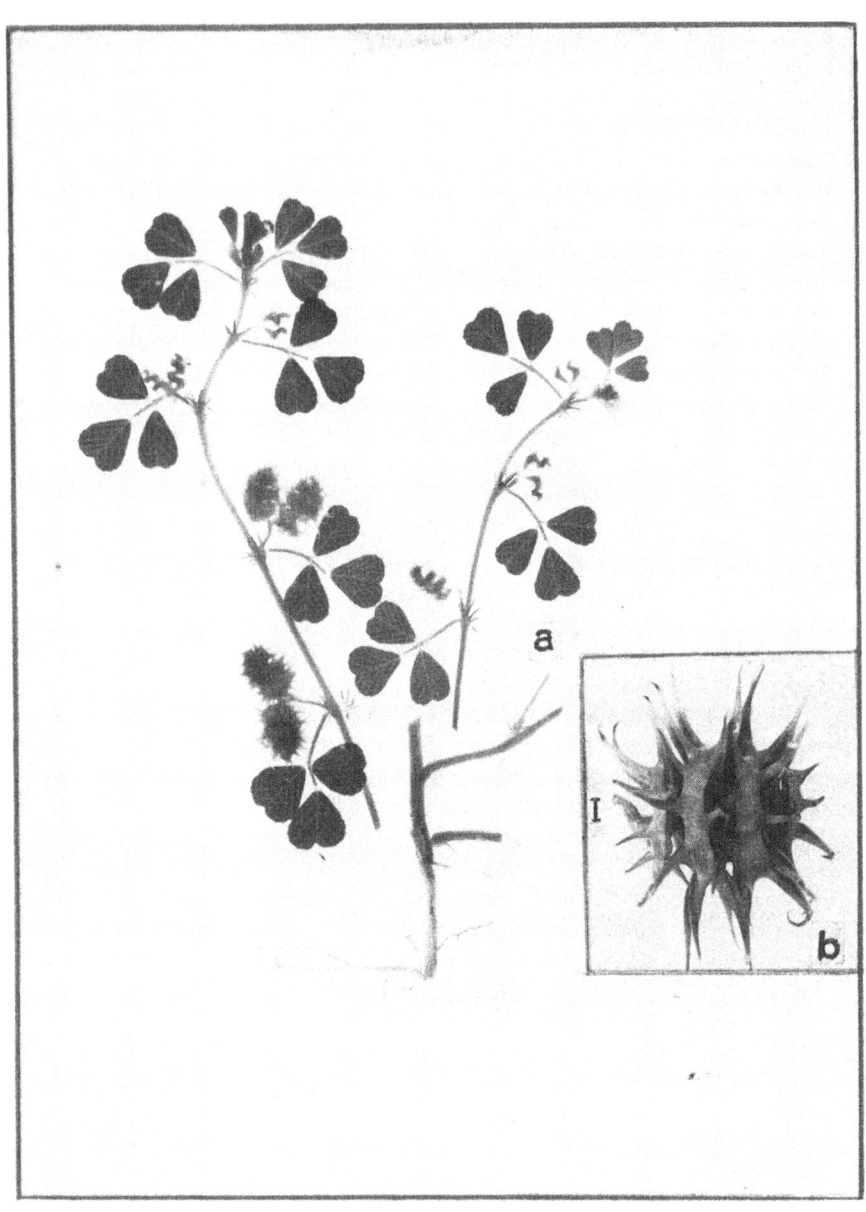

*Fig. 62,  M. praecox.* Parts of herb (a), and pod (b).

Plants up to 50 cm long, branches arising from the base, decumbent, ascending, branching secondarily throughout their length, sparsely covered with simple hairs. Stipules deeply toothed, laciniate. Leaflets usually glabrous on the upper side, 8-12 mm x 5-13 mm, obcordate, obovate; margin slightly serrate in its apical part; midrib without or with a minute terminal tooth. Peduncle 1 to 2-flowered, shorter than the corresponding petiole, without a terminal cusp. Florets small, 3-4 mm long. Pedicel shorter than the calyx tube; bract longer than the pedicel. Calyx slightly longer than half of the floret; teeth shorter than the tube. Corolla yellow; standard roundish (breadth may be slightly greater than length); wings shorter than the keel. Young pod emerges from the calyx, then turns sideways. Mature pod straw-colored to dark brown, glabrous (rarely covered sparsely with simple and glandular hairs), cylindrical to oval, spiny. Coils 2.5-4, loose, 2.5-4.2 mm in $\phi$, turning clockwise. On coil face 8-10 strongly curved veins, anastomosing near the lateral vein; between it and the dorsal plate a narrow, often deep groove which may be observed more readily from the face-side of the coil (the dorsal plate may cover it from the edge-side). Spines 10-12 in each row, grooved, hooked, 1.5-3 mm long, inserted at 90°-130° to the coil face. Seeds light to brownish-yellow, 2-2.8 mm x 1-1.3 mm, 1-2 in each coil, separated. Seed weight 2.6 g/1000 for the largest seeds. Radicle about half of the seed length. $2n = 14$ (Lesins & Lesins, 1962).

*Habitat and Distribution.* Growing on dry, rocky hillsides. A north-Mediterranean species. Nègre (1959) does not mention it from North Africa, nor did we find it there. We collected it on Cyprus, Crete, in Greece, Italy, on Sardinia and in southern Spain. Generally it is not abundant in any of its growing sites.

*Distinguishing M. praecox from Other Taxa.* Most often *M. praecox* is confused with *M. coronata*, especially because *M. praecox* has forms (UAG No. 957) with small pods and seeds (half the size of most other accessions); also, in some accessions the dorsal suture (plate) is wide and completely covers the lateral groove, making the pod appear like *M. coronata*. A good distinguishing character is usually the number of florets per peduncle: in *M. praecox* there are only 1-2, in *M. coronata* there are many (with an exception). If live material is available a decision may be reached by counting chromosomes: *M. praecox* has $2n = 14$, *M. coronata* $2n = 16$.

47. *Medicago coronata* (L.) Bartalini, Catal. Piante Siena : 61 (1776). Syn. *M. polymorpha* var. *coronata* L., Sp. Pl. 780 (1753). Figs. 63; 5,-9.

Plants 25-60 cm long, branches ascending to erect. Vegetative parts covered with simple or simple and glandular hairs. Stipules with 3-5 small teeth in their basal part. Leaflets 7-12 mm x 5-12 mm, obcordate, obovate, truncate; margin coarsely serrate in its apical 1/3, with a triangular apical tooth. Peduncle usually with many florets (up to 17), rarely with few (2-4) florets; several times longer than the corresponding petiole, with a short cusp. Florets 3-5 mm long, gathered in a compact raceme. Pedicel $\pm$ the length of the calyx tube; bract usually shorter than the pedicel. Calyx 1.8-2 mm long,

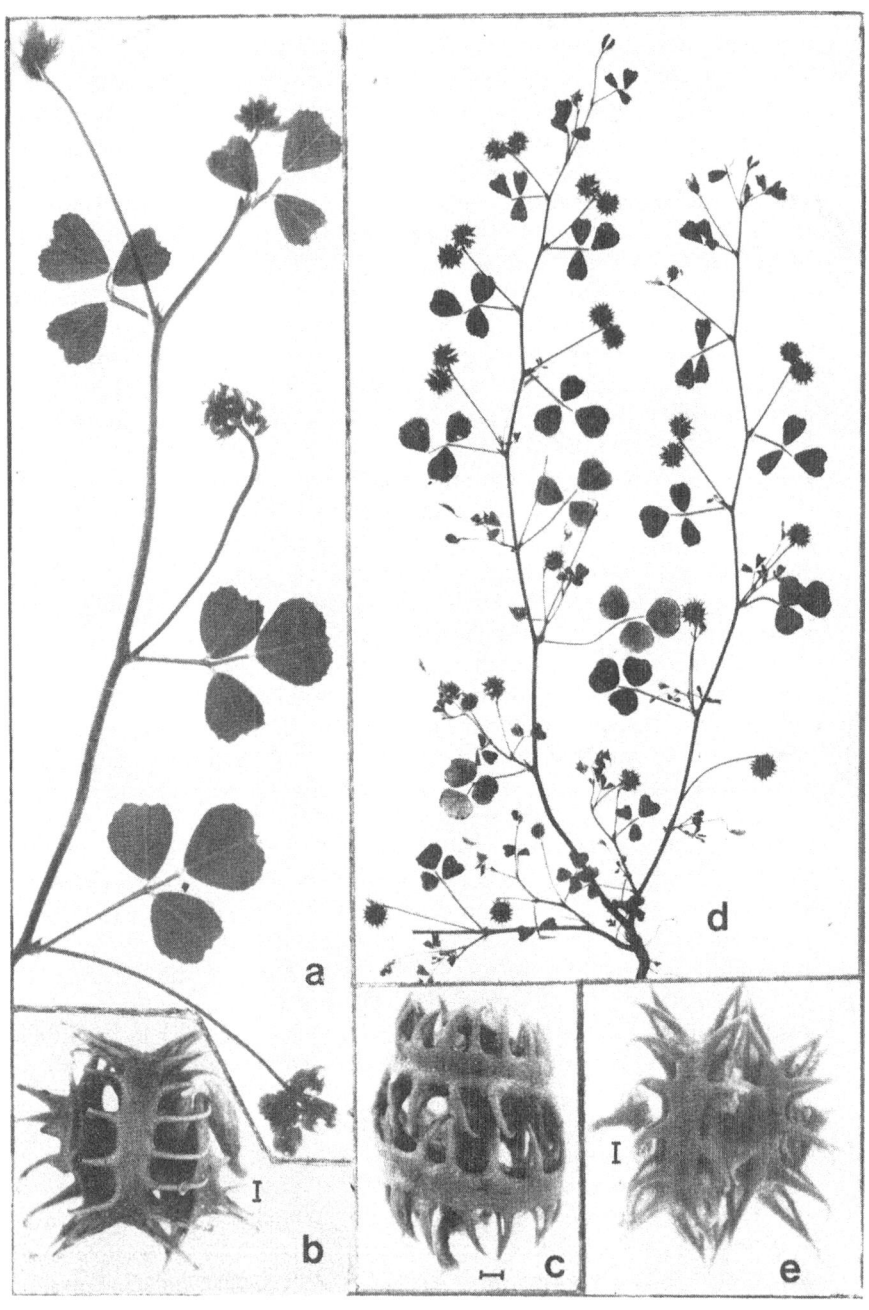

*Fig. 63, M. coronata.* Branch (a), and pods (b) and (c) prevalent in the species. Plant (d), and pod (e) of f. *pauciflora*, at (d) note 1-2-podded racemes, and at (e) dorsal plate not completely covering lateral grooves. Magn (a) 1.6; (d) 2/3 nat. size.

covered with simple or simple and glandular hairs; teeth narrow, usually shorter than the tube (rarely longer). Corolla lemon yellow; standard broadly obovate to obcordate; wings about the length of the keel. Young pod emerges straight from the calyx, then bends sideways. Mature pod greenish-brown to dark brown, glabrescent, or covered with simple or simple and glandular hairs, spiny. Coils 1-3, loose, 2.5-4 mm in $\phi$, turning clockwise. On coil face 6-8 strongly curved veins, slightly branching before entering the lateral vein. The edge of the coil 0.6-0.7 mm wide, the dorsal suture often in a shallow depression in the middle of the edge; margins of the edge protruding over the facial plane, therewith usually covering the lower-lying lateral veins; between lateral veins and the dorsal plate deep grooves not readily observable from the edge-side. Spines 11-15 in each row, 1-2.7 mm long (longest often hooked), inserted usually at about 90° to the facial plane, giving a crownlike appearance to the coil edge. Seeds light yellow, small, 2-2.5 mm x 1 mm, may be bow-shaped, 1-2 in each coil, not separated. Seed weight about 0.7 g/1000. Radicle longer than half of the seed. $2n = 16$.

*Habitat and Distribution.* Growing on rocky hillsides, often in limestone soils. According to Heyn (1963), *M. coronata* is a North-Mediterranean species spreading from Portugal to Iraq and Egypt. Nègre (1959) does not mention it for North Africa, nor did we find it there. We collected it in Lebanon, Turkey, Greece and on Cyprus.

*Variation in M. coronata and Distinguishing Characteristics.* Heyn (l.c.) mentions variations involving simple and glandular hairs, size of pods and florets, length of spines, number of florets in a raceme, and ratio of peduncle to petiole. Since intermediate forms and combinations of different characters were frequent, no formal subdivision was proposed. We found a form (UAG No. 1303, Fig. 63,-d) on Cyprus with consistently few, 2-4, florets per raceme. Other characteristic differences were noted on pods (Fig. 63,-e): 1) spines inserted to the coil face at about 135° vs. 90°, and 2) dorsal plate not completely covering the grooves to the lateral veins vs. completely covering them in the typical *M. coronata* (Fig. 63,-b,c). We collected this form between Lefkonikos and Boghaz and farther on the road to Cape Andreas. In crosses with the many-flowered type, the $F_1$ plants had intermediate number of florets (around 6) per raceme; the pollen viability was somewhat lower than in parents, viz. 85% vs. 98% in parents. We propose to call the few-flowered form f. *pauciflora.*

The pods of *M. coronata* resemble most closely those of *M. praecox*. The f. *pauciflora* is, in fact, almost indistinguishable from *M. praecox*. Hence, Heyn (1963) reported that intermediate forms occur between the two species. Morphological details disclose some of the differences: upper side of leaflets is hairy in *M. coronata*, vs. glabrous or sparsely hairy in *M. praecox*; stipules serrate in *M. coronata* vs. deeply toothed or laciniate in *M. praecox*; leaflet's midrib ending in a triangular tooth in *M. coronata* vs. tooth absent or tiny in *M. praecox*; peduncle longer than petiole in *M. coronata* vs. shorter in *M. preacox*, and seeds in pod not separated in *M. coronata* vs. separated by a thin wall in *M. praecox*. Above all, $2n = 16$ is in *M. coronata* vs. $2n = 14$ in

195

*M. praecox.* Moreover, the two species could not be hybridized (unpubl.).
The pods of f. *pauciflora* also resemble those of *M. laciniata.* One of the
distinguishing features here is the venation of coil face: in *M. coronata* the
veins, though strongly curved, are not S-shaped, as they are in *M. laciniata.*

48. *Medicago polymorpha* Linnaeus, Sp. Pl.:779 (1753). Syn. *M. hispida*
Gaertn., Fruct. et Semin. 2:349 (1791); *M. nigra* Krock., Fl. Siles. 2,2:244
(1790); *M. lappacea* Desr. in Lam., Encycl. Meth. Bot. 3:637 (1792); *M.
denticulata* Willd., Sp. Pl. 3:1414 (1802); *M. apiculata* Willd., Sp. Pl. 3:1414
(1802). Figs. 64; 3,-8; 5,-35.

Plants 20-70 cm long, branches decumbent to ascending, arising from the
base and secondarily along the main branches. Vegetative parts glabrate,
underside of leaves sparsely hairy. Stipules laciniate. Leaflets 11-20 mm x
10-20 mm, obovate, sometimes emarginate at the apex; margin vaguely
serrate in its apical 1/3-1/2; midrib ending in a thin tooth. Peduncle 1 to
6-flowered, longer than, equal to, or shorter than the corresponding petiole,
with or without a cusp. Florets 4-6 mm long. Pedicel shorter than the calyx
tube; bract usually longer than the pedicel. Calyx equal to or slightly longer
than half of the floret; teeth ± the length of the tube. Corolla yellow;
standard broadly obovate (5:4), emarginate; wings longer than the keel.
Young pod protruding sideways from the calyx. Mature pod ash-grey to
black, discoid, short to long-cylindrical, or conical-truncate, spiny, tuberculate or spineless, usually glabrous. Coils 1.5-7, not tightly appressed, 3.5-8
mm in $\phi$, turning clockwise. On coil face 6-10 well discernible veins, curved,
anastomosing freely before entering the lateral vein; between it and the
dorsal suture a marked groove traversed by roots of spines or tubercles.
Spines, if present, 0.5-4 mm long, up to 18 in each row, grooved, inserted at
90° (on end coils) to 180° to the face of the coil, the longest ones usually
hooked. Seeds light yellow to brownish, 2.5-4 mm x 1.5-2.2 mm, 1-2 in
each coil, separated. Seed weight 2.2-5.8 g/1000. Radicle approximately
half the seed length, tip of radicle somewhat protruding. $2n = 14$.
*Habitat.* Growing in various habitats. Competes more successfully with
grasses in moister soils and in cooler climates than most of the annual
*Medicago* species.
*Distribution.* Originally a Mediterranean species, *M. polymorpha* has now
spread all over the world; its boundaries are drawn only by low winter
temperatures destroying imbibed seeds, and spring and fall frosts killing
sprouted seedlings. The species is naturalized and constitutes a valuable part
of pasture stands in Australia, South America and in the southern U.S.A.
*Variation Within Species. M. polymorpha* is a rather variable species. Urban
(1873) divided it according to the diameter of the coils, and in turn, to the
number of coils per pod, and further, to the length of the spines. Heyn
(1963) recognized three varieties: *brevispina* Heyn, pods spineless or tubercled; *polymorpha*, coils spined, 5-8(10) mm in $\phi$, number of coils 4-6; and
*vulgaris* Shin., spined, 2.5-4.5 mm in $\phi$, number of coils 1.5-3.5.

We investigated (unpubl. data) some characters which have been used in

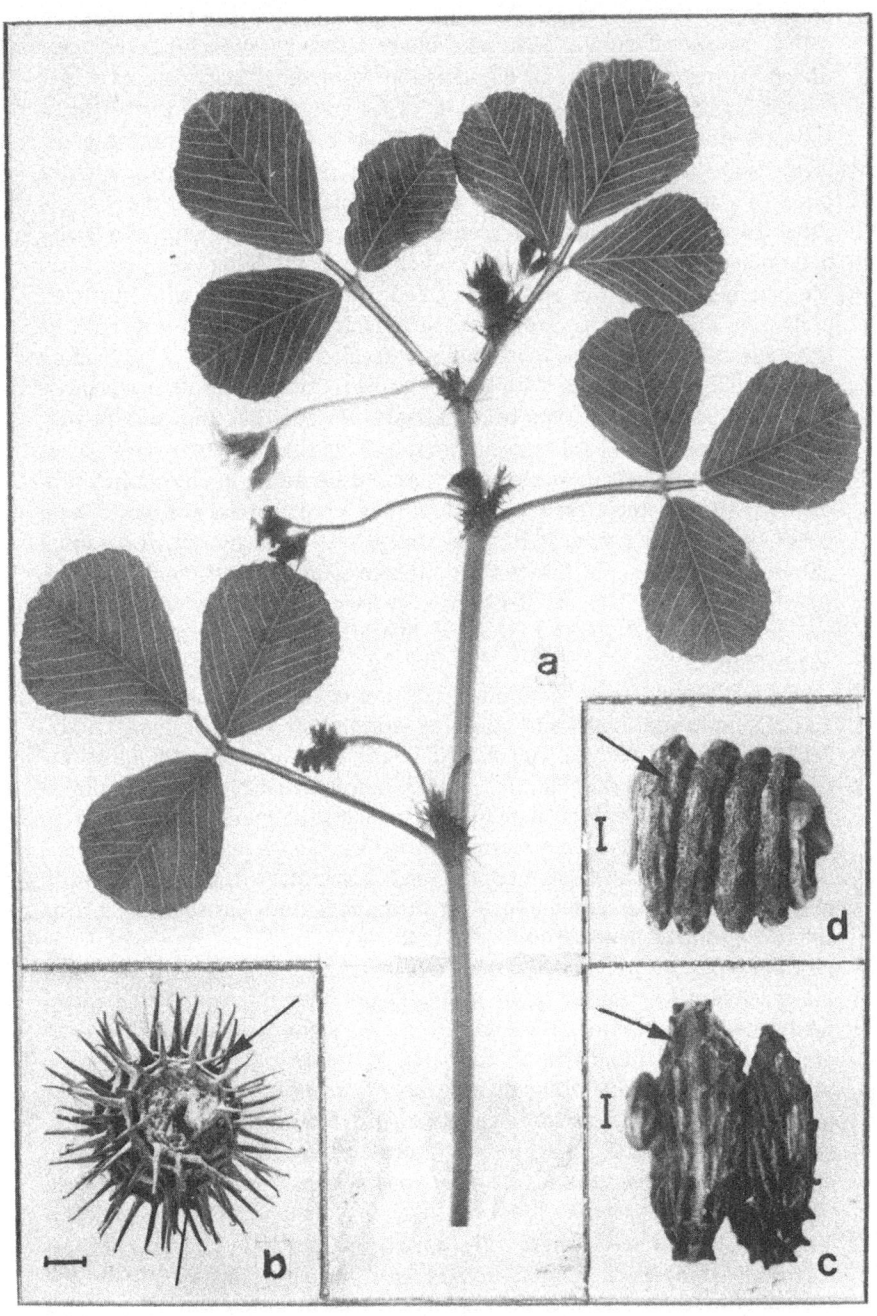

*Fig. 64, M. polymorpha.* Branch (a), and types of pods: spiny (b), tuber-cled (c), and spineless (d); arrows point to grooves between dorsal suture and lateral vein.

delimiting taxonomic ranks. Although *M. polymorpha* is a self-pollinating species as are all annual *Medicago*, intercrossing occasionally takes place under natural conditions. In an accession from the Canary Islands (UAG No. 2239), where spiny and spineless types grew together, 7 seeds from a spineless pod gave 6 spineless and 1 spiny-pod plants. Very probably the spiny plant had originated from pollen transferred by insects from a spiny plant to a spineless one, since the progeny from the spiny-pod plant segregated 43 spiny to 11 spineless plants. Thus, a single gene with two alleles determined the spiny vs. spineless pod character, the latter being recessive. The number of coils was also investigated in two accessions with tubercled pods. One had pods with more than three coils, the other had three coils or less. $F_2$s of the cross between them segregated in 170 plants with more than 3 coils to 10 plants with 3 coils or less. A two-gene involvement is plausible, the fewer-coil type being recessive. In addition, the same $F_2$ progeny segregated in 171 fully green plants to 9 light green, semidwarfed ones. Dwarfing of light greens was probably caused by deficient chlorophyll production; though their seed production was poor, several progenies were raised, all of which were light green and dwarfed. Obviously, for normal chlorophyll synthesis in the two parental accessions two different pathways (governed by two different genes) have evolved. $F_2$ plants which happened not to possess any of the two active factors failed in normal chlorophyll production.

Fryer (1930), Simon & Simon (1965), and Gillies (in Lesins & Gillies, 1972) found variations in chromosome morphology in *M. polymorpha*. Gillies' accessions, UAG Nos. 409 and 401, both with short-spined pods belong to Heyn's variety *brevispina*, but their chromosome morphology was shown to be distinctly different. It may be assumed that under spatial isolation, chromosome mutations and rearrangements by chance take place. On interbreeding of different-origin material some impairment in progeny viability may appear, such as noted above for chlorophyll-deficient segregants from the two normally green forms.

*Distinguishing M. polymorpha from Other Species.* We found that half of the specimens we received (as *M. hispida*) from different sources were incorrectly identified (Lesins & Lesins, 1962). The spiny forms of *M. polymorpha* had been confused with *M. laciniata, M. minima, M. praecox* and some other species. *M. polymorpha* differs from *M. laciniata* in that: in *M. laciniata* radial veins on pod face are sigmoid, slightly branching, not freely anastomosing, whereas in *M. polymorpha*, though curved, they are not sigmoid, and anastomose freely; in *M. laciniata* wing petals are shorter than the keel, whereas in *M. polymorpha* they are longer; in *M. laciniata* the seed radicle is long, 2/3 of the seed length (Fig. 5,-21), and seeds in the pod are not partitioned, whereas in *M. polymorpha* radicle is half or slightly more of the seed length (Fig. 5,-35), and seeds are partitioned by a thin wall. Finally, if a specimen has the laciniate-lyrate leaves characteristic of *M. laciniata* (Fig. 60,-a) then no further proof is needed for a decision. *M. polymorpha* may be distinguished from *M. minima* as follows: in *M. minima* the stipules are entire or slightly serrate and the upper side of the leaf is hairy, while in *M. polymorpha* stipules are incised, and the upper side of the leaf is glabrous or nearly so; in *M. minima* the lateral vein on the pod face is located 1/3-2/5 of the radius from the dorsal suture, radial veins are sigmoid and sparsely branched, whereas in *M. polymorpha* the lateral vein is closer to the dorsal suture, 1/6-1/5 of the radius, radial veins are not sigmoid, anastomosing freely; in *M. minima* the seeds are very small (Fig. 5,-26), not partitioned, whereas in *M. polymorpha* they are distinctly larger and partitioned. In general the pods of *M. minima* are also smaller than those of *M. polymorpha*. The spineless, tubercled and short-spined *M. polymorpha* may be confused with some varieties of *M. tornata*, especially because tubercled pods of *M. polymorpha* tend to become hard, particularly on the edges, resembling to some extent those of *M. tornata*. The most useful morphological character for distinguishing the two species is the groove on the coil face between the lateral vein and the dorsal suture. In *M. polymorpha* this groove is always discernible, though sometimes shallow (Fig. 64,-d), whereas in *M. tornata* no clearly discernible groove is present, though a slight depression below the dorsal suture sometimes may be found. Although at a first glance pods of *M. arabica* may be confused with those of *M. polymorpha*, the coil edge of *M. arabica* has 3 grooves and 4 ridges, a unique character in the genus.

If living material is available, there is no difficulty in distinguishing *M. polymorpha* from other species as its 2*n* number is 14, whereas in resembling species it is 2*n* = 16, with a single exception of *M. praecox*. There is no difficulty in discerning *M. praecox* (2*n* = 14) from *M. polymorpha*, however, as the pods of the former have a rather wide edge covering almost completely the grooves between it and the lateral veins.

*49.* *Medicago arabica* (L.) Hudson, Fl. Angl.: 288 (1762). Syn. *M. polymorpha* var. *arabica* L., Sp. Pl.:780 (1753). *M. maculata* Sibth., Fl. Oxon.: 232 (1794). Figs. 65; 3,-3; 5,-2.

Plants 40-65 cm long, profusely branching from the base, decumbent, sparsely covered with diffuse simple, or simple and multicellular glandular hairs. Stipules large, deeply toothed. Leaflets large, wide, 11-23 mm x 14-32 mm, obcordate to obovate, usually with an anthocyanin-colored patch in the middle; glabrous on upper side; margin in its apical half slightly toothed, midrib ending in a small tooth. Peduncle ± the length of the corresponding petiole, (1)2 to 5-flowered, with a terminal cusp. Florets 5-6.5 mm long. Pedicel shorter than the calyx tube; bract as long as the pedicel, often longer. Calyx 2.5-3 mm long, sparsely covered with appressed hairs; teeth ± the length of the tube. Corolla bright yellow; standard wide, oval; wings shorter than the keel. Young pod tightly coiled, protruding sideways through the calyx teeth (Fig. 3,-3). Mature pod cylindrical to spherical, flat at both ends, straw-colored to brownish, glabrous, spiny. Coils 3.5-6(7), turning clockwise, 4-6 mm in $\phi$. Coil face with 5-8 veins, forming elongated cells near and parallel to the edge, then entering the lateral vein. Lateral

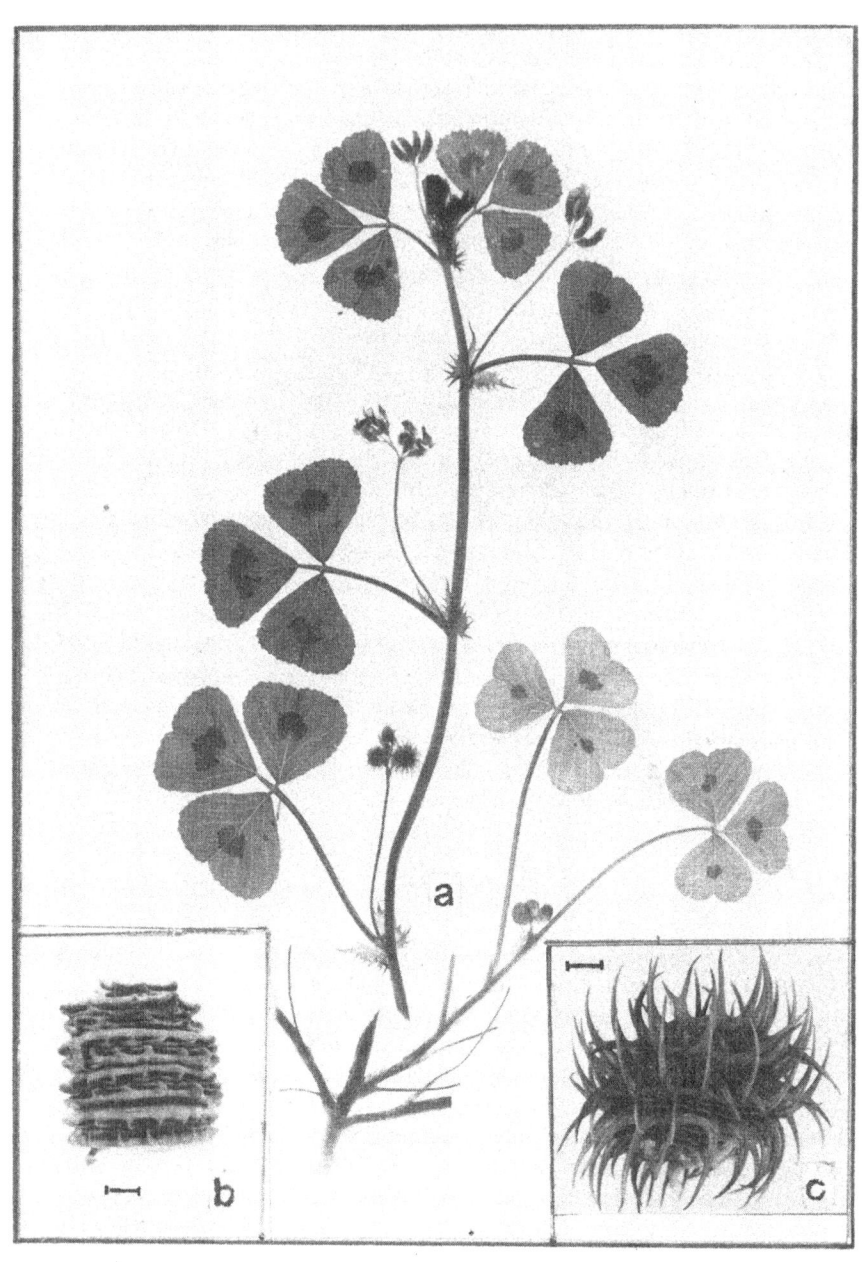

*Fig. 65,  M. arabica.* Parts of herb (a), and types of pods: short-spined (b), and long-spined (c); note three grooves on the coil edge.

veins prominent, higher than the dorsal suture which lies in a furrow with elevated sides; between these and the lateral veins are grooves bridged by roots of the spines; thus, the edge of the coil consists of 4 ridges and 3 grooves. Spines 13-15 in each row, 1.5-3.5 mm long, thin and soft, arching over the sides of the coils; the tubercled form (Fig. 65,-b) is rather rare. Seeds yellow to brownish, 2.5-3.5 mm x 1.2-1.5 mm, 1-3 in each coil, separated. Seed weight about 1.7 g/1000. Radicle longer than half of the seed, its tip protruding. $2n = 16$.

*Habitat and Distribution.* M. arabica prefers grassy, moist places (Casellas, 1962; Carbonell, 1962), unlike most other *Medicago* species.

It is distributed on both sides of the Mediterranean Sea from the Canary Islands to Asia Minor. Heyn (1963) has not found it in Israel, and indicates that it has never been found in Arabia. For that reason Willdenow (1802) has preferred the epithet *maculata* rather than *arabica*. We collected the species in Turkey, Greece, Italy, on Crete, Sicily, Capri, Corsica, Sardinia, in France (Nice, Cannes, Biarritz) and in Holland (temporally adventitious near wool factories). In North Africa it probably grows in the coastal region (Heyn, l.c.); inland it is rare, we saw it only once at Mt. Bargou, Tunisia.

*Variation Within Species.* Leaves with four and even five leaflets, along with the normal three leaflets, are often observed on the same plant. This seems characteristic of *M. arabica*. Urban (1873) mentions a variety *heptacycla* with seven pod coils. Heyn (l.c.) does not subdivide the species into formal subdivisions; however, she indicates a number of variations including leaflets without an anthocyanin patch (rare), and pods without spines (not seen by us). The anthocyanin patch appears to be influenced by light conditions and the age of the plant. Thus on one occasion a supposedly patchless type developed clear patches under our greenhouse conditions. Recently, Latschaschvili (1969) described as *M. talyschensis* a form with rounded pods, and short appressed spines. In our collection we have an accession (No. 2095) from Calabria, Italy, corresponding to the above description. The four ridges and three grooves on the coil edge leave no doubt that it belongs to *M. arabica*.

50. *Medicago lanigera* Winkler et Fedtschenko ex Fedtschenko, Bull. Jard. Bot. Petersbg. 5:41 (1905). Figs. 66; 5,-22.

Small plants, 5-25 cm long, branches arising from the base, decumbent. Vegetative parts densely lanate. Stipules usually with 2-4 large teeth, ending in a long terminal tooth. Leaflets 8-15 mm x 6-9 mm, obovate; margin in its 1/3-2/3 apical part coarsely serrate, midrib ending in a wide tooth. Peduncle 1(2?)-flowered, shorter than or equal to the corresponding petiole, with a cusp. Florets small, up to 4 mm long. Pedicel longer than the calyx tube; bract much shorter than the pedicel. Calyx longer than half of the floret, densely covered with long hairs; teeth equal to or slightly longer than the tube. Corolla yellow; standard oval; wings slightly shorter than the keel. Young pod protruding sideways from the calyx. Mature pod densely covered with long (5-6 mm), thin, many-celled hairs, giving the pod the appear-

*Fig. 66*, *M. lanigera*. Branch (a), and pod (b).

ance of a small ball of cotton, spineless. Coils 3-4, soft-walled, approximately 6 mm in $\phi$, turning clockwise; on coil face (after removing hairs) are seen 8-10 S-shaped veins which at 2/3 from the centre unite in a thicker lateral vein, from which they reradiate to the dorsal suture. Seeds yellow, 3.5-4 mm x 2 mm, 1-2 in each coil, not separated. Seed weight about 3.3 g/1000. Radicle up to 2/3 of the seed length; at outset it bends inwards following the concavity formed by cotyledons (Fig. 5,-22), subsequently the tip protrudes slightly outwards. $2n = 16$ (Lesins and Lesins, 1961).

*Habitat.* Rocky slopes, at an altitude of 600-900 m.

*Distribution.* Endemic to Central Asiatic region of the USSR.

*Distinction from Other Species.* The species cannot be misidentifed because of the unique cover of cottony hairs on the pods, Fedtschenko (l.c.) put it in a separate section *Lanigerae*. We follow Heyn's (1963) classification and consider it as belonging to the section *Leptospirae*. The chromosomes were found to be 2.8-3.6 $\mu$ long, thus fairly uniform, in which they differ from those of most other annual *Medicago* species (Lesins & Lesins, l.c.).

*51.* **Medicago disciformis** de Candolle, Cat. Hort. Monsp.:124 (1813). Figs. 67; 5,-12.

Plants 15-70 cm long, few branches arising from lower part of the main stem, decumbent. Vegetative parts covered with simple and glandular hairs, the younger parts appearing more glandular-hairy. Stipules entire, or with 2-4 small teeth toward the base. Leaflets small, 5-14 mm x 4-10 mm, obovate (lower ones may be obcordate), occasionally truncate; margin serrate in the apical 1/3; midrib ending in a small triangular tooth. Peduncle 1 to 3-flowered, distinctly longer than the petiole, with a terminal cusp. Florets 5-7 mm long. Pedicel shorter than, rarely equal to, or longer than the calyx tube; bract shorter than the pedicel. Calyx 2.5-3.5 mm long, covered with glandular and simple hairs (sometimes with glandular hairs only); teeth ± the length of the tube. Corolla yellow; standard broadly obovate to obcordate; wings shorter than the keel. Young pod glabrate, or glandular-hairy, emerges in a loose spiral from the calyx, then bends sideways. Mature pod cylindrical, disk-shaped, straw-colored to brown, spiny, glabrous to sparsely covered with glandular hairs. Coils 5-8, loose, 4.5-7 mm in $\phi$, turning clockwise, apical coil spineless. On face of the coil 10-16 S-shaped veins, entering a veinless zone which about 1/3 the width of the coil radius. Spines 0.5-6 mm long, 14-16 in each row, inclined to the basal part of the pod (Fig. 67,-c). Between lateral veins and the dorsal suture a deep, narrow groove bridged by the roots of the spines. Seeds light yellow, 2.5-3.5 mm x 1.5-1.8 mm, 1-2 in each coil, not separated. Seed weight about 3.6 g/1000. Radicle longer than half of the seed. $2n = 16$.

*Habitat.* In dry soils, on rocky hillsides.

*Distribution.* From Turkey to Spain along the northern coast and on isles of the Mediterranean, more abundantly represented in the eastern part. We collected it on Cyprus, the Aegean islands, in Greece and Italy. The short spined form (Fig. 67,-b) is rare, distributed in the eastern part of the species area. We collected it at Kytherea on Cyprus.

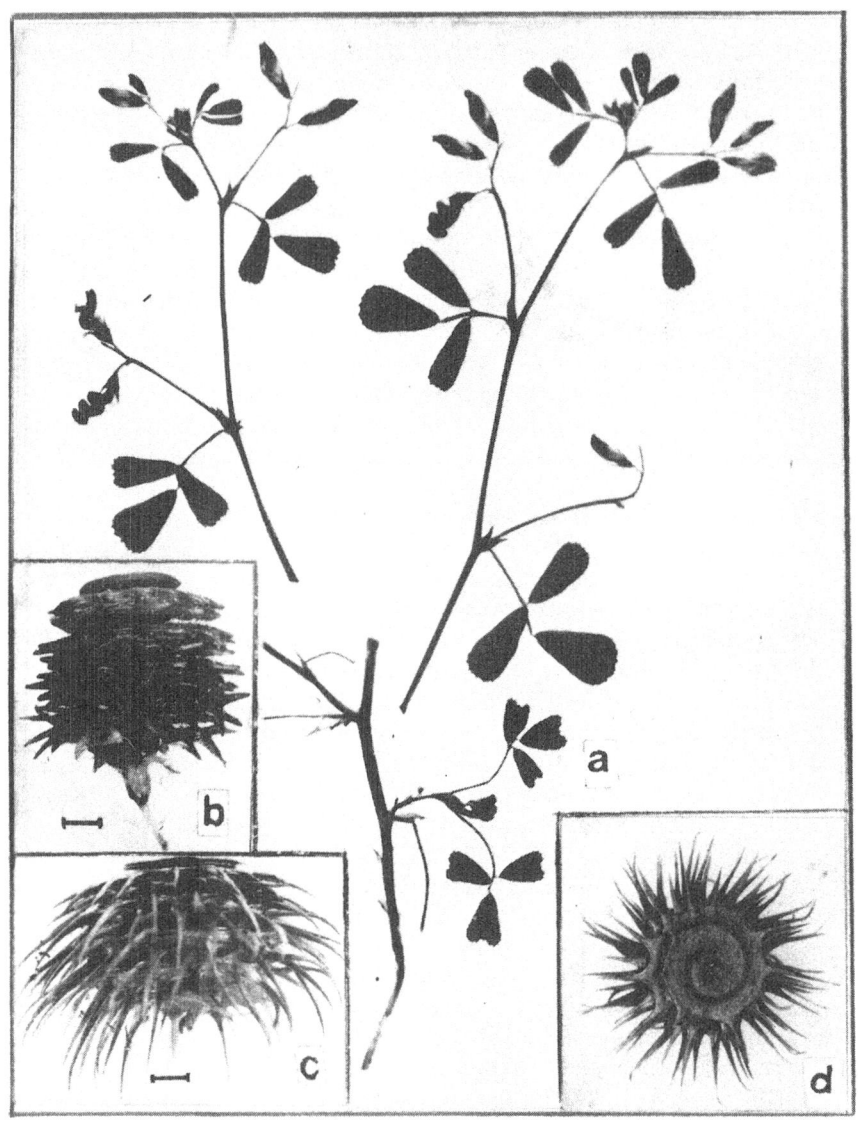

*Fig. 67, M. disciformis.* Parts of herb (a), and types of pods: short-spined
(b), long-spined (c); in side view note spines directed to the basal part of the
pod; and in apical view (d) note spineless apical coil.

204

*Relationship to Other Species.* Heyn (1963) pointed to the resemblance between *M. disciformis* and *M. tenoreana.* On looking over the two species, one indeed notes a number of similar features, the outstanding morphological character in common being the veinless zone in the outer part of the coil face. Differential characters between the two species are discussed under *M. tenoreana.*

Our attempts to hybridize the two species did not succeed. A number of seeds produced all turned out to be of self-origin. It was of interest to note that under identical greenhouse conditions the leaves of *M. disciformis* were of a distinctly less intense green color than those of *M. tenoreana.* The character could be used to identify selfs in early plant development where *M. tenoreana* was used as the maternal parent. These observations, however, involved only two accessions from each species.

52. *Medicago tenoreana* Seringe in DC., Prodr. Syst. Nat. 2:180 (1825). Syn. *M. cancellata* Ten., Fl. Nap.:44 (1811) non M.B. (1808). Figs. 68; 6,-57.

Plants 20-35(50) cm long, decumbent, branching from the base. Vegetative parts covered with diffuse, simple hairs, upper sides of leaves less densely haired. Stipules usually shallow-toothed with acuminate teeth. Leaflets 8-14 mm x 7-10 mm, obovate; 1/3 of apical margin finely serrate; midrib ending in a slightly larger tooth. Peduncle 1 to 4-flowered, longer than the corresponding petiole, with a distinct terminal cusp. Florets 4-7 mm long. Pedicel shorter than the calyx tube; bract slender, equal in length to the pedicel. Calyx hairy, more than half the length of the floret; teeth slender, longer than the tube. Corolla yellow; standard roundish; wings slightly shorter than the keel. Young pod covered with simple, or simple and glandular hairs (glands often purple), emerging straight from the calyx tube then protruding sideways through the teeth. Mature pod cylindrical, glabrate, brownish, spiny. Coils 4-6, loose, 5-6.5 mm in $\phi$, turning clockwise, coil edge wide (0.5 mm) without grooves. On coil face about 10 S-shaped veins, branching slightly before entering the veinless zone, which 1/3 the width of the coil radius; between it and the coil's edge a groove traversed by the roots of spines. Spines 2-3 mm long, grooved, some hooked, inserted at 90°-130° to the face of the coil, somewhat slanted opposite to the direction of coiling. Some accessions have a spineless apical coil. Seeds light yellow to brownish, 2.5-3 mm x 1-1.3 mm, 1-2 in each coil, not separated. Seed weight about 1.6 g/1000. Radicle longer than half of the seed. $2n = 16$.

*Habitat.* We found *M. tenoreana* growing on dry, rocky hill slopes.

*Distribution.* The species is growing on northwestern Mediterranean coasts and islands. We collected it on Sicily (San-Martino, Palermo), on the western coast of the Adriatic Sea (Gargano), and in southern France (Nice, Mt. Baron). In no site did we find it in abundance. Though Carbonell (1962) and Casellas (1962) assume that the species is well represented in N. Africa, Nègre (1959) does not list it there, nor did we find it in Tunisia, Algeria and Morocco.

*Distinguishing M. tenoreana from Other Taxa.* Pods of *M. tenoreana* have

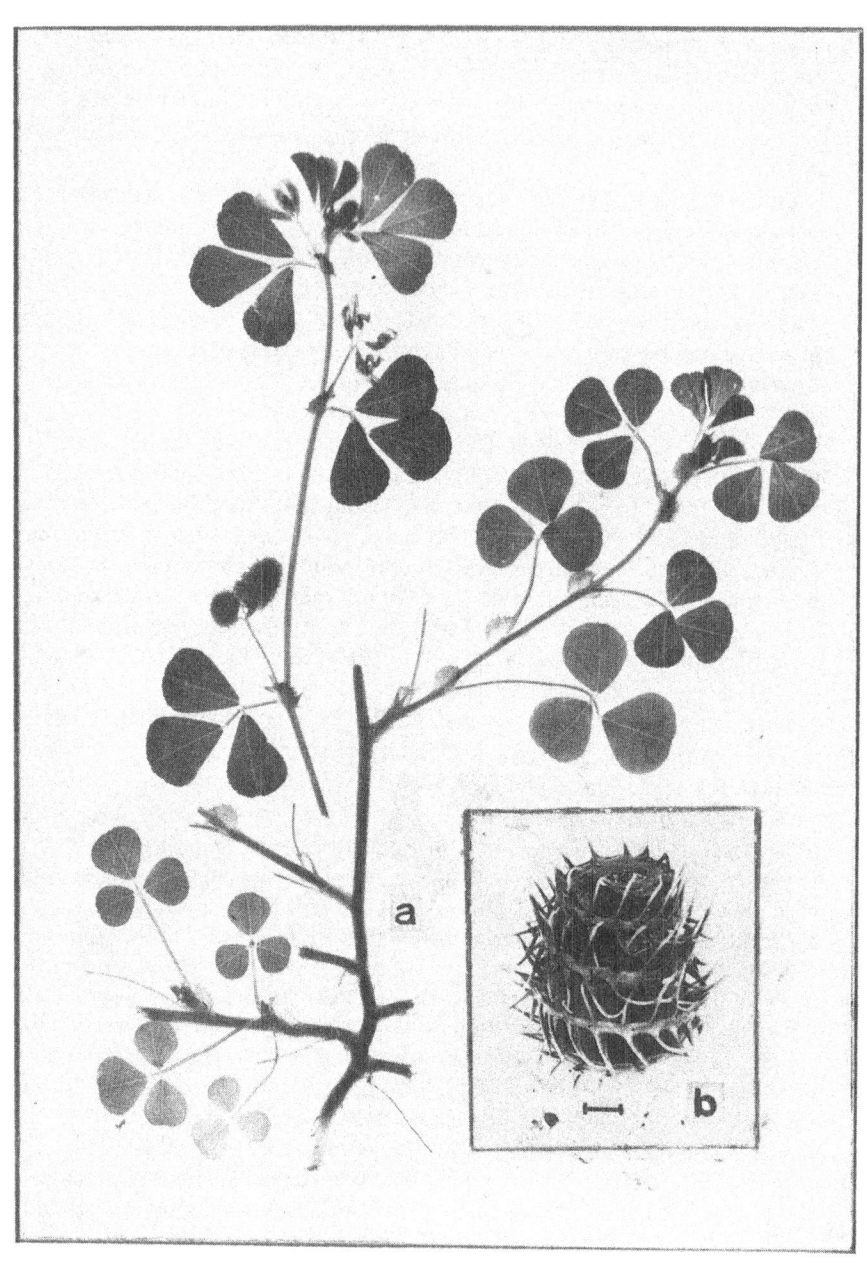

*Fig. 68, M. tenoreana.* Parts of herb (a), and pod (b).

206

similarities in the first place with those of *M. disciformis*. However, the coil edge in *M. disciformis* has grooves formed by the outer roots of the spines and the dorsal suture, whereas in *M. tenoreana* the edge is smooth (outer roots of the spines are well down on the face of the coil). The spines of *M. disciformis* are inclined to the basal end of the pod, whereas in *M. tenoreana* the spines point to both ends; hence in *M. disciformis* the apical spineless coil is very clearly displayed (since spines of the other coils point away from the apex), whereas in *M. tenoreana* the apical coil when spineless is inconspicuous, being partly covered by spines from the adjacent coil. Still, it is probable that the two taxa are related: both have veinless zones on the pod face, and both have radicles distinctly longer than half (about 2/3) of the seed, their tips protruding off the main body (Figs. 5,-12; 6,-57). Our attempts to cross the two species, however, failed (see also *M. disciformis*). There is some resemblance in general pod appearance of *M. tenoreana* to *M. polymorpha*, *M. minima* and *M. laciniata*. None of these three, however, have the veinless zone on coil face; moreover, *M. polymorpha* has a $2n = 14$ chromosome complement.

SECTION INTERTEXTAE Urban, Verh. bot. Ver. Brand. 15:48 (1873).

Pods with soft walls. Coils not tightly appressed, turning clockwise, spiny; coil face without lateral vein or veinless zone; facial veins well discernible, entering directly into the base of spines. Radicle half the length of the seed or shorter. Seeds black or red-brown. $2n = 16$, (32?).

Key to species of section *Intertextae*:

| | | |
|---|---|---|
| 1 | Pods, including spines, covered profusely with glandular, many-celled hairs | *M. ciliaris* |
| — | Pods glabrous, or with simple hairs, or with simple and sparse glandular hairs | 2 |
| 2 | Pods with (6)8-11 coils. Florets large, 8.5-10 mm long. Seeds large, 13-17 g/1000 | *M. intertexta* |
| — | Pods with (3)5-7 coils. Florets smaller, 5-7.5 mm long. Seeds smaller, 7—10 g/1000 | 3 |
| 3 | Pods short-cylindrical, disc-shaped; seeds black | *M. muricoleptis* |
| — | Pods barrel-shaped, roundish; seeds black or reddish-brown | *M. granadensis* |

The species of this section are morphologically well delineated from all others of the genus, especially in pod appearance. It was also found that its pollen, cube-shaped when expanded (Fig. 8,-8), was different from that of any other *Medicago* (Lesins & Lesins, 1963). We further observed that the *Intertextae* were extremely susceptible to alfalfa mosaic virus, so that under greenhouse conditions plants often died from this disease. It may be added that members of this section prefer heavy, moist soils, and exceed all other annual *Medicago* in vigour of growth.

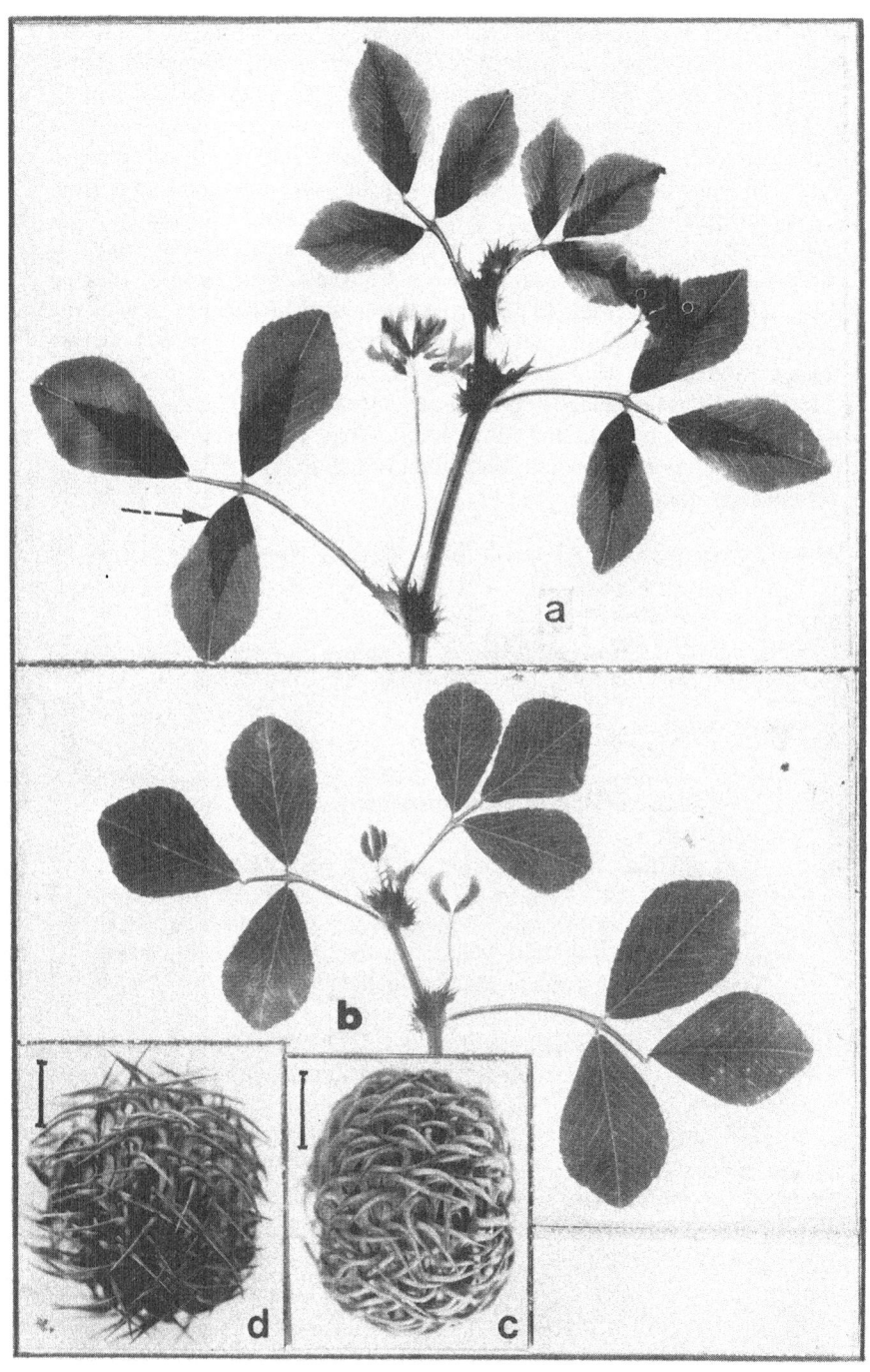

*Fig. 69, M. intertexta.* Branches: with leaflets having basal anthocyanin fleck (arrow at a), and without fleck (b). Types of pods: with interlocked appressed spines (c), with less appressed spines (d).

53. *Medicago intertexta* (L.) Miller, Gard. Dict. ed. 8: *Medicago* no. 4 (1768).
Syn. *M. polymorpha* var. *intertexta* L., Sp. Pl.: 780 (1753). Figs. 69; 5,-20.

Plants 35-100 cm long, branches quadrangular, decumbent to ascending, arising from the base; glabrescent or sparsely covered with simple, or with simple and glandular hairs. Stipules incised to 1/3 of their length, forming 13-18 slender teeth. Leaflets large, 14-29 mm x 10-20 mm, broadly to narrowly obovate; upper sides usually glabrous, sometimes with a red blotch in their basal part; margin in its 1/2-3/4 apical part serrate, with small, not closely spaced teeth; midrib ending in a small terminal tooth. Peduncle 3 to 11-flowered, ± the length of the corresponding petiole, with or without a terminal cusp. Florets 8.5-10 mm long. Pedicel as long as the calyx tube; bract equal to or shorter than the pedicel. Calyx 3-5 mm long; teeth ± the length of the tube. Corolla yellow to orange yellow; standard broadly ovate; wings longer than the keel. Young pod rises in a loose spiral from the calyx, later bends sideways. Mature pod greyish to dark brown, roundish to elongate, glabrous, spiny. Coils (6)8-11(12), loose, 10-13 mm in $\phi$, turning clockwise. Veins 6-9, forming a coarse net in the outer half of the coil face, from which one vein enters one root of each spine. Spines 16-19 in a row, 2.5-6 mm long, grooved to ± half their length, inserted at 90°-120° to the facial plane, arching and usually interlocking with spines of adjacent coils. Seeds brownish-black to black, 4.5-6 mm x 3-3.5 mm, 1-2 in each coil, separated. Seed weight 13.9-17.4 g/1000. Radicle shorter than half the length of the seed. $2n = 16$, (32?).

*Habitat and Distribution.* Growing mainly in heavy, moist soils. Distributed from the Canary Islands over Portugal, Spain to Italy and its islands. We collected it also in Algeria and Morocco. A west-Mediterranean species.

*Variation Within Species.* Urban (1873) divided the species into long-spined (*aculeata*) and short-spined (*tuberculata*) groups. Under long-spined he listed: 1) var. *decandollei* Urb., pods lentil-shaped, peduncle 2 to 3-flowered, coils 6-8, 2) var. *panormitana* Urb., pods spheroid, peduncle 6 to 10-flowered, coils 7-9, and 3) var. *echinus* Urb., pods round to oval, penduncle 7 to 10-flowered, coils 7-9. In this latter variety he discerned f. *variegata* with a purple blotch in the leaves, and f. *pilifera* with glandular hairs on the pedicels. Heyn (1963) discerned: 1) var. *intertexta* with large leaflets, spherical or ovoid pods, 2) var. *decandollei* Urb. with small leaflets and discoid pods, and 3) var. *ciliaris* Heyn with multi-cellular hairs on pods. Urban's varieties *echinus* and *panormitana* and Heyn's var. *intertexta*, we consider here as *M. intertexta*. Urban's var. *decandollei* we include in *M. muricoleptis*, and Heyn's var. *ciliaris* as *M. ciliaris*.

We could not confirm the $2n = 32$ chromosome number (Fernandes & Santos de Fatima, 1961) on material supposedly collected from the site where such plants were growing.

*Relationship to Other Species.* We carried out some crossing experiments (Lesins et al., 1971) and found that *M. intertexta* could be hybridized without difficulties with: 1) *M. ciliaris* and 2) *M. muricoleptis*. In 1) the characters investigated were: red patch vs. no patch in basal part of leaflets,

209

and hairy vs. glabrous pods. The characters in each pair segregated in the $F_2$ in a monogenic pattern (3:1). In 2) we investigated red patch vs. no patch, and seed weight 15 g vs. seed weight 8 g. The characters segregated normally with one and two genes involved, respectively. In addition to morphological differences of parents, in $F_1$ plants of both crosses, pollen fertility was between 40 and 50%; hence we consider the three taxa as separate species. Crossing *M. intertexta* with *M. granadensis* did not yield any offspring; consequently, we consider the latter somewhat more distantly related to *M. intertexta*, than *M. ciliaris* and *M. muricoleptis* (see these).

54. *Medicago ciliaris* (L.) Krocker, Fl. Siles. 2,2:244 (1790). Syn. *M. polymorpha* var. *ciliaris* L., Sp. Pl.: 780(1753). Figs. 70; 3,-2; 5,-7.

Plants 35-55 cm long, branches prostrate to ascending, arising from near the base, covered sparsely with simple hairs. Stipules finely incised to 1/4-1/2 of their length. Leaflets 14-25 mm x 9-22 mm, obovate to elliptical; apical 1/2 to 2/3 of the margin serrate; midrib ending in a small, thin tooth. Peduncle 1 to 4-flowered, equal to or longer than the corresponding petiole,

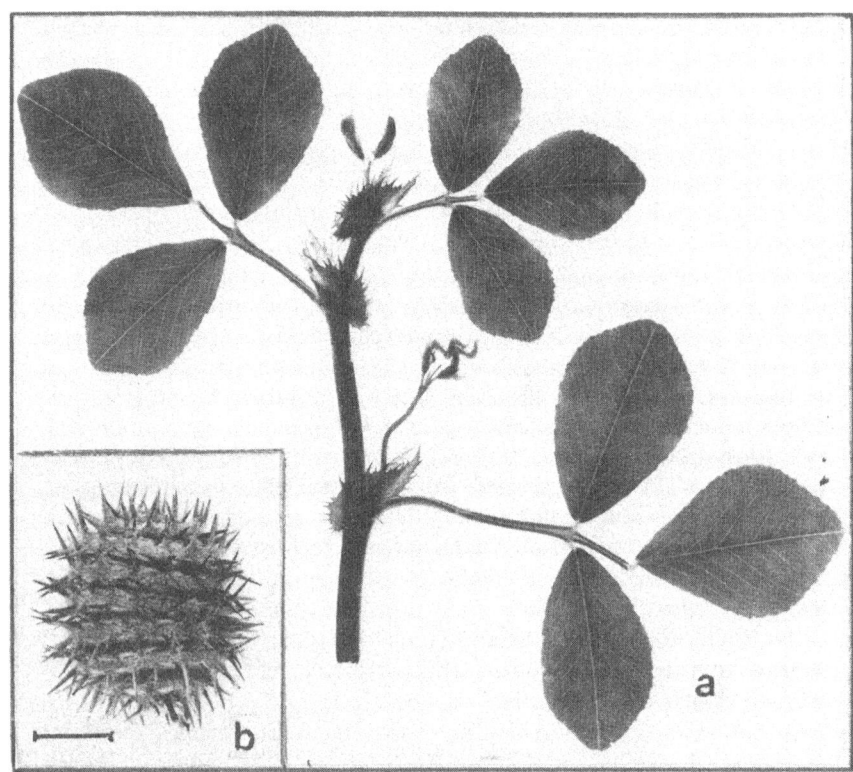

*Fig. 70, M. ciliaris.* Branch (a), and pod (b).

with a distinct terminal cusp. Florets 7.5-8 mm long. Pedicel equal to or longer than the calyx tube; bract shorter than the pedicel. Calyx 3.5-4 mm long, sparsely covered with simple hairs; teeth equal to or shorter than the tube. Corolla bright yellow; standard broadly ovate; wings longer than the keel. Young pod emerges from the calyx in a loose spiral, then bends sideways; covered densely with multicellular, glandular hairs, glands often purple. Mature pod spherical to oval, yellowish-grey to light brown, covered with multicellular hairs, also on spines. Coils 9-10, not firmly appressed, turning clockwise, 9-11 mm in $\phi$. Veins 6-8, anastomosing in the outer half of the coil face, one vein entering one root of each spine. Spines usually straight, 15-18 in each row, inserted at about 135° to the face of the coil (90° on end-coils), 2-3 mm long, grooved to about half of their length; spines of two adjacent coils crossing. Seeds brownish-black to black, 4.8 mm x 3 mm, 1-2 in each coil, separated. Seed weight about 13 g/1000. Radicle less than half the length of the seed. $2n = 16$.

*Habitat and Distribution.* Growing in heavy, loamy soils. Distributed from the Canary Islands to Israel on both sides of the Mediterranean Sea. We collected it in Lebanon, Spain, N. Africa and on the Mediterranean islands of Cyprus, Crete, Sicily Malta and Sardinia.

*Relationship to Other Taxa.* Urban (1873) mentions, as one of the characters distinguishing *M. ciliaris* from *M. intertexta*, the former's dorsal suture being wider than the width of the zone spanned by the roots of the spines.

*M. ciliaris* could be intercrossed readily with *M. intertexta* and also with *M. muricoleptis* (Lesins et al., 1971). In this latter cross, pollen fertility in $F_1$s was about 35%, and in 30% of meiotic metaphase cells some irregularities were found. The segregation of pod hairs in $F_2$s, however, followed a monogenic, 3:1, ratio, hairiness being dominant. We retained the taxon as a separate species, though Heyn (1963) considered it a variety of *M. intertexta*, noting that pod hairiness was too variable a character for delineating species boundaries. Agreeing with Heyn that generally pod hairiness is a variable character, we find that the occurence of such multicellular, glandular hairs on pods and on spines, to the extent as in *M. ciliaris*, is not common in other annual *Medicago*. Ponert (1973) ranked *M. ciliaris* as a subspecies of *M. intertexta*. Our attempts to hybridize *M. ciliaris* with *M. granadensis* were unsuccessful (unpubl.).

55.   *Medicago muricoleptis* Tineo, Pl. Sic. Rar. Pug. 1:18 (1817). Syn. *M. decandollii* Tin. ex Guss., Fl. Sic. Syn. 2:369 (1844). Figs. 71; 5,-28.

Plants 15-70 cm long, branches decumbent to ascending, arising from near the base. Vegetative parts glabrate. Stipules incised, having 8-11 slender teeth. Leaflets 10-17 mm x 7-15 mm, oblong, obovate; apical 2/3 of leaflet margin serrate; midrib ending in a small tooth. Peduncle 2 to 6-flowered, longer or equal in length to the corresponding petiole, with or without a small terminal cusp. Florets 5-7.5 mm long. Pedicel equal to or shorter than the calyx tube; bract ± the length of the pedicel. Calyx 3-4 mm long, glabrous or with appressed hairs; teeth ± the length of the tube. Corolla bright yellow; standard broadly ovate; wings about the length of the keel. Young pod glabrous (very rarely with glandular hairs), rises up in the calyx, then bends sideways. Mature pod greyish to dark brown, discoid, or short-cylindrical with smaller end-coils, spiny. Coils 3-6.5, loose, 8-12 mm in $\phi$, turning clockwise. On coil face 7-10 slightly curved veins, anastomosing soon after emergence from the ventral suture, one vein entering one root of each spine. Spines slender, 16-24 in a row, 1-2 mm long, inserted at 90°-180° to the face of the coil, often pointing in different directions. Seeds

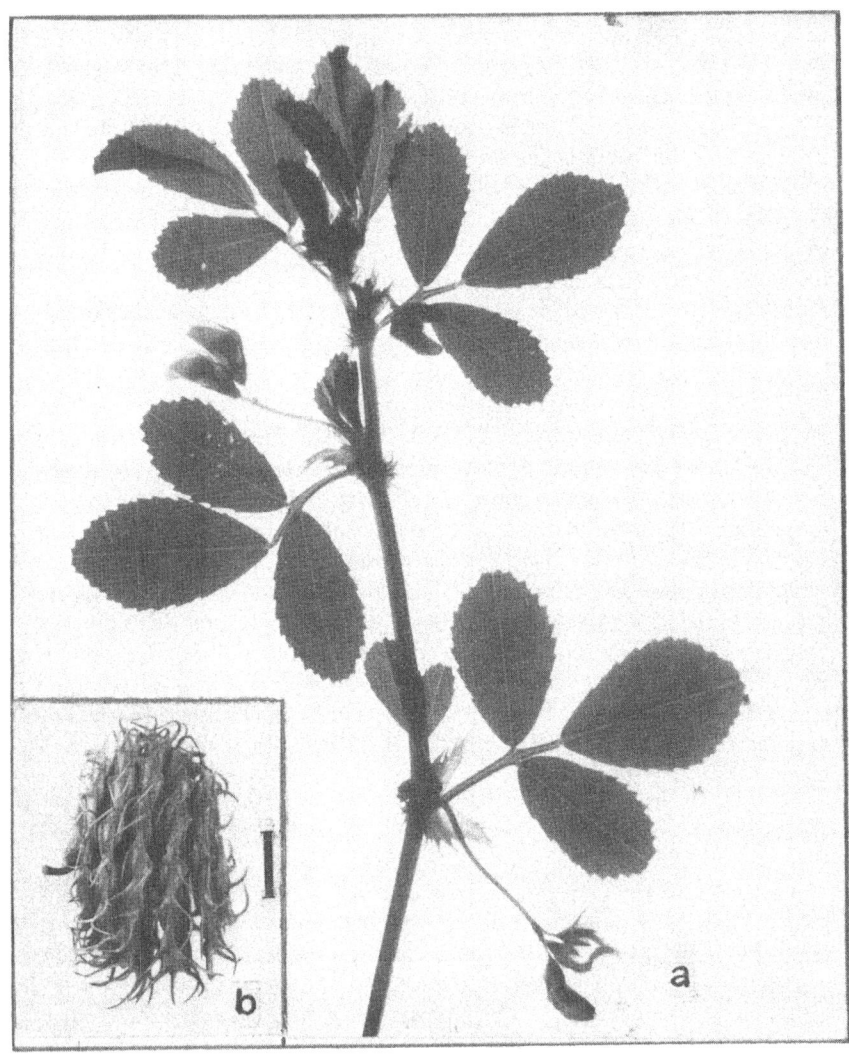

*Fig. 71,  M. muricoleptis.* Branch (a), and pod (b).

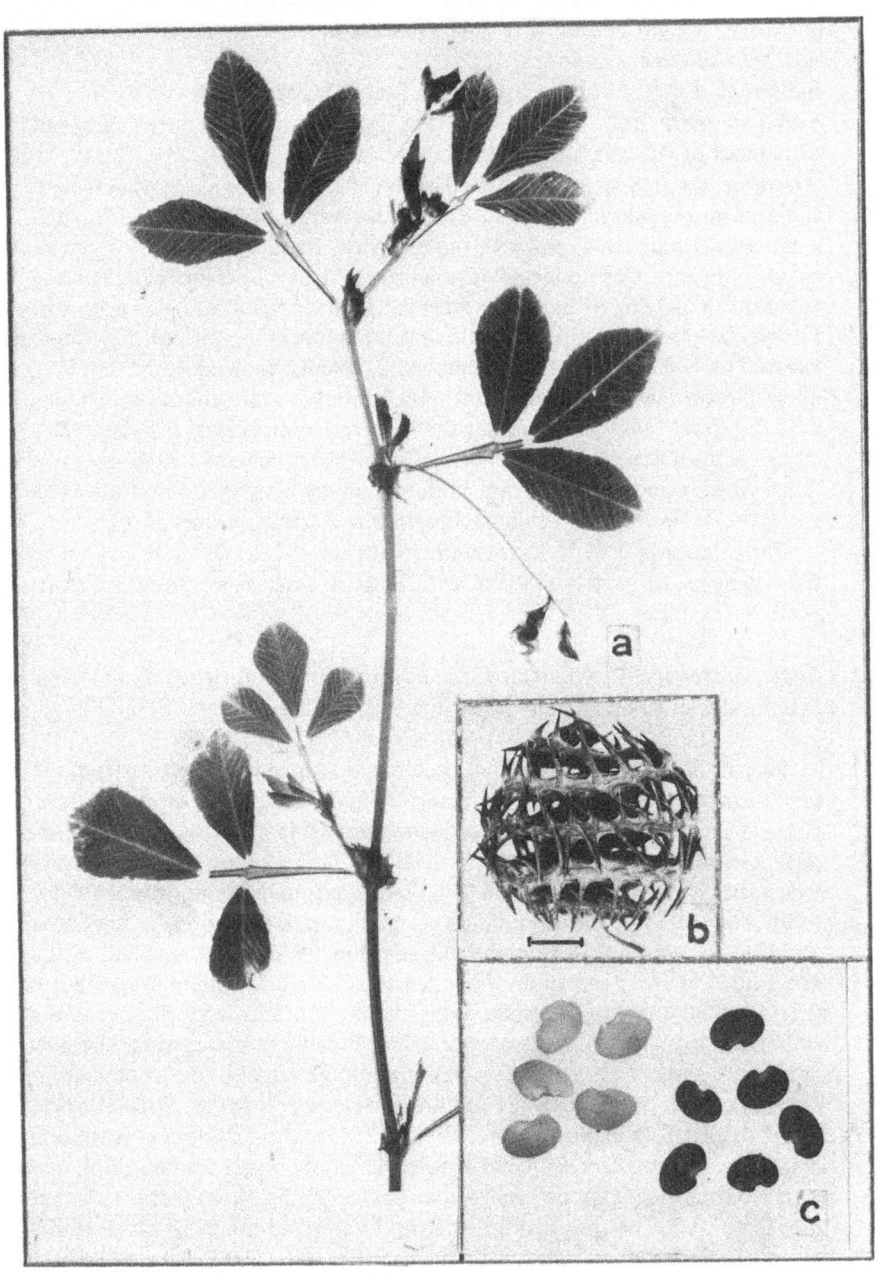

*Fig. 72*, *M. granadensis*. Branch (a), pod (b), brown and black seeds (c).

black, 4-5 mm x 2-2.5 mm, 1-3 in each coil, not separated, or by a thin wall only. Seed weight about 7.5 g/1000. Radicle half the length of the seed or less. $2n = 16$ (Lesins & Lesins, 1965).

*Habitat and Distribution.* Growing in heavy, moist soils. Distributed in Sicily and southern Italy. We collected it at Collesano near Palermo, Sicily, and 17 km east of Potenza in southern Italy.

*Variation Within Species.* Neither Urban (1873) nor Heyn (1963) notes any variation in the taxon. Heyn and Urban, however, list under *M. intertexta* as a variety *decandollei*, types with discoid pods; we include them in *M. muricoleptis*. From seeds collected in a single spot at Collesano near Palermo (accession UAG No. 768), plants with the following pod forms were grown: 1) pods barrel-shaped, up to 11 mm in $\phi$, up to 6.5 coils, with prickly spines inserted at 135° to the facial plane, and 2) discoid pods, up to 13 mm in $\phi$, up to 5.5 coils, with thinner, softer spines, inserted more obliquely, bending over the facial plane; we consider the latter variation typical for *M. muricoleptis*. A third variation (accession UAG No. 2149) collected at Potenza had much smaller pods, up to 8 mm in $\phi$, resembling more in shape those of *M. granadensis*. However, hybridization with the latter was unsuccessful.

The relationship of *M. muricoleptis* with other *Intertextae* shows that a free exchange of genetic material with both *M. intertexta* and *M. ciliaris* is possible (Lesins et al., 1971).

56. *Medicago granadensis* Willdenow, Enum. Hort. Reg. Bot. Berol.:803 (1809). Syn. *M. galilaea* Boiss., Diagn. Pl. Nov. 1, 9:10 (1849). Figs. 72; 5,-18; 7,-2; 8,-2.

Plants 35-70 cm long, branches decumbent to ascending, arising from the lower part of the main stem, quadrangular in cross section. Vegetative parts glabrate or covered with glandular hairs, upper side of leaves glabrous. Stipules incised to 1/3 of their length. Leaflets 15-23 mm x 7-14 mm, obovate to elliptical; margin serrate in its 3/4 apical part; midrib ending in a narrow tooth. Peduncle 2 to 6-flowered, longer than the corresponding petiole, with a distinct terminal cusp. Florets 5-6.5 mm long. Pedicel ± the length of the calyx tube; bract ± the length of the pedicel. Calyx 2.5-3.5 mm long; teeth ± the length of the tube. Corolla orange yellow (in bud stage the tip of the standard often with a violet tinge); standard broadly obovate; wings equal to or slightly shorter than the keel. Young pod emerges in loose spiral from the calyx, then bends sideways. Mature pod straw-colored to greyish-brown, barrel-shaped to roundish, spiny. Coils 5-7, loose, 8-10 mm in $\phi$, turning clockwise. On coil face 8-14 radial veins, anastomosing in its outer 1/3, one vein entering one root of each spine. Spines 18-20, short (approx. 2 mm long), grooved ± halfway toward the tip, inserted to the facial plane at 90° or less (bending downwards over the facial plane). Seeds black or brown, 3.3-4 mm x 2-2.5 mm, 1-3 in each coil, separated by thin, spongy tissue. Seed weight about 9.4 g/1000. Radicle less than half the length of the seed. $2n = 16$.

*Habitat and Distribution.* We collevted the species in red. heavy soils in fallow fields. Heyn (1963) considers *M. granadensis* an East-Mediterranean species. We agree, as we found it only in southern Turkey (Maras, Gaziantep, Mersin).

*Variation Within Species.* Urban (1873) considered *M. granadensis* and *M. galilaea* as two separate species giving as distinguishing features for *granadensis*: pods with simple hairs, grooves in spines halfway toward the tip, and spines appressed to the coils (90° angle). For *galilaea* he noted: pods sparsely glandular hairy, grooves in spines almost to the tip, and spines not appressed to the coils (angle more than 90°). Heyn (l.c.), studying many specimens, found that the characters assigned by Urban to the two taxa occur in different combinations, hence it was impossible to draw a dividing line. Accordingly, she lists *M. galilaea* as a synonym of *M. granadensis*. We found in hybridization experiments that the accessions bearing the two names behaved identically (Lesins et al., 1971). This then, and Heyn's observations, was why we consider them to be a single taxon. It may be noted that in *M. granadensis* the black and brown seed color (UAG Nos. 866 and 867, respectively) in hybridization tests segregated in a 3:1 ratio, black being dominant over brown (Lesins et al., l.c.).

*Relationship to Other Species.* The similarity in characters of pods (spines, veins), seeds, length of radicle, and shape of pollen, leaves no doubt that *M. granadensis* is closely related to the other three species of the section *Intertextae*. However, it was found that there was a very strong interfertility barrier between it and the *M. intertexta* — *M. ciliaris* (Lesins et al., l.c.). Recently we found that *M. granadensis* did not hybridize with the third species, *M. muricoleptis* (unpubl.), either.

# REFERENCES

In addition to citations provided under species authorship, the pagination is given here for the entire genus if the original publication (or a Xerox copy of it) was available. Otherwise, the particular species or incomplete set of *Medicago* [sp.] is put as the title of the reference. Translations of titles are enclosed in brackets, and so are explanatory remarks. If the main text has been published in Cyrillic alphabets, this is noted in parentheses.

Åkerberg, E. & Lesins, K. 1949. Insects pollinating alfalfa in central Sweden. *Annal. Royal Agr. Coll. Sweden* 16:630-643.

Allioni, C. 1785. Medicago. *Fl. Pedem.* 1:314-316.

Arcangeli, G. 1876. Sopra una nuova specie del genere Medicago. *Nuovo Giorn. Botan. Ital.* 8:5-8.

Armstrong, J.M. & Gibson, D.R. 1941. Inheritance of certain characters in the hybrid of Medicago media and M. glutinosa. *Sci. Agr.* 22:1-10.

Armstrong, J.M. & White, W.J. 1935. Factors influencing seed-setting in alfalfa. *J. Agr. Sci.* 25:161-179.

Atwood, S.S. & Grun, P. 1951. Cytogenetics of alfalfa. *Bibliogr. Genet.* 14:133-188.

Balabaev, G.A. 1934. [Yellow lucernes of Siberia]. *Bull. Appl. Botan. Genet. Plant Breed.*, Ser. 7(1):113-123 (Russ.).

Balbis, G.B. 1801. Medicago glomerata. *Elenco Piante Torino* :93.

Barnes, D.K. 1972. A system for visually classifying Alfalfa flower colour. *U.S. Dept. Agr., Agr. Handbook.* No. 424.

Bartalini, B. 1776. Medicago [sp.]. *Catal. Piante Siena* :60-61.

Bastard, T. 1814. Medicago striata. In Desvaux, *Jour. Bot. Appl.* 3:19.

Bauhin, C. 1623. II. Trifolium fyl. luteum... *Pinax Theatri. Bot.* :330.

Baum, B.R. 1968. A clarification of the generic limits of Trigonella and Medicago. *Can. J. Bot.* 46:741-749.

Bentham, G. 1826. Medicago. *Catal. Pl. Indig. Pyrenées* :98-105.

Bertoloni, A. 1819. Medicago sphaerocarpos. *Amoen. Ital.* :91-92.

Besser, W.S.J.G. 1809. Medicago procumbens. *Prim. Fl. Gallic.* 2:127.

Boissier, P.E. 1843. Medicago rotata. *Diagn. Pl. Orient.*, Ser. 1, 2:23-24.

Boissier, P.E. 1849. Medicago galilaea. *Diagn. Pl. Orient.* Nov., Ser. 1, 9:10-11.

Boissier, P.E. 1856a. Medicago noëana. *Diagn. Pl. Orient.*, Ser. 2, 2:10.

Boissier, P.E. 1856b. Medicago blancheana. *Diagn. Pl. Orient.*, Ser. 2, 5:75-76.

Boissier, P.E. 1872. Medicago. *Fl. Orient.* 2:91-105.

Bolton, J.L., Goplen, B.P. & Baenziger, H. 1972. World distribution and historical developments. In Hanson (ed.), *Alfalfa Science and Technology* :1-34. Amer. Soc. Agr., Madison, Wis.

Bordzilowsky, E.I. 1909. [On some Caucasian plants]. *Prot. zased. Kievsk. Obschch. Estestv.* 1907-8:24-26 (Russ.).

Bordzilowsky, E.I. 1915. [To the flora of Caucasus]. *Zapiski Kievsk. Obschch. Estestv.* 25:86-87 (Russ.).

Borges, O.L., Stanford, E.H. & Webster, R.K. 1976. Sources and inheritance of resistance to Stemphylium leafspot of alfalfa. *Crop Sci.* 16:458-461.

Britton, N.L. & Brown, A. 1913. Medicago. *Illustr. Fl. U.S.A., Can., Brit. Possess.* 2:350-352. Scripner's, New York.

Burkill, I.H. 1894. On the fertilization of some species of Medicago L. in England. *Proc. Cambr. Philos. Soc.* 8:142-153.

Carbonell, J.B. 1962. *Las Mielgas y Carretones Españoles*, Inst. Nac. Invest. Agr., Madrid.

Carmignani, V. 1810. Sulle medich. tornata e turbinata. *Giorn. Sci. Litt. Pisano* 12(N.32):43-48.

Casellas, J. 1962. El género Medicago L. en España. *Collectanea Bot.* 6(1-2):183-291.

Chmelar, J. 1975. *Generis Salix Iconographia*, Inst. Bot. Forest., Brno.

Clement, Jr., W.M. 1962. Chromosome numbers and taxonomic relationships in Medicago. *Crop Sci.* 2:25-28.

Clement, Jr., W.M. 1963. Chromosome relationships and diploid hybrid between Medicago sativa L. and M. dzhawakhetica Bordz. *Can. J. Genet. Cytol.* 5:427-432.

Cooper, R.L. & Elliott, F.C. 1964. Flower pigments in diploid alfalfa. *Crop Sci.* 4:367-372.

Coutinho, A. & Santos, A. 1943. Novas contribuicoes para a cariologia de género Trigonella L. *Agron. Lusit.* 5:349-361.

Cummings, M.B., Jenkins, E.W. & Dunning, R.G. 1936. Sterility in Pears. *Vermont Res. Sta. Bull.* 408. U.S.A.

Davidov, B. 1902. Medicago glandulosa. *Österr. Bot. Zeitschr.* 52:493.

Davis, P.H. 1970. Medicago. *Fl. Turkey* 3:483-511. Univ. Press., Edinburgh.

Davis, P.H. & Heywood, V.H. 1963. *Principles of Angiosperm Taxonomy*. Oliver and Boyd, Edinburgh.

De Candolle, A.P. 1813. Medicago [sp.]. *Cat. Pl. Horti Monsp.* :123-125.

Desrousseaux, M. 1792. Luserne, Medicago. In Lamarck, *Encycl. Meth. Bot.* 3:627-638.

Döll, J.C. 1843. Medicago. *Rhein. Fl.* :802-803.

Drower, M.S. 1969. The domestication of the horse. In Ucko and Dimbleby (eds.), *Domest. and Exploit. Plants, Anim.* :471-478. London.

Duby, J.E. 1828. Medicago. *Bot. Gall.* :123-127.

Durham, J.W. 1975. Tertiary Period. In *Encyclopedia Brittannica* 18:151-160.

Durieu de Maisonneuve, M.C. 1845. Medicago [sp.]. In Duchartre, *Revue Botan.* :365-366.

Durieu de Maisonneuve, M.C. 1873. Medicago constricta. *Extr. Compt. Rend. Soc. Linn. Bord.* 29:15-18.

Fedtschenko, B. 1905. [New species of Turkestan flora]. *Bull. Jard. Bot. Petersbg.* 5:41-44 (Russ.).

Fedtschenko, B. 1940. Medicago ciscaucasica. *Bot. Inst. Acad. Sci., Bot. Mater. Herb.* 8:176-177 (Russ.).

Fernandes, A. & Santos de Fatima, M. 1971. Contribution a la connaissance cytotaxinomique des spermatophyta du Portugal. IV. Leguminosae. *Bol. Soc. Broter.* 45:177-225.

Fiori, A. 1925. Medicago. *Nuova Fl. Analit. Ital.* 1:828-840.

Font Quer, P. 1924. Medicago arborea L. var. citrina. *Mem. Mus. Cience Nat. Barcelona*, Secc. Bot. 1(2):7.

Fryer, J.R. 1930. Cytological studies in Medicago, Melilotus and Trigonella. *Can. J. Res.* 3:3-50.

Gaertner, J. 1791. Medicago. *Fruct. Semin. Plant.* 2:348-350.

Ghimpu, M.V. 1930. Etude cariologique des especes Medicago. *Arch. Anat. Microsc.* 26:207-215.

Girshman, R. 1954. *Iran* [Engl. translation]. Penguin Books Ltd., Harmondsworth, Middlesex.

Gillies, C.B. 1971. Pachytene studies in 2n = 14 species of Medicago. *Genetica* 42:278-298.

Gillies, C.B. 1972a. Pachytene chromosomes of perennial Medicago species. I. Species closely related to M. sativa. *Hereditas* 72:277-288.

Gillies, C.B. 1972b. Pachytene chromosomes of perennial Medicago species. II. Distantly related species whose karyotypes resemble M. sativa. *Hereditas* 72:289-302.

Gillies, C.B. 1972c. Pachytene chromosomes of perennial Medicago species. III. Unique karyotypes of M. hybrida Trautv. and M. suffruticosa Ram. *Hereditas* 72:303-310.

Greuter, W. 1970. Medicago heyniana. *Candollea* 25(2):189-192.

Grossheim, A.A. 1919. [Survey of Crimean-Caucasian Medicago]. *Notes (Zapiski) Tiflis Bot. Gard.* 1:1-53 (Russ.).

Grossheim, A.A. 1925. [New data on Caucasian lucernes]. *Reports (Vestnik) Tiflis Bot. Gard.* 4:143-147 (Russ.).

Grossheim, A.A. 1945a. Trigonella. In Komarov (ed.), *Fl. USSR* 11:102-129 (Russ.).

Grossheim, A.A. 1945b. Medicago. In Komarov (ed.), *Fl. USSR* 11:129-176 (Russ.).

Grossheim, A.A. 1945c. Medicago borealis. In Komarov (ed.), *Fl. USSR* 11:391 (Russ.).

Gruner, L. 1867. Medicago meyeri. *Bull. Soc. Nat. Mosc.* 40:416-418.

Gunn, C.R., Skrdla, W.H. & Spencer, H.C. 1978. Classification of Medicago sativa L. using legume characters and flower colors. *U.S. Dept. Agr., Agr. Res. Serv., Techn. Bull.* 1574.

Gussone, G. 1844. Medicago decandollii. *Fl. Sic. Syn.* 2:369.

Hagerup, O. 1932. On pollination in the extremely hot air at Timbuctu. *Dansk bot. Ark.* 8(1):1-20.

Hagerup, O. 1951. Pollination in the Faroes — in spite of rain and poverty in insects. *Dan. Biol. Medd.* 18(15):1-48.

Hayek, A. 1927. Medicago. *Prodr. Fl. Penins. Balc.* 1:834-843.

Hendry, G.W. 1923. Alfalfa in history. *J. Amer. Soc. Agr.* 15:171-176.

Heyn, C.C. 1963. The annual species of Medicago. *Scripta Hierosolymitana* 12:1-154.

Heyn, C.C. 1970. Medicago [annuals]. In Davis, *Flora of Turkey* 3:494-510.

Heszky, L. 1972. Role played by parts of flower in the tripping mechanism of alfalfa. *Acta Agron. Acad. Sci. Hung.* 21:186-190.

Howe, W.L. & Gorz, H.J. 1960. Feeding preferences of the Cowpea Aphid among species of Melilotus. *Annal. Entom. Soc. Amer.* 53-696.

Hsü, K.J. 1973. When the Mediterranean dried up. *Sci. Amer.* 227(6):27-36.

Hsü, K.J. 1978. History of the Mediterranean salinity crisis. In *Initial Reports, Deep Sea Drilling Project*, Vol. 42A. (in press).

Hsü, K.J., Ryan, W.B.F. & Cita, M.B. 1973. Late miocene desiccation of the Mediterranean. *Nature* 242:240-244.

Hsü, K.J., Montadert, L. & Bernoulli, L., *et al.* 1977. History of the Mediterranean salinity crisis. *Nature* 267:399-403.

Hudson, W. 1762. Medicago. *Fl. Angl.* 1st. ed. 287-288.

Ignasiak, T. & Lesins, K. 1972. Carotenoids in petals of Medicago falcata. *Phytochem.* 11:2581-2583.

Ignasiak, T. & Lesins, K. 1973. Carotenoids in petals of some perennial Medicago species. *Biochem. Syst.* 1:97-100.

Ignasiak, T. & Lesins, K. 1975. Carotenoids in petals of perennial Medicago species. *Biochem. Syst. and Ecol.* 2:177-180.

Jacquin, N.J. 1770. Medicago prostrata. *Hort. Bot. Vindob.* 1:39.

Jordan, A. 1854. Medicago [sp.]. In Schultz, *Arch. Fl. Fr. Allem.* :312-316.

Karelin, G.S. 1839. Medicago lessingii [nom. nud.]. *Bull. Soc. Nat. Mosc.* 1:150.

Kasimenko, M.A. 1951. Trigonella. In Larin (ed.), [*Forage Plants of Meadows and Pastures, in USSR*] 2:549-560 (Russ.). Selhozgiz, Leningrad.

Kihara, H. 1972. Right- and left-handedness in plants. *Seiken Zihô* 23:1-37.

Kisliakov, P.V. 1927. [To the question of forage plants of Apsheron]. *Bull. Appl. Bot. (Genet.) Plant Breed.* 17(4):235-251 (Russ.).

Klokov, M.V. 1948. [The Black Sea yellow lucernes]. *Bot. J.* 5:44-49 (Ukr.).

Koperzhinsky, V.V. 1946. [The mechanics of automatic opening of alfalfa flowers]. *Proc. Lenin Acad. Agr. Sci. USSR* 3-4:1-3 (Russ.).

Kotov, M.I. 1940. Medicago erecta. *Bot. J.* 1(2):276-279 (Ukr.).

Kožuharov, S. 1965. The species of genus Medicago L. (Lucerne) in Bulgaria [Engl. summ.]. *Bulg. Acad. Sci., Reports (Izvestiya) Bot. Inst.* 15:119-188 (Bulg.).

Krocker, A.J. 1790. Medicago nigra. *Fl. Siles.* 2, 2:244.

Kunkel, G. 1972. Enumeración de las plantas vasculares de Gran Canaria. *Monogr. Biol. Canar.* 3:1-86.

Lamarck, J.B. et De Candolle, A.P. 1805. Luserne, Medicago. *Fl. Franc.* 4:539-549.

Larkin, R.A. & Graumann, H.O. 1954. Anatomical structure of the alfalfa flower and an explanation of the tripping mechanism. *Bot. Gaz.* 116:40-52.

Latschaschvili, I. 1962. [Perennial Medicago in Daghestan]. *Reports (Vestnik) Bot. Soc. Georg. SSR.* 1:19-20 [Russ. summ.].

Latschaschvili, I. 1969. Medicago talyschensis. *Georg. SSR, Acad. Sci., Inst. Bot. Notes (Zametki) Plant. Sist. Geogr.* 26:76-77 (Russ.).

Ledebour, C.F. 1843. Medicago. *Fl. Ross.* 1:523-530.

Lesins, K. 1950. Investigations into seed setting of Lucerne at Ultuna, Sweden, 1945-1949. *Annal. Royal Agr. Coll. Sweden* 17:441-483.

Lesins, K. 1952. Some data on the cytogenetics of alfalfa. *J. Heredity* 43:287-291.

Lesins, K. 1954. Procedure to facilitate chromosome counts in difficult plant material. *Stain Techn.* 29:261-264.

Lesins, K. 1955. Techniques for rooting cuttings, chromosome doubling, and flower emasculation in alfalfa. *Can. J. Agr. Sci.* 35:58-67.

Lesins, K. 1956a. Somatic flower colour mutations in alfalfa. *J. Heredity* 47:171-179.

Lesins, K. 1956b. Interspecific hybrids between alfalfa, Medicago sativa L., and M. dzhawakhetica Bordz. *Agron. J.* 48:583.

Lesins, K. 1956. Cytogenetic study on a tetraploid plant at the diploid chromosome level. *Can. J. Bot.* 35:181-196.

Lesins, K. 1959. Note on a hexaploid Medicago: M. cancellata M.B. *Can. J. Genet. Cytol.* 1:133-134.

Lesins, K. 1961a. Interspecific crosses involving alfalfa. I. Medicago dzhawakhetica (Bordz.) Vass. x M. sativa L. and its peculiarities. *Can. J. Genet. Cytol.* 3:135-152.

Lesins, K. 1961b. Interspecific crosses involving alfalfa. II. Medicago cancellata M.B. x M. sativa L. *Can. J. Genet. Cytol.* 3:316-324.

Lesins, K. 1961c. Mode of fertilization in relation to breeding methods in alfalfa. *Ztschr. Pflanzenz.* 45:31-54.

Lesins, K. 1962. Interspecific crosses involving alfalfa. III. Medicago sativa L. x M. prostrata Jacq. *Can. J. Genet. Cytol.* 4:14-23.

Lesins, K. 1968. Interspecific crosses involving alfalfa. IV. Medicago glomerata x M. sativa with reference to M. prostrata. *Can. J. Genet. Cytol.* 10:536-544.

Lesins, K. 1969. Relationship of taxa in genus Medicago as revealed by hybridization. IV. M. hybrida x M. suffruticosa. *Can. J. Genet. Cytol.* 11:340-345.

Lesins, K. 1970. Interspecific crosses involving alfalfa. V. Medicago saxatilis x M. sativa with reference to M. cancellata and M. rhodopea. *Can. J. Genet. Cytol.* 12:80-86.

Lesins, K. 1971. Interspecific hybrids involving alfalfa. VI. Ineffectiveness of alloploidy in induction fertility in Medicago pironae x M. daghestanica hybrids. *Can. J. Genet. Cytol.* 13:437-442.

Lesins, K. 1972. Interspecific hybrids involving alfalfa. VII. Medicago sativa x M. rhodopea. *Can. J. Genet. Cytol.* 14:221-226.

Lesins, K. 1976. Alfalfa, lucerne. In Simmonds (ed.), *Evolution of Crop Plants* :165-168. Longman, New York.

Lesins, K., Åkerberg, E. & Böjtös, Z. 1954. Tripping in alfalfa flowers. *Acta Agr. Scand.* 4(2):239-256.

Lesins, K. & Erac, A. 1968a. Relationship of taxa in the genus Medicago as revealed by hybridization. I. M. striata x M. littoralis. *Can. J. Genet. Cytol.* 10:263-275.

Lesins, K. & Erac, A. 1968b. Relationship of taxa in the genus Medicago as revealed by hybridization. III. M. aschersoniana x M. laciniata. *Can. J. Genet. Cytol.* 10:777-781.

Lesins, K. & Gillies, C.B. 1968. Relationship of taxa in genus Medicago as revealed by hybridization. II. M. pironae x M. daghestanica with reference to M. sativa. *Can. J. Genet. Cytol.* 10:454-459.

Lesins, K. & Gillies, C.B. 1972. Taxonomy and cytogenetics of Medicago. In Hanson (ed.), *Alfalfa Science and Technol.* :53-86. Amer. Soc. Agr., Madison, Wis.

Lesins, K. & Lesins, I. 1958. Some improvements in paper chromatography technique for sap-soluble plant pigments. *Proc. Genet. Soc. Can.* 3:44-46.

Lesins, K. & Lesins, I. 1960. Sibling species in Medicago prostrata Jacq. *Can. J. Genet. Cytol.* 2:416-417.

Lesins, K. & Lesins, I. 1961. Some little-known Medicago species and their chromosome complements. *Can. J. Genet. Cytol.* 3:7-9.

Lesins, K. & Lesins, I. 1962. Trueness-to-species in seed samples of Medicago, with a note on 2n=14 species. *Can. J. Genet. Cytol.* 4:337-339.

Lesins, K. & Lesins, I. 1963a. Some little-known Medicagos and their chromosome complements. II. Species from Turkey. *Can. J. Genet. Cytol.* 5:133-137.

Lesins, K. & Lesins, I. 1963b. Pollen morphology and species relationship in Medicago L. *Can. J. Genet. Cytol.* 5:270-280.

Lesins, K. & Lesins, I. 1963c. Medicago saxatilis M.B., a second high-ploidy species in the genus. *Can. J. Genet. Cytol.* 5:348-350.

Lesins, K. & Lesins, I. 1964. Diploid Medicago falcata L. *Can. J. Genet. Cytol.* 6:152-163.

Lesins, K. & Lesins, I. 1965. Little-known Medicagos and their chromosome complements. III. Some Mediterranean species. *Can. J. Genet. Cytol.* 7:97-102.

Lesins, K. & Lesins, I. 1966. Little-known Medicagos and their chromosome complements. IV. Some mountain species. *Can. J. Genet. Cytol.* 8:8-13.

Lesins, K., Lesins, I. & Gillies, C.B. 1970. Medicago murex with 2n = 16 and 2n = 14 chromosome complements. *Chromosoma* 30:109-122.

Lesins, K., Sadasivaiah, R.S. & Singh, S.M. 1976. Relationship of taxa in genus Medicago as revealed by hybridization. VIII. Section Rotatae. *Can. J. Genet. Cytol.* 18:345-355.

Lesins, K. & Singh, S.M. 1973. Relationship of taxa in genus Medicago as revealed by hybridization. VII. M. tornata complex. *Can. J. Genet. Cytol.* 15:321-325.

Lesins, K., Singh, S.M., Baysal, I. & Sadasivaiah, R.S. 1975. An attempt to breed hexaploid alfalfa. *Ztschr. Pflanzenz.* 75:192-204.

Lesins, K., Singh, S.M. & Erac, A. 1971. Relationship of taxa in the genus Medicago as revealed by hybridization. V. Section Intertextae. *Can. J. Genet. Cytol.* 13:335-346.

Lilienfeld, F.A. 1959. Investigations in Medicago laciniata (L.) Mill. *Seiken Zihô* 10:1-10.

Lilienfeld, F.A. 1962. Plastid behavior in reciprocally different crosses between two races of Medicago truncatula Gaertn. *Seiken Zihô* 13:3-38.

Lilienfeld, F.A. & Kihara, H. 1956. Dextrality and sinistrality in plants. *Proc. Acad. Jap.* 32:620-632.

Linnaeus, C. 1738. Medicago. *Hort. Cliffort.* :376-378.

Linnaeus, C. 1753. Trigonella. Medicago. *Spec. Pl.* :776-781.

Linnaeus, C. 1754. Medicago. *Gen. Pl.* 5th. ed. :339.

Loesener, T. 1930(1931). Ignatius Urban. *Ber. Deutsch. Bot. Ges.* 48:(205)-(225).

Loiseleur-Deslongchamps, J.L.A. 1810. Medicago littoralis. *Not. Fl. France* :118-119.

Lowe, R.T. 1868. Medicago. *Man. Fl. Madeira* 1:156-167.

Lubenetz, P.A. 1953. [Species of cultivated and wild lucernes and their breeding value]. *Bull. Appl. Bot. Genet. Plant Breed.* 30(2):3-155 (Russ.).

Lubenetz, P.A. 1972. Lucerne-Medicago. *Bull. Appl. Bot. Genet. Plant Breed.* 47(3):3-68 (Russ.).

Maire, R. 1933. M. sativa ssp. faurei. *Bull. Soc. Hist. Nat. l'Afrique Nord* 24:208.

Mariani, A. 1963. Determinazione del numero cromosomico di alcune specie di Medicago. *Caryologia* 16:139-142.

Marschall von Bieberstein, F.A. 1808. Medicago. *Fl. Taur.-Cauc.* 2:223-228.

Marschall von Bieberstein, F.A. 1819. Medicago sp. *Fl. Taur.-Cauc.* Suppl. to 3:516-518.

Martyn, T. 1792. Medicago varia. *Fl. Rust.* 3:86-87.

Mayr, E. 1970. *Populations, Species and Evolution.* Belknap Press, Cambridge, Mass.

McComb, J.A. 1974. Annual Medicago species with particular reference to those occurring in Western Australia. *J. Royal Soc. West. Austral.* 57(3):81-96.

McComb, J.A. & Andrews, R. 1973. Sequential softening of hard seeds in burrs of annual medics. *Austral. J. Exp. Agr.* 14:68-75.

McKee, R. 1918. Glandular pubescence in various Medicago species. *J. Amer. Soc. Agr.* 10:159-162.

McKee, R. & Ricker, P.L. 1913. Nonperennial Medicagos. *U.S. Dept. Agr., Bur. Pl. Ind., Bull.* 267:7-37.

Miller, P. 1768. Medicago [Nos. 1, 2, 3, 4, 5] *Gard. Dict.* 8th. ed.

Murbeck, S. 1897. De la flore du Nord-Ouest de l'Afrique. *Act. Reg. Soc. Physiogr.* Lund, 33:61-63.

Nègre, R. 1954. Deux nouvelles luzernes pour la flore marocaine. *Soc. Sci. Nat. Phys. Maroc, Compt. Rend.* Mensuelles No. 7:175-176.

Nègre, R. 1956. Les luzernes du Maroc. *Travaux l'Inst. Sci. Cherif. Maroc,* Ser. Bot. 3:xxii, 1-120.

Nègre, R. 1959. Révision des Medicago d'Afrique du Nord. *Bull. Soc. Hist. Nat. l'Afrique Nord* 50:267-314.

Newton, W.C.F. & Pellew, C. 1929. Primula kewensis and its derivatives. *J. Genet.* 20:405-467.

Noulet, J.B. 1837. Luzerne de Pourret. *Fl. Sous-Pyren.* :151.

Oakley, R.A. & Garver, S. 1917. Medicago falcata, a yellow-flowered alfalfa. *U.S. Dept. Agr. Bureau Pl. Ind., Bull.* 428:1-70.

Oldemeyer, R.K. 1956. Distant relatives of cultivated alfalfa, Medicago ruthenica and M. platycarpa. *Agron. Jour.* 48:583-584.

Oostroom, S.J. & Reichgelt, T.J. 1958. Het geslacht Medicago in Nederland en België. *Acta Bot. Neerl.* 7:90-123.

Pedersen, M.W., Barnes, D.K., Sorensen, E.L., et al. 1976. Effects of low and high saponin selection in alfalfa on agronomic and pest resistance traits and the interrelationship of these traits. *Crop Sci.* 16:193-199.

Persoon, C.H. 1807. Medicago. *Synop. Plant.* 2:356-357.

Plinius Secundus, G. (Pliny). *Natural History* [Transl. Rackham] Vol.4:175-179. Harvard Univ. press, London, 1938.

Ponert, J. 1973. Neue taxonomische Kombinationen, Kategorien und Taxa vor allem der turkischen Arten [Medicago]. *Feddes Repert.* 83:639-640.

Post, G.E. 1888. Medicago shepardi. *J. Linn. Soc. Bot.* 24:425.

Post, G.E. 1896. Medicago. *Fl. Syria, Palest., Sinai,* 1st. ed. :226-231.

Post, G.E. 1932. Medicago. *Fl. Syria, Palest., Sinai,* 2nd. ed. :320-329.

Pourret, P.A. 1788. Trigonella hybrida. *Mem. Acad. Toulouse* 3:335.

Presl, J.S. and Presl, C.G. 1822. Medicago. *Delic. Prag.* :44-46.

Proctor, M. & Yeo, P. 1973. *The Pollination of Flowers.* Collins, London.

Prodan, J. 1923. Medicago romanica. *Fl. Rom.* 1:617.

Raven, P.H., Kyhos, D.W. & Hill, A.J. 1965. Chromosome numbers of spermatophytes, mostly Californian. *Aliso* 6:105-113.

Reichenbach, H.G.L. 1832. Medicago. *Fl. Germ. Excurs.* 2:501-505.

Retzius, A.J. 1779. Medicago obscura. *Obs. Bot.* 1:24.

Roberts, E.H. 1975. Problems of long-term storage of seed and pollen for genetic resources conservation. In Frankel and Hawkes (eds.), *Crop Genetic Resources for Today and Tomorrow* :269-296. Cambridge Univ. Press, London.

Romell, L.-G. 1954. Växternas spridningsmöjligheter. In Skottsberg (ed.), *Växternas Liv,* 2nd. ed. 8:30-199. Norden, Malmö.

Rouy, G. 1899. Medicago. *Fl. France* 5:5-44.

Schmalhausen, I.F. 1895. Medicago. *Fl. Sred. Juž. Ross.* 1:224-228 (Russ.).

Schmiedeknecht, M. & Lesins, K. 1968. Weitere Untersuchungen zur Resistenz von Medicago-Arten gegen Pseudopeziza medicaginis (Lib.) Sacc. *Theoret. Appl. Genet.* 38:188-194.

Schröck, O. 1943. Beobachtungen an einem Bastard zwischen Luzerne und Gelbklee. *Züchter* 15:4-10.

Schulz, O.E. 1901. Monographie der Gattung Melilotus. *Bot. Jahrb.* 29:660-735.

Scofield, C.S. 1908. The botanical history and classification of alfalfa. *U.S. Dept. Agr., Bureau Pl. Ind., Bull.* 131(2):11-19.

Scott-Moncrieff, R. 1936. A biochemical survey of some Mendelian factors for flower colour. *J. Genet.* 32:117-170.

Seringe, N.C. 1825. Medicago. In de Candolle, *Prodr. Syst. Nat.* 2:171-181.

Shade, R.E., Thompson, T.E. & Campbell, W.R. 1975. An alfalfa weevil larval resistance mechanism detected in Medicago. *J. Econ. Entom.* 68:399-404.

Sibthorp, J. 1794. Medicago. *Fl. Oxon.* :232.

Simon, J.P. 1965. Relationship in annual species of Medicago. II. Interspecific crosses between M. tornata (L.) Mill. and M. littoralis Rhode. *Austr. J. Agr. Res.* 16:51-60.

Simon, J.P. 1969. Serological studies in Medicago, Melilotus, Trigonella and certain other genera. *Bot. Gaz.* 130(2):127-141.

Simon, J.P. & Simon, A. 1965. Relationship in annual species of Medicago. *Austr. J. Agr. Res.* 16:37-50, 6 pl.

Singh, S.M. & Lesins, K. 1972. Relationship of taxa in the genus Medicago as revealed by hybridization. VI. M. laciniata x M. sauvagei. *Can. J. Genet. Cytol.* 14:823-828.

Sinskaya, E.N. 1945. On the diploid species of yellow alfalfa. *Compt. Rend. Acad. Sci. URSS.* 48(4):281-282.

Sinskaya, E.N. 1948. Medicago difalcata. *Bull. Appl. Bot. Genet. Pl. Breed.* 28(1):29 (Russ.).

Sinskaya, E.N. 1950. Medicago. *Flora Cultiv. Pl. USSR.* 13:7-195 (Russ.). [Engl. transl. 1961].

Sinskaya, E.N. & Maleeva, Z.P. 1959. [Ploidy in perennial species of lucerne]. *Bot. J.* 44:1103-1113 (Russ.).

Širjaev, G. 1928. Medicago falcata var. tirnensis. *Izvest. Bulg. Bot. Druzhestvo* :44.

Širjaev, G. 1928. *Gen. Trigonella* 1/1. Fac. Sci., Univ. Masaryk, Brno.
Širjaev, G. 1933. *Gen. Trigonella* 1/6. Fac. Sci., Univ. Masaryk, Brno.
Širjaev, G. 1934. *Gen. Trigonella* 2/1. Fac. Sci., Univ. Masaryk, Brno.
Širjaev, G. 1935. Die Entwicklungsgeschichte der Gattung Trigonella. *Assoc. Russe Sci. Prague, Bull.* 2(9):135-162.
Širjaev, G. 1938. In Handel-Mazzetti, H.: Kleine Beiträge zur Kenntnis der Flora von China [M. edgeworthi]. *Österr. Bot. Zeitschr.* 87:123-124.
Southworth, W. 1928. Influences which tend to affect seed production in alfalfa and an attempt to raise a high seed producing strain by hybridization. *Scient. Agric.* 9:1-28.
Sprague, E.W. 1959. Cytological and fertility relationships of Medicago sativa, M. falcata, and M. gaetula. *Agron. J.* 51:249-252.
Stearn, W.T. 1966. *Botanical Latin.* Nelson, London.
Stebbins, G.L. 1974. *Flowering Plants.* Belknap Press, Cambridge, Mass.
Stevenson, G.A. 1969. An agronomic and taxonomic review of the genus Melilotus. *Can. J. Plant Sci.* 49:1-20.
Straw, R.M. 1955. Hybridization, homogamy, and sympatric speciation. *Evolution* 9:441-444.
Sumnevicz, G. 1932. De speciebus nonnulis asiaticis generis Medicaginis L. *Animadv. System. Herb. Univ. Tomsk* 1-2:1-15 (Russ.).
Tenore, M. 1811. Medicago cancellata [= M. tenoreana]. *Fl. Nap.* :44.
Tenore, M. 1815. Medicago agrestis. *Cat. Horti Bot. Neapol., Append.* 1:66-67.
Tineo, V. 1817. Medicago muricoleptis. *Pl. Sic. Rar. Pug.* 1:18.
Torsell, R. 1936. Influence of external factors on seed setting in Lucerne. *Lantbr. Högsk. Annal.* 3:191-241.
Torssell, R. 1943. Odling, förädling och fröodling av blåluzern. *Sveriges Ustädesför. Tidskr.* 53:275-284.
Torssell, R. 1948. Different methods in the breeding of lucerne. *Svalöf 1886-1946* :237-248.
Trabut, L. 1917. Origine hybride de la luzerne cultivée. *Compt. Rend. Acad. Sci.* (Paris) 164:607-609.
Trautvetter, E.R. 1841. Über die mit Trifolium verwandten Pflanzengattungen. *Acad. Imp. St. Petersburg, Bull. Sci.* 8:267-272.
Troitzky, N. 1923. Medicago dzhawakhetica var. timofeewii. *Report (Vestnik) Bot. Gard. Tiflis,* New Ser. 1:90-91 (Russ.).
Troitzky, N. 1928. [A contribution to the question of the role played by hybridization in the formation of species]. *Bull. Appl. Bot. Genet. Plant Breed.* 19(2):213-231 (Russ.).
Tschechow, W. 1932. [Karyo-systematic analysis of the tribe Trifolieae D.C. (Family Leguminosae Juss.)]. *Bull. Appl. Bot. Genet. Plant Breed.,* Ser. 2(1):119-143 (Russ.).
Tutin, T.G. 1968. Medicago. In *Fl. Europ.* 2:153-157.
Urban, I. 1873. Prodromus einer monographie der gattung Medicago L. *Verh. bot. Ver. Brandenb.* 15:1-85, 2 pl.
Urban, I. 1877. [Remarks in Medicago falcata and M. sativa hybrids and on seed characteristics of several Medicago species]. *Verh. bot. Ver. Brandenb.* 19:125-134.
Vassilczenko, I.T. 1949a. [Lucerne — the best forage plant]. *Acta Inst. Bot. Komarovii,* Ser. 1(8):1-240 (Russ.).

Vassilczenko, I.T. 1949b. Medicago gunibica. *Acad. Sci. USSR, Not. Syst. Herb. Inst. Bot.* 11:100 (Russ.).

Vassilczenko, I.T. 1952. [New Trigonella species for cultivation]. *Acad. Sci. USSR*, Ser. Sci.-popular (Russ.). Acad. Nauk, Moscow.

Vassilczenko, I.T. [Synopsis of the species of the genus Trigonella]. *Acta Inst. Bot. Komarovii*, Ser. 1/10:124-269 (Russ.).

Velenovsky, J. 1893. Medicago rhodopea. *Sitzungsber. Böhm. Ges. Wiss.* 37:21-22.

Visiani, R. 1855. Medicago pironae. *Ind. Sem. Hort. Bot. Patav.* :365.

Webb, P.B., and Berthelot, S. 1836. Medicago. *Hist. Nat. I. Canaries*, 1st. ed. :59-66, pl. 56.

Webb, P.B., and Berthelot, S. 1850. Medicago canariensis Benth. *Phytographia Canariensis*, 2nd. ed., 2:pl. 56.

Willdenow, C.L. 1802. Medicago. *Sp. Plant.* 3:1403-1419.

Willdenow, C.L. 1809. Medicago. *Enumer. Pl. Horti Reg. Bot. Berol.* :800-805.

Willdenow, C.L. 1814. Medicago applanata [nom. nud.]. *Enumer. Pl. Horti Reg. Bot. Berol., Suppl.* :52.

Woods, J. 1850. Medicago cuneata. *Tourist's Flora* :84.

Wulfen, F.X. 1786. Medicago carstiensis. In Jacquin, *Collect. ad Botan.* :86-87.

# INDEX